AF803513

5G and Beyond Wireless Communications

This book provides a thorough introduction of 5G and B5G wireless networks, as well as cutting-edge technologies that aid in network design and development. This book also covers machine learning techniques for advanced communications.

5G and Beyond Wireless Communications: Fundamentals, Applications, and Challenges discusses the newest technologies for 5G and future networks, including CR networks, D2D networks, UAV-assisted communications, RIS-assisted communications, and ML for communication networks. Additionally, it discusses using antenna systems for advanced communications networks. It also explores various security issues and their solutions, as well as power and interference management and machine learning for optimization of network parameters. The book also examines the design of 5G antennas from a materials perspective, and a thorough analysis of the materials utilized to create innovative antennas for advanced communication network is discussed. The book concludes by discussing the advancement of ML-based communication networks and their future opportunities and challenges.

This book will be helpful for researchers and master students who want to focus their research work in the area of next-generation advanced wireless communications.

Materials, Devices, and Circuits: Design and Reliability

Series Editor: Shubham Tayal, K. K. Paliwal, Amit Kumar Jainy

This series is designed to illustrate the many new and exciting challenges in the expansive and interdisciplinary field of materials science, device engineering, reliability, device-circuit co-design and their applications. The scope of this series is broad, with reference works and textbooks offering insight into all aspects of materials development, device fabrication, circuit analysis and their reliability. The titles in this series deliver authoritative information to professionals, researchers, and students across material and device engineering as well as other scientific disciplines. Each volume offers a comprehensive approach covering fundamentals, technologies, their evolution advancements and applications and includes real-world examples where appropriate.

Device Circuit Co-Design Issues in FETs
Edited by Shubham Tayal, Billel Smaani, Shiromani Balmukund Rahi, Samir Labiod, Zeinab Ramezani

Human-Machine Interface Technology Advancements and Applications
Edited by Edited By Ravichander Janapati, Usha Desai, Shrirang Ambaji Kulkarni, Shubham Tayal

Negative Capacitance Field Effect Transistors
Physics, Design, Modeling and Applications
Edited by Young Suh Song, Shubham Tayal, Shiromani Balmukund Rahi, Abhishek Kumar Upadhyay

Laser Applications in Manufacturing
Edited by Pankaj Kumar, Manowar Hussain, Amit Kumar Jain, Sunil Pathak

5G and Beyond Wireless Networks
Technology, Network Deployments and Materials for Antenna Design
Edited by Indrasen Singh, Shubham Tayal, Niraj Pratap Singh, Vijay Shankar Tripathi, and Ghanshyam Singh

5G and Beyond Wireless Communications
Fundamentals, Applications, and Challenges
Edited by Abhijit Bhowmick, Yogesh Kumar Choukiker, Indrasen Singh, and Arumugam Nallanathan

For more information about this series, please visit: www.routledge.com/Materials-Devices-and-Circuits/book-series/MDCDR

5G and Beyond Wireless Communications

Fundamentals, Applications, and Challenges

Edited by Abhijit Bhowmick,
Yogesh Kumar Choukiker, Indrasen Singh, and
Arumugam Nallanathan

CRC Press
Taylor & Francis Group
Boca Raton London New York

CRC Press is an imprint of the
Taylor & Francis Group, an **informa** business

Designed cover image: © Shutterstock

MATLAB® and Simulink® are trademarks of The MathWorks, Inc. and
are used with permission. The MathWorks does not warrant the accuracy of the text or exercises in this book. This book's use or discussion of MATLAB® or Simulink® software or related products does not constitute endorsement or sponsorship by The MathWorks of a particular pedagogical approach or particular use of the MATLAB® and Simulink® software.

First edition published 2025
by CRC Press
2385 NW Executive Center Drive, Suite 320, Boca Raton FL 33431

and by CRC Press
4 Park Square, Milton Park, Abingdon, Oxon, OX14 4RN

CRC Press is an imprint of Taylor & Francis Group, LLC

© 2025 selection and editorial matter, Abhijit Bhowmick, Yogesh Kumar Choukiker, Indrasen Singh, and Arumugam Nallanathan; individual chapters, the contributors

Reasonable efforts have been made to publish reliable data and information, but the author and publisher cannot assume responsibility for the validity of all materials or the consequences of their use. The authors and publishers have attempted to trace the copyright holders of all material reproduced in this publication and apologize to copyright holders if permission to publish in this form has not been obtained. If any copyright material has not been acknowledged please write and let us know so we may rectify in any future reprint.

Except as permitted under U.S. Copyright Law, no part of this book may be reprinted, reproduced, transmitted, or utilized in any form by any electronic, mechanical, or other means, now known or hereafter invented, including photocopying, microfilming, and recording, or in any information storage or retrieval system, without written permission from the publishers.

For permission to photocopy or use material electronically from this work, access www.copyright.com or contact the Copyright Clearance Center, Inc. (CCC), 222 Rosewood Drive, Danvers, MA 01923, 978-750-8400. For works that are not available on CCC please contact mpkbookspermissions@tandf.co.uk

Trademark notice: Product or corporate names may be trademarks or registered trademarks and are used only for identification and explanation without intent to infringe.

ISBN: 978-1-032-62256-9 (hbk)
ISBN: 978-1-032-62502-7 (pbk)
ISBN: 978-1-032-62503-4 (ebk)

DOI: 10.1201/9781032625034

Typeset in Times
by Apex CoVantage, LLC

Contents

About the Editors

Abhijit Bhowmick received his BE (hons) degree in electronics and telecommunication engineering in 2002 from Burdwan University, West Bengal, India, and MTech degree in telecommunication engineering in 2009 and PhD in December 2016 from NIT Durgapur. He worked for Cubix Control System Pvt. Ltd. from 2004 to 2006. He joined the Department of Electronics and Communication Engineering, Bengal College of Engineering and Technology, Durgapur, as a lecturer in 2006. Thereafter, he joined VIT University, Vellore, TN, India, in the School of Electronics Engineering (SENSE) in June 2016 and is currently serving as an associate professor there. His research interests include cognitive radio networks, focusing on spectrum sensing and spectrum sharing issues, cooperative communications in cognitive radio networks, energy harvesting in wireless network, D2D communication, and physical layer security issues in wireless networks, and UAV-assisted communication.

Yogesh Kumar Choukiker received his BE degree in 2007 and MTech and the PhD degrees from the National Institute of Technology, Rourkela, Orissa, India, in 2009 and 2014, respectively, both in electronics engineering. He is specialized in antenna and wave propagation. In particular, his research interest is microstrip antenna (fractal antennas) in the area of wireless application, MIMO environment, UWB, and reconfigurable fractal. He completed the ISRO RESPOND project in 2013, which is based on optimization techniques. At present, he has been working as a professor in Vellore Institute of Technology, Vellore, from January 2014. He was also a visiting research scholar in San Diego State University, California, USA, in his PhD.

Indrasen Singh is working as an assistant professor (senior grade 2) in the Department of Embedded Technology under the School of Electronics Engineering (SENSE), Vellore Institute of Technology (VIT), Vellore, Tamil Nadu, India. He received his BTech and MTech degrees in electronics and communication engineering from Uttar Pradesh Technical University, Lucknow, India, in 2006 and 2010, respectively. He obtained his PhD degree in electronics and communication engineering from then National Institute of Technology Kurukshetra, Haryana, India, in 2019. He has qualified GATE examination many times. He has more than 15 years of teaching/research experience in the various reputed technical institutes/universities. He is supervising BTech, MTech, and PhD students. He is an editorial board member for AJECE, Science Publishing Group, USA. He is the reviewer of many SCI/SCOPUS-indexed international journals and served as TPC member and reviewer in different conferences. He has published more than 25 research papers in national/international journals/conferences of repute, and many are under review. His research interests are in the area of cooperative communication, stochastic geometry, modelling of wireless networks, heterogeneous networks, millimetre wave communications, device-to-device communication, machine learning, and 5G/6G communication.

Arumugam Nallanathan is a professor of wireless communications and the founding head of Communication Systems Research (CSR) group in the School of Electronic Engineering and Computer Science at Queen Mary University of London (QMUL) from September 2017. He was with the Department of Informatics at King's College London from December 2007 to August 2017, where he was a professor of wireless communications from April 2013 to August 2017. He was an assistant professor in the Department of Electrical and Computer Engineering, National University of Singapore, from August 2000 to December 2007. His areas of interest are communication, stochastic geometry, modelling of wireless networks, heterogeneous networks, millimetre wave communications, device-to-device communication, and 5G/6G communication.

Contributors

Ahmad, Saif
Department of ECE
Integral University
Lucknow, UP, India

Alam, Munshi Yusuf
Department of Computer Science and Engineering
Budge Budge Institute of Technology
Kolkata, WB, India

Annepu, Visalakshi
School of Computer Science and Engineering
VIT-AP University
Amaravati, Andhra Pradesh, India

Aparna, E.
National Institute of Technology
Warangal, Telangana, India

Bagadi, Kalapraveen
School of Electronics Engineering
VIT-AP University
Amaravati, India

Baranwal, Alok
Department of Electronics and Communication Engineering
National Institute of Technology
Durgapur, WB, India

Bhattacharyya, Budhaditya
Department of CE (SENSE), VIT
Vellore, TN, India

Challa, Naga Raju
Department of Electronics and Communication Engineering
Bapatla Engineering College
Bapatla, India

Chandra, Kamireddy Vijay
VNR Vignana Jyothi Institute of Engineering and Technology
Bachupally, Hyderabad, India

Chattopadhyay, Sudipta
Department of Electronics and Telecommunication Engineering
Jadavpur University
Kolkata, WB, India

Dhar Roy, Sanjay
Department of Electronics and Communication Engineering
National Institute of Technology
Durgapur, WB, India

Gaur, Pooja
Department of ECE, Integral University
Lucknow, UP, India

Ghosh, Priyanka
Department of Electronics and Communication Engineering
National Institute of Technology
Durgapur, WB, India

Godugu, Kiran Kumar
School of Electronics Engineering
VIT-AP University
Amaravati, Andhra Pradesh, India

Hota, Lopamundra
Department of Computer Science and Engineering
National Institute of Technology
Rourkela, India

Kavitha, Kankanala
School of Electronics Engineering
VIT-AP University
Amaravati, Andhra Pradesh, India

Khan, Mohd Javed
Department of ECE
Integral University
Lucknow, UP, India

Kingsly, Saffrine
Symbiosis Institute of Technology
Symbiosis International (Deemed University)
Pune, India

Kondepogu, Vijayakumar
Department of CE (SENSE), VIT
Vellore, TN, India

Kumar, Arun
Department of Computer Science and Engineering
National Institute of Technology
Rourkela, India

Kumar, Vaegae Naveen
Department of CE (SENSE), VIT
Vellore, TN, India

Kundu, Sumit
Department of Electronics and Communication Engineering
National Institute of Technology
Durgapur, WB, India

Malathi A., Christina Josephine
Department of CE (SENSE), VIT
Vellore, TN, India

Nandi, Subrata
Department of Computer Science and Engineering
National Institute of Technology
Durgapur, WB, India

Nayak, Biraja Prasad
Department of Computer Science and Engineering
National Institute of Technology
Rourkela, India

Parbat, Gayatree
Department of Computer Science and Engineering
National Institute of Technology
Durgapur, WB, India

Poonkuzhali, R.
Department of CE (SENSE), VIT
Vellore, TN, India

Prasad, Binod
Department of Electrical and Electronic Engineering
Atal Bihari Vajpayee Indian Institute of Information Technology and Management (ABV-IIITM)
Gwalior, Madhya Pradesh, India

Priyadarshini, Prangya
Department of Computer Science and Engineering
National Institute of Technology
Rourkela, India

Rajesh, G.
SENSE, VIT,
Vellore, TN, India

Ram, Gopi
National Institute of Technology
Warangal, Telangana, India

Rao, Yamarthi Narasimha
School of Electronics Engineering
VIT-AP University
Amaravati, Andhra Pradesh, India

Shamina, S. S.
SENSE, VIT, Vellore
Chennai, India

Singh, Indrasen
Department of CE (SENSE), VIT
Vellore, TN, India

Sona, Deepika Rani
Department of CE (SENSE)
VIT
Vellore, TN, India

Srivastava, Vishwas
Department of Electrical and Electronic Engineering
Atal Bihari Vajpayee Indian Institute of Information Technology and Management (ABV-IIITM)
Gwalior, Madhya Pradesh, India

Subbaraj, Sangeetha
College of Engineering
Anna University
Chennai, India

Thakur, Chandrima
Department of Electronics and Telecommunication Engineering
Jadavpur University
Kolkata, India

Thomas, Benedict
Vellore Institute of Technology
Vellore, India

Thomas, Shweta B
Department of CE (SENSE)
VIT
Vellore, TN, India 632014

Vappangi, Suseela
School of Electronics Engineering
VIT-AP University
Amaravati, Andhra Pradesh, India

Vishnu, P. R.
National Institute of Technology
Warangal, Telangana, India

1 A Comprehensive Research Look on Opportunities and Challenges in Advanced Wireless Communication

Vijayakumar Kondepogu and Budhaditya Bhattacharyya

1.1 HISTORIC VIEW OF WIRELESS COMMUNICATION

The world of wireless mobile communication has undergone a remarkable transformation over the past few decades, evolving from analogue systems with limited capabilities to sophisticated digital networks that enable seamless connectivity and a plethora of advanced services. This journey has been driven by technological advancements, societal needs, and the ever-increasing demand for seamless communication. From the nascent stages of its emergence, strategies involving wireless communication have ruled engineering and technological innovations. Amongst all the other innovations, the mobile phone, or portable option of communication between two remote users, became a hugely popular and accepted engineering marvel [1]. Starting with the first generation (1G), it has developed and grown into the sixth generation (6G) in performing communication or transmission in a wireless manner [2]. The first generation is specified as the first generation in cellular systems that was introduced in 1980s. This system relied and performed on an analogue basis, where the 'frequency-division multiplexing access' (FDMA) model has been utilized. Dominated by analogue technologies like advanced mobile phone system (AMPS) and Nordic Mobile Telephone (NMT) [3], it suffered heavily with respect to limited capacity, poor voice quality, and susceptibility to interference. However, they laid the foundation for the future of mobile communication, allowing individuals to make and receive calls without the constraints of wired connections. The second generation (2G) of wireless mobile communication systems marked a significant leap forward, introducing digital technology and paving the way for the widespread adoption of mobile phones. The Global System for Mobile Communications (GSM) [4] emerged as the dominant standard, offering improved voice quality, increased capacity, and the ability to send and receive text messages (SMS). 2G also introduced the concept of data services, albeit at relatively slow speeds. It is further

DOI: 10.1201/9781032625034-1

distinguished into two multiplexing methods as 'code-division multiple access' (CDMA) and 'time-division multiple access' (TDMA).

Subsequently, the third generation (3G) of wireless mobile communication systems revolutionized the mobile landscape, enabling multimedia applications and high-speed data transmission. The Universal Mobile Telecommunications System (UMTS) and its successors, such as High-Speed Packet Access (HSPA) [5] and HSPA+, provided data rates in the range of hundreds of kilobits per second, allowing users to access the internet, stream videos, and download files at much faster speeds.

The fourth generation (4G) of wireless mobile communication systems supported by secure-based IP communication saw a dramatic shift in the usage of mobile devices. Represented by Long-Term Evolution (LTE) [6] and LTE-Advanced, it further enhanced data rates to reach speeds of up to several gigabits per second [7]. The increase in bandwidth enabled the proliferation of mobile internet applications, including fast video streaming, online gaming, and cloud-based services. 4G also paved the way for the Internet of Things (IoT) in a true sense, connecting a vast array of devices and sensors to the internet, thus starting a new revolution in the form of 'smart device'–based applications. The terms *smart healthcare*, *smart city*, *smart automations*, etc. were coined after the advent of successful implementation of 4G services.

The fifth generation (5G) [8] of wireless mobile communication systems is currently rolling out, promising up to 10 Gbps, which is almost 10 to 100× improvement over 4G and 4.5G network. The tremendously low latency (of around 1 ms) for 100 times more number of connected devices per unit area makes it a true candidate of global recognition. It also promised a 99% availability of network, with cent percentage coverage. One of the biggest features that 5G provides is 90% reduction in network energy usage, which, as envisioned, will make battery life of low-powered IoT devices to around ten years and more. 5G offers primarily three enablers in the form of:

- Enhanced mobile broadband (eMBB)
- Massive machine type communications (mMTC)
- Ultra-reliable low-latency communications (URLLC)

All these enablers are based on the demands from users of different quarters. eMBB has made its name in *faster, on-the-go* solutions as it has promised to provide a seamless, on-the-go HD movie viewing experience, immersive augmented reality (AR) [9], and virtual reality (VR). Services provided to the most densely populated metropolitan centres at a DL speed of 1 Gbps and 300 Mbps have broaden the horizon of end user satisfaction. This plethora of users requires connections all the time. This has been well handled using mMTC – which supports almost 1 md/km^2 with low data rate and little energy consumption. Most of the IoT-based applications and growth of such infrastructure have been heavily rolled out based on the amicable mMTC feature. The third wing, in the form of ultra-reliable low-latency communications (URLLCs), which boasts of *staying connected with low lag*, provides a potent solution for 1 ms or less end-to-end latency. URLLC is the backbone to support various mission-critical communications services like autonomous cars and remote surgery. These are situations where data transfer needs to have as little lag as possible, and connections need to be as reliable as possible. URLLC targets around 99.999% reliability and single-digit millisecond latency, a 75% improvement over 4G LTE.

The 5G technologies, such as cognitive radio (CR) technologies, device-to-device (D2D) communications, and reconfigurable intelligent surfaces (RIS)–assisted communications are explained as follows.

CR technology [10]. CR is a wireless transmission model that can change its environment. CR can identify which transmission channels are in use and forward it to vacant channels while preventing the occupied channels. CR is employed in 5G wireless communications. The significant features of CR systems are adaptability, cognition, and awareness. *Adaptability* means CR has the capacity to allocate the system parameters without any changes in the device hardware. *Cognition* means it displays the capacity to execute the data-attained environment that must be employed for creating optimal device performance. Finally, *awareness* means that the model has the capacity to estimate and monitor the sector if the spectrum is occupied or free. The significant limitations as faced by CR can be summarized in the subsequent discussion:

- The CR technology has high network complexity and security issue
- The CR technology has a limited frequency range and has standardization issues.
- The CR technology has environmental adaptability issues and is susceptible to unknown risks.

D2D communication [11]. It is the process of making direct communication among 5G-enabled systems. In the 5G network, the installed systems are grouped together via specific connections for rapid data transmission from one system to another. The D2D communication is partitioned into two, namely, out-band D2D and in-band D2D. Out-band communication is utilized in the unlicensed spectrum. It prevents interference from cellular candidates, yet interference among the systems exists, which results in a more critical issue of managing the interference. In-band communication is employed in the licensed cellular spectrum. It offers spectral efficacy because of sharing the licensed spectrum among cellular candidates and D2D. However, it doesn't guarantee effective resource utilization because of rigid resource allocation. The problems faced in D2D communications are explained in what follows.

- It has a high risk of security breaches and data theft.
- Inference management is one of the biggest weaknesses in D2D communication.
- It may face numerous technical challenges and congestion.

RIS-assisted communication [12]. RISs have the capacity to drive electromagnetic (EM) waves in particular direction. The RISs are artificial surfaces created by the EM material. RISs can manage the EM signal's propagation by tuning the magnetic and electric attributes of the surface. RISs can enhance the wireless link's performance by driving EM waves. This permits the receivers to be illuminated. The usage of energy has become a significant concern to 5G. RISs have developed as a new paradigm with enhanced power for the reduction of energy consumption. RISs are integrated with various low-cost factors, such as PIN diode. RISs improve spatial multiplexing by the attained power through beamforming. Several issues faced in RIS-aided communication are given here.

- It has high cost and design complexity.
- It does not support digital processing and is restricted by path loss.
- It does not perform better than relay.

The evolution of the standards is depicted in Figure 1.1.

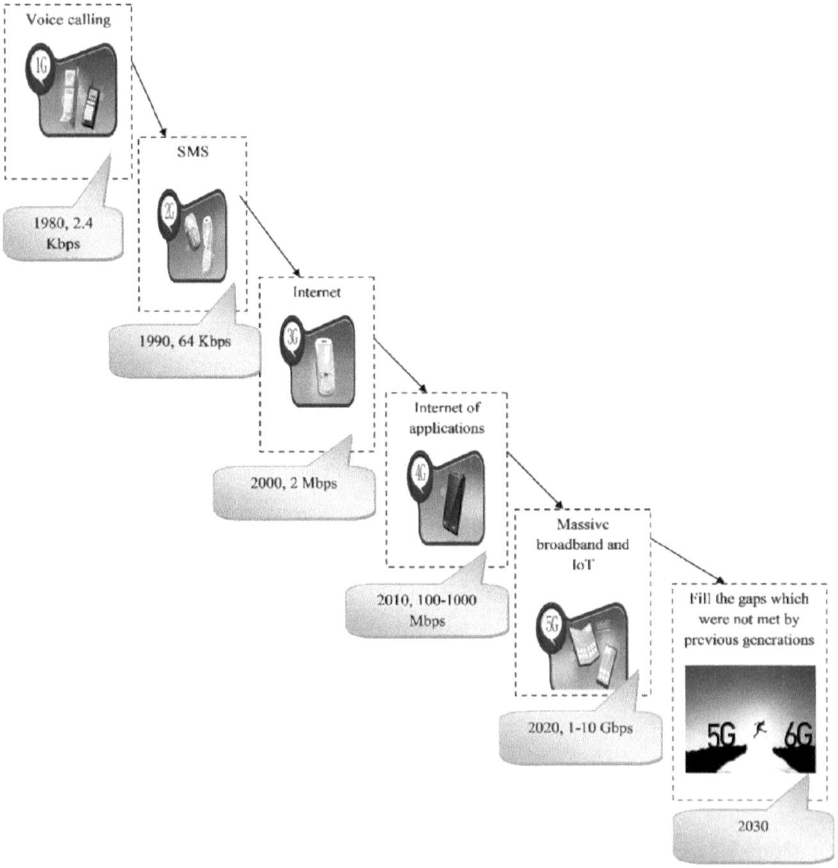

FIGURE 1.1 Timeline picture of the wireless communication system.

1.2 BLOCK DIAGRAM OF THE WIRELESS COMMUNICATION SYSTEM

The three major elements included in communication systems are 'transmitter, receiver, and channel', which are depicted in Figure 1.2. A typical transmitter is responsible for encoding and modulating the information to be transmitted, while the receiver is responsible for demodulating and decoding the received signal.

The transmitter comprises of the source encoder, which primarily collects information in the form of voice, data, or video and converts it into a digital format. The security of these transmitted data is ensured by several encryption algorithms. These include algorithms like chaotic encryption scheme [13], secret key agreement (SKA) protocols [14], and physical layer network coding (PLNC) [15], which all exhibit good performance by generating unpredictable and secure encryption keys to encode the data. SKA protocols establish shared secret keys between communicating parties without exchanging them over a vulnerable channel [16]. Along with this, cooperative jamming methods deliberately introduce multiple friendly transmitters introducing interference into the channel, thus disrupting eavesdroppers' ability to intercept the desired signal. In fact, this is established to keep maximum interference while minimizing impact on the legitimate communication. The conventional spread spectrum-based technique or adaptive beamforming also helps in making directed transmission to the receiver, avoiding unnecessary interventions of intruders.

The channel encoder adds redundancy to the digitally secured data even further to protect it from errors caused by the transmission channel. The data finally gets modulated in modulator block using standard schemes like QPSK, QAM modulations, before conversion to an analogue signal for transmission over the radio channel. This channel is also referred to as the medium to pass the signal from the source to the

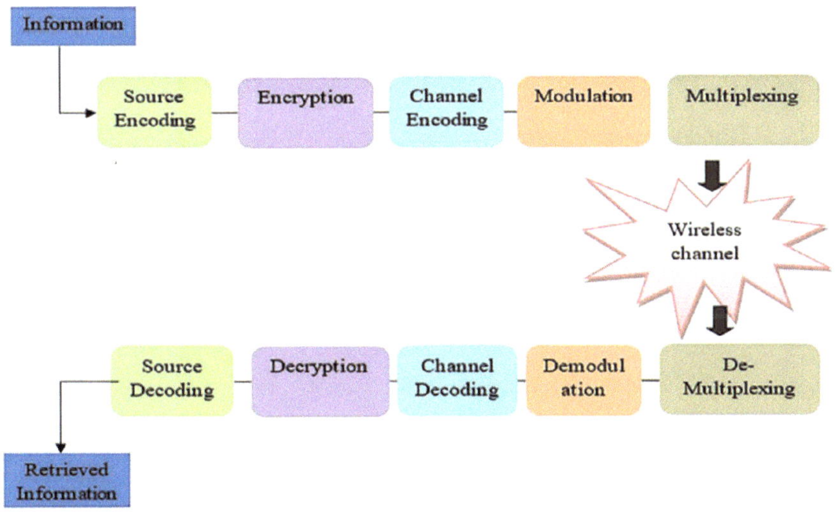

FIGURE 1.2 Block diagram view of the wireless communication system.

destination. Since it has wireless nature, the wireless channel is taken and used. It has the characteristics of being unpredictable and random to achieve the better process [17]. In some cases, however, the channel is vulnerable to distortion, scattering, interference, etc., and the result is that the signal is received with such errors. The main intent of a receiver is to fetch the signal from the wireless channel, and it is also used to reconstruct the source signal. It is a 'vice versa' process of transmission path, in which it constitutes de-multiplexing, demodulation, channel and source decoding, and also decryption. With the initial block of receiving side, the transmitted signal is applied for de-multiplexing, which separates the signal and that given for demodulation. It is further followed by the decoding process, where the redundant bit information gets declined. Since the message is encrypted form, the decryption block is helped to retrieve the original information or data or signal.

1.3 CHANNEL ESTIMATION TECHNIQUES

Channels are the critical part in a wireless system in transferring data from the source to the receiver. In the communication system, the transmitting and receiving devices can be fixed or mobile. Still, the channels are affected by multipath and its fading characteristic. Therefore, the channel estimation plays a pivotal part during data or signal transmission, and it also helps increase the system efficiency. Prior to transmission, channel estimation takes place. In order to estimate the channels in the system, it is based on varying the time and the channel behaviour. The types of channel estimation are depicted in the Figure 1.3.

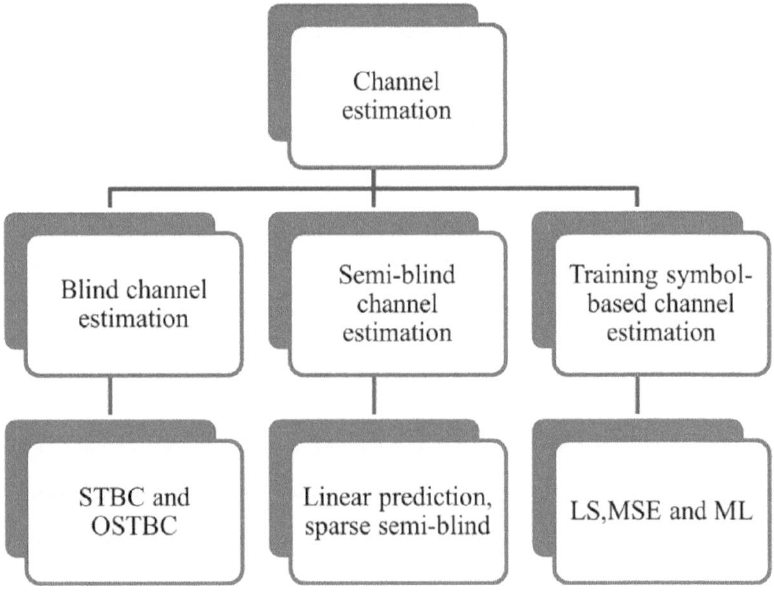

FIGURE 1.3 Types of channel estimation techniques.

1.3.1 Blind Channel Estimation

In this type of estimation, the pilot symbols are not considered at both the transmitter and the receiver. The receiver module is used to estimate the channels without using the pilot symbols. Thus, to effectively determine the channels, the receiver can use the large amount of information by influencing the characteristics of the channel [13]. The main advantages of this method are to reduce overhead, as it contains pilots. Nevertheless, the computation complexity gets increased as it retrieves the signal characteristics in massive collection. Since it is am earlier model, it is unable to give promising results.

STBC. It is the most used technique in the wireless communication network in transferring multiple numbers of information based on multiple antennas to enhance system performance. It combines all the information copies of the original signal, formed in the corresponding manner to recover the best possible information.

OSTBC. It is one of the most promising approaches to exploit spatial diversity in the wireless network [18]. It offers full diversity and minimizes the complexity of the optimal decoder to a simple linear receiver.

1.3.2 Semi-Blind Channel Estimation

By implying the technique name, it is developed by getting the objectives from both training and blind-based approach. Rather than other methods, this technique has noted that the pilots present at the transmitter side assist in estimating the channel at the receiver side. During this process, the channel could be assessed by the pilot as well as data and pilot symbols. Once the channel is estimated in the system, its respective coefficient gets back to the transmitter source. When the feedback symbols match with the original source, the signal could be further detected by the receiver easily. Similarly, the channel coefficients are represented in symmetry form along with the pilot symbols, which drive rapid prediction of symbols. Hence, the prime key point of using this technique is to appropriately track the channel behaviour during the transmission.

Linear prediction [19]. It provides two favours, like depending on whether the transmitted symbol is modelled as unknown deterministic or as uncorrelated random sequence.

Sparse semi-blind [20]. In most of the practical situations, the wireless channel is actually the sparse channel, which is observed in cross talk, digital television transmission, network echo cancellation, and communication channels.

1.3.3 Training Symbol-Based Channel Estimation

The training symbols are also termed as the pilots taken for channel estimation process. While using the pilot symbols, it can cause overhead issue, which can mislead the communication. It uses the earlier method to estimate the channel, where the time domain features are interpolated by applying the least square methods.

LS. The main aim of the channel LS is to reduce the square distance between the received and the original signal.

MSE [21]. It is the most popularly utilized channel estimation approach. Also, it is applicable to any multicarrier model.

ML [22]. The ML-based technique is to estimate the discrete time channel parameter. It is also useful for coherent demodulation. It combines unbiased channel estimation and offers better convergence of overall receiver than the conventional technique.

1.4 TRAITS AND CHARACTERISTICS

- In the mobility state, it could make the communication and send the information. As it is wireless, its related devices can be used at any places.
- This system becomes so feasible and convenient as it does not have any wires for connection, so that it eases the process of accessing the information and talking to another person.
- It is growing faster and aids in sending information rapidly. Thus, it makes the system bigger, where it supports multiple devices and handles more users. Therefore, it ensures better scalability.
- It is also less cost-effective than wired systems, which do not require any installation. Thus, it reduces the complexity level.
- Security and privacy of data are a significant factor, owing its data handling to the open nature of system. It is also very helpful in easily connecting people who are at different places.

1.5 DIVERGENT APPLICATIONS

Due to the evolutionary development of the communication system, it is used for various applications that are explained as follows:

1. **Vehicle-based application.** Transmitting news, weather reports, road or traffic condition, and other major information passed via digital broadcasting services. When it comes to personal and remote area communication [14], GPS has been used to identify the location or position of vehicles. Information has to be exchanged between two vehicles to prevent accidents, traffic jams, etc. In this application, the vehicular-based data can be transmitted from the source, such as buses, trains, and trucks.
2. **Business and IT organizations.** In recent times, Wi-Fi has become an inevitable source of supply of the data that have been used for various purposes. Majorly, this wireless-based communication can be installed in several organizations to utilize the data effectively and for business purposes also. This set-up is done in homes, buildings, stores, warehouses, etc. without the need for any physically connected wires.

3. **Infotainment.** Since it is wireless in nature, it has the potential to yield information whenever required at any respective location. It can be accessed in outdoor areas also. For entertainment, ticket booking is done by paying cash, information on which is then sent to the provider. Similarly, for education, online studies have evolved, where students participate in learning the skills.
4. **Mobile communication and IoT.** For long-range communication, we people use mobile or cell phones and satellites to talk globally. As the name IoT implies, multiple things like gadgets are interconnected over a network applied for industries, home appliances, and agriculture sectors.
5. **WSN and local area network.** Small-size sensor nodes are present in this network to fetch information such as pressure, temperature, and other data. It is then passed to the central system. Due to such beneficial traits, the WSN has been extensively utilized in environmental and traffic monitoring, whereas in local network, the devices are connected directly to the internet in homes and some public places without the use of wires.
6. **Wireless power transfer (WPT).** It sends electrical power without using cords, which have been employed in electric vehicles and medical industries. While in satellite communication satellites are used for sending and receiving information, it can be applied in TV and weather broadcasting.

1.6 PRIME TECHNICAL CHALLENGES IN WIRELESS COMMUNICATION

Albeit it generates multiple advantages, still it faces shortcomings such as in deploying the communication system more effectively. Hence, different challenges to the system are pointed out in what follows.

- ***Security, privacy, and secrecy.*** The communication system should carry confidential information from one place to another. Since it is a wireless-based service, third users or intruders can easily enter the network to mislead the communication process [15]. Here, data security gets affected. While the information is broadcast, the transmitted data is more vulnerable to eavesdropping. It also leads to privacy and secrecy issues. Owing to this, data integrity gets reduced.
- ***Resource and spectrum utilizations.*** Scarcity of resources and spectrum is the major cause of the decreased efficiency of communication systems. When using larger bandwidth, the coverage area with RF bands must be smaller, as it heavily depends on the propagation loss. Therefore, the challenge is in how the system significantly alters the structure based on multiple user scenarios. Sometimes, the radio spectrum is also stringent in doing the transmission. Handling these functions makes the system more intricate.
- ***Communication environment.*** The growing demand for wireless services is another concern. Distributing antenna arrays, many base stations, and spectrum resources are included in the system. The challenges fall under

how it uses the potential performance that gains implementation cost and complexity.

- ***Improving energy efficiency.*** Enormous sensor nodes do their function of sending information from source to destination. But some sensors have limited resources or energy for performing communication [23]. Some nodes have adequate energy, but it is not involved in the process. Thus, making the nodes more energy-efficient is quite a challenging task.
- ***Dynamic topology.*** Future networks are envisaged with the distinct topology with intense network blocks or parameters. The tricky part is where users stumble to make a connection within the structure. Owing to the dynamic entry of devices upon leaving or entering, the entire resources vary. It can cause performance degradation.

Some of the technical challenges are pointed in subsequent section.

- ***Spectrum limitations.*** Since it is limited with such agreements, communication services become less effective [24]. To overcome this, the spectrum must be enhanced and utilized properly.
- ***Limited energy.*** It not only sends information over the channel but also notices charging or energy capacity of devices. The amplifier and transmission power must be improved, yet it is critical.
- ***Multipath propagation.*** The channel is primarily used to make a connection between the transmitter and the receiver [25]. During this process, the channel may fade due to less resources or some line-of-sight issue.
- ***Noise and interference limited.*** The quality of transmission in the system gets affected when the channel is susceptible to different types of noises. It can be evaluated by the SNR. Based on value, less signal power may lead to bad link quality. Signal or channel interference is a major challenge.
- ***Channel estimation.*** As channels are the prominent part in communication systems, estimating channels is a quite challenging concern. Over the past years, different estimation techniques [26, 27] have been developed and deployed, as discussed in the preceding sections.

1.7 CONCLUSION

This book chapter has provided the historic growth of wireless communication, its types, applications, and characteristics. It has also explained the general architecture of a communication system, where it has included the multiple blocks for transmission. Also, it explained channel estimation techniques to perform for better communication over the wireless network. Additionally, different types of channel estimation approaches, like blind, semi-blind, and training-based methods, were discussed. This chapter also explained the significant traits and characteristics of the communication system to make the system more effective. The development of communication system used in multiple applications was also discussed. Finally, the most effective challenges were noted to evade system inefficiency and performance degradation in the system.

REFERENCES

[1] H. Tataria, M. Shafi, A. F. Molisch, M. Dohler, H. Sjöland and F. Tufvesson, "6G wireless systems: vision, requirements, challenges, insights, and opportunities," Proceedings of the IEEE, Vol. 109, No. 7, pp. 1166–1199, 2021.
[2] X. You, C.-X. Wang, J. Huang, X. Gao, Z. Zhang, M. Wang, Y. Huang, C. Zhang, Y. Jiang, J. Wang, M. Zhu, B. Sheng and D. Wang, "Towards 6G wireless communication networks: vision, enabling technologies, and new paradigm shifts," Science China Information Sciences, Vol. 64, 2021.
[3] A. A. A. Solyman and K. Yahya, "Evolution of wireless communication networks: from 1G to 6G and future perspective," International Journal of Electrical and Computer Engineering, Vol. 12, No. 4, pp. 3943–3950, 2022.
[4] Q. Bi, G. L. Zysman and H. Menkes, "Wireless mobile communications at the start of the 21st century," IEEE Communications Magazine, Vol. 39, No. 1, pp. 110–116, 2001.
[5] K. Tachikawa, "A perspective on the evolution of mobile communications," IEEE Communications Magazine, Vol. 41, No. 10, pp. 66–73, 2003.
[6] E. T. Tchao, J. D. Gadze and J. O. Agyapong, "Performance evaluation of a deployed 4G LTE network," Networking and Internet Architecture, Vol. 9, No. 3, pp. 165–178, 2018.
[7] N. Shaik and P. K. Malik, "A comprehensive survey 5G wireless communication systems: open issues, research challenges, channel estimation, multi carrier modulation and 5G applications," Multimedia Tools and Applications, Vol. 80, pp. 28789–28827, 2021.
[8] S. A. Mahmod, "5G wireless technologies - future generation communication technologies," International Journal of Computing and Digital Systems, Vol. 6, No. 3, pp. 139–147, 2017.
[9] C.-Lin, S. Han, Z. Xu, Q. Sun and Z. Pan, "5G: rethink mobile communications for 2020+," The Royal Society Philosophical Transactions A, Vol. 374, No. 2062, pp. 1–13, 2016.
[10] M. N. Hindia, F. Qamar, H. Ojukwu, K. Dimyati, A. M. Al-Samman and I. S. Amiri, "On platform to enable the cognitive radio over 5G networks," Wireless Personal Communications, Vol. 113, pp. 1241–1262, 2020.
[11] S. Hashima, B. M. ElHalawany, K. Hatano, K. Wu and E. M. Mohamed, "Leveraging machine-learning for D2D communications in 5G/beyond 5G networks," Electronics, Vol. 10, No. 2, p. 169 (1–16), 2021.
[12] M. Nemati, J. Park and J. Choi, "RIS-assisted coverage enhancement in millimeter-wave cellular networks," IEEE Access, Vol. 8, pp. 188171–188185, 2020.
[13] S. Zubiya Hashmi and M. Vijaya Lakshmi, "Different channel estimation techniques used in MIMO-OFDM communication systems-a survey," Journal of Emerging Technologies and Innovative Research, Vol. 6, No. 6, pp. 150–254, 2019.
[14] N. Bogdanov, J. Jansons, K. Martinsons and E. Petersons, "Estimating the speed of an integrated wireless network for transportation applications," Automatic Control and Computer Sciences, Vol. 48, pp. 274–281, 2014.
[15] C. Liang, Q. Zhang, J. Ma and K. Li, "Research on neural network chaotic encryption algorithm in wireless network security communication," EURASIP Journal on Wireless Communications and Networking, Vol. 2019, No.151, pp. 1–10, 2019.
[16] Z. R. M. Hajiyat, A. Sali, M. Mokhtar and F. Hashim, "Channel coding scheme for 5G mobile communication system for short length message transmission," Wireless Personal Communications, Vol. 106, pp. 377–400, 2019.
[17] C.-H. Kang, H.-J. Yang and H.-K. Song, "Cooperative communication system with multiple relays for performance improvement in wireless communication system," Journal of Ambient Intelligence and Humanized Computing, Vol. 10, pp. 3461–3467, 2019.
[18] J. Via and I. Santamaria, "Correlation matching approaches for blind OSTBC channel estimation," IEEE Transactions on Signal Processing, Vol. 56, No. 12, pp. 5950–5961, 2008.

[19] F. Wan, W.-P. Zhu and M. N. S. Swamy, "Linear prediction based semi-blind channel estimation for MIMO-OFDM system," 2007 IEEE International Symposium on Circuits and Systems, New Orleans, LA, USA, pp. 3239–3942, 2007.
[20] P. De, "Semi-blind sparse channel estimation using reduced rank filtering," IEEE Transactions on Wireless Communications, Vol. 17, No. 3, pp. 1418–1431, 2018.
[21] S. K. Bandari, V. M. Vakamulla and A. Drosopoulos, "Training based channel estimation for multitaper GFDM system," Mobile Information Systems, Vol. 2017, pp. 1–10, 2017.
[22] M. Kobayashi, J. Boutros and G. Caire, "Successive interference cancellation with SISO decoding and EM channel estimation," IEEE Journal on Selected Areas in Communications, Vol. 19, No. 8, pp. 1450–1460, 2001.
[23] M. Bhatnagar, "Design and analysis of spectral and energy efficient wireless communication networks," CSI Transactions on ICT, Vol. 8, pp. 181–184, 2020.
[24] N. C. Kapsalis and P. G. Cottis, "Channel allocation to multiservice wireless networks: a collocation—based approach to enhance spectrum utilization," Wireless Personal Communications, Vol. 74, pp. 359–371, 2014.
[25] L. V. Kuz'min, A. V. Grinevich and M. D. Ushakov, "Analysis of the link budget of a wireless communication channel with a multipath propagation in ultra-wideband systems," Journal of Communications Technology and Electronics, Vol. 64, pp. 787–796, 2019.
[26] R. Govil, "Different types of channel estimation techniques used in MIMO-OFDM for effective communication systems," International Journal of Engineering Research & Technology, Vol. 7, No. 7, 2018.
[27] B. Ahlawat, "Channel estimation and performance evaluation of STBC with channel coding," International Journal of Engineering Research & Technology, Vol. 1, No. 3, pp. 1–6, 2012.

2 Opportunities and Challenges in Advanced Wireless Communication

Dr Shweta B Thomas, Dr S. Sangeetha, and Dr Benedict Thomas

2.1 INTRODUCTION TO WIRELESS COMMUNICATION

Wireless communication stands as one of the most dynamic and vibrant domains within the communication field today. Although it has been an object of study since the 1960s, the past decade has witnessed an unprecedented surge in research activities within this realm. This phenomenon can be attributed to a convergence of several compelling factors. First and foremost, there has been an explosive increase in the demand for wireless connectivity without physical tethering. This demand has been primarily fuelled by the proliferation of cellular telephony, yet it is expected to be soon overshadowed by the burgeoning applications of wireless data. Secondly, the remarkable advancements in very large-scale integration (VLSI) technology have paved the way for the efficient implementation of sophisticated signal processing algorithms and coding techniques in compact, low-power form factors. Thirdly, the resounding success of second-generation (2G) digital wireless standards, with the IS-95 code-division multiple access (CDMA) standard as a prominent example, has provided tangible evidence that sound concepts from communication theory can yield significant practical benefits [1].

This surge in research activity over the past decade has contributed to the development of a far more diverse array of perspectives and tools for effective wireless channel communication. The landscape in this field continues to evolve rapidly, promising even more exciting advancements in the near future. Two fundamental challenges are described in wireless communication that distinguish it from wire line communication: fading and interference. Let's examine each of these challenges in more detail:

1. **Fading**
    - **Multipath fading.** In wireless communication, signals often take multiple paths to reach the receiver due to reflections, diffractions, and scatterings. As a result, these multiple signal paths can interfere constructively or destructively at the receiver, causing the received signal strength to fluctuate rapidly over time and space. This phenomenon is known as multipath fading.
    - **Path loss.** Wireless signals also experience attenuation as they travel through space. This attenuation, known as path loss, occurs due to

DOI: 10.1201/9781032625034-2

factors like distance, free-space loss, and obstacles in the signal path. The signal strength diminishes with increasing distance from the transmitter.
- **Shadowing.** Obstacles such as buildings, trees, and terrain can create shadow regions where the signal strength is significantly reduced. Shadowing is another aspect of fading that introduces spatial variations in signal strength.

To deal with fading, wireless communication systems employ techniques such as diversity reception (e.g., using multiple antennas), error correction coding, and adaptive modulation and coding to adapt to changing channel conditions.

2. **Interference**

- **Co-channel interference.** In wireless networks, multiple devices often share the same frequency bands. When multiple transmitters transmit signals in the same frequency band simultaneously, they can interfere with each other. This interference can occur within the identical cell (intra-cell intervention) or between cells (inter-cell interference).
- **Interference between users.** In cellular networks, multiple handlers interconnect with the main station (uplink), or vice versa (downlink). Interference can arise when numerous operators communicate at the identical time and share the same resources. This is a substantial challenge in the arrangement of wireless systems.
- **Interference management.** Methods like frequency reprocess, power control, and progressive signal processing algorithms are recycled to mitigate intervention and advance the total system routine.

Addressing these challenges is crucial in the strategy of wireless communication systems. Advanced technologies like MIMO (multiple-input, multiple-output), beamforming, interference cancellation, and arranging processes have been established to progress the reliability and effectiveness of wireless communication networks. Additionally, the use of higher-frequency bands, like mmWave, and the deployment of small cells and massive MIMO are some of the strategies employed to improve wireless communication in the face of fading and interference.

2.2 WIRELESS COMMUNICATION NETWORKS

Over the past few years, there has been a substantial surge in global mobile data traffic. Recent data composed by the International Telecommunication Union (ITU) reveals projections indicating a significant growth in worldwide mobile statistics circulation. By 2025, it is anticipated to reach a remarkable 607 exabytes (EB) per month, with expectations soaring to 5,016 EB by 2030 [2]. In terms of data consumption per user, the estimate stands at around 39 EB by 2025 and an astonishing 257 EB by 2030. Furthermore, it is projected that nearly 70% of the world's population will be subscribed to mobile services by 2025. Within this 70% statistic, it is anticipated that approximately 60% will make use of mobile internet services.

This surge in data traffic has led to escalating demands for a range of services, spanning from comprehensive coverage to ultra-high-speed wireless communication, characterized by unparalleled reliability and minimal latency. The remarkable surge in data traffic can be attributed to a diverse range of devices, encompassing individual processors (CPUs), PCs, tablets, smart mobile phones, sensor devices, and Internet of Everything (IoE) devices and solicitations. All these devices predominantly demand records volume, with a particular emphasis on video content, as opposed to conventional voice traffic. Additionally, we foresee a significant rise in the global count of internet operators, mobile phone users, machine-to-machine (M2M) networks, and interconnected devices in the foreseeable future.

To visualize these trends, please refer to Figure 2.1, which offers a comparison of internet connectivity figures in billions across four distinct categories: internet operators, mobile phone subscribers, machine-to-machine (M2M) associates, and associated devices. This comparison spans from 2018 to 2023 [3]. As depicted in Figure 2.1, projections indicate that the amount of internet users will touch 5.3 billion by 2023. Furthermore, the data indicates a nearly twofold increase in the count of machine-to-machine (M2M) networks and associated devices during the same year. For a comprehensive comparison of the whole quantity of linked devices across six distinct regions from 2018 to 2023, please refer to Figure 2.2 [3]. Remarkably, the development in the quantity of linked devices in the Asia Pacific countries (APAC) is anticipated to surge to 13.5 billion devices.

These numbers underscore the remarkable expansion of internet connectivity, which holds significant implications for various sectors, including conveyance facilities, well-being care, organization, and applications in domestic and soldierly contexts. Consequently, the existing wireless communication infrastructure faces challenges in accommodating the ongoing surge in data traffic and the emergence of new applications.

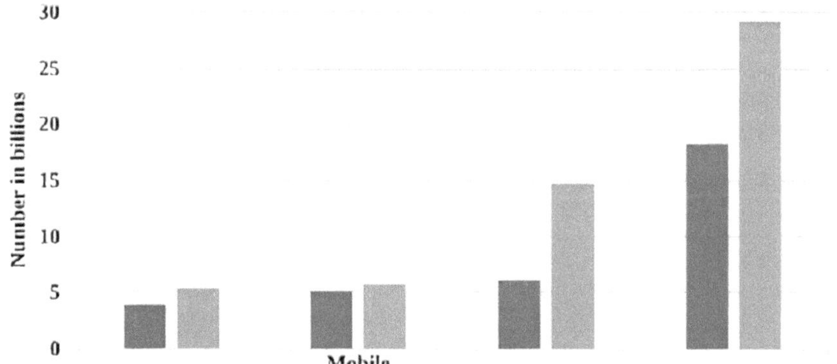

	Internet users	Mobile subscribers	M2M connections	Connected devices
2018	3.9	5.1	6.1	18.3
2023	5.3	5.7	14.7	29.2

FIGURE 2.1 The combined count of Wi-Fi operators, mobile users, machine-to-machine (M2M) networks, and interconnected devices.

Source: [4].

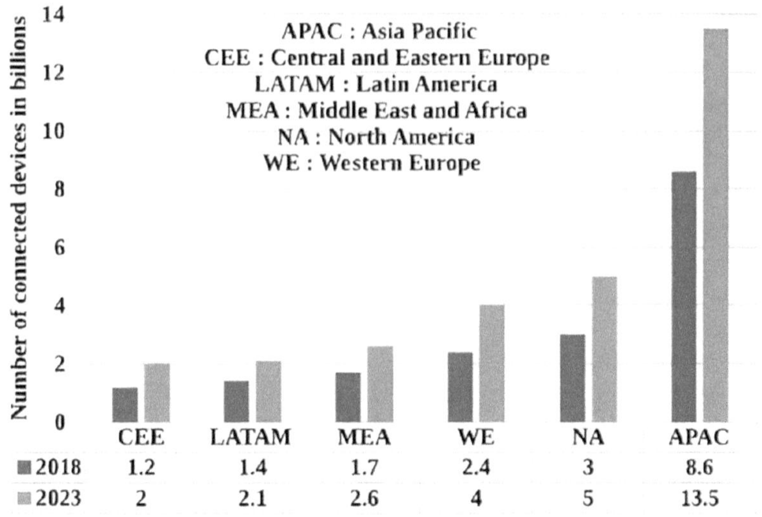

FIGURE 2.2 The count of associated devices in billions across six distinct areas.

Source: [4].

In response to the diverse and extensive requirements of internet connections, including statistics degrees, consistency, dormancy, cleverness, safety, and energy effectiveness, the subsequent group of wireless communication schemes, known as sixth-generation (6G) nets, is being introduced. Nevertheless, the successful implementation of 6G wireless communication networks will demand a series of advanced solutions at the physical layer. These solutions encompass innovative modulation techniques, progressive multi-access methods, energy garnering, edge calculating, the exploration of novel spectral groups, the combination of earthly and non-earthly public services, the deployment of cell-free enormous MIMO (multiple-input, multiple-output), blockchain, and quantum technologies, together with the addition of artificial intelligence and machine learning techniques.

2.2.1 Cellular Network Progression from 1G to 6G

To gain a comprehensive understanding of the potential advancements that 6G wireless networks may bring, it is essential to provide a concise overview of the development of mobile phone communication networks, spanning from the inception of the first generation (1G) to the current fifth generation (5G). So far, there have been five distinct groups of mobile communication schemes, each distinguished by varying standards, techniques, and capabilities, as detailed in Table 2.1. The time frames associated with each mobile generation in Table 2.1 are projected based on references [5, 6]. Notably, each generation of mobile communication networks has witnessed significant development approximately every decade.

TABLE 2.1
Key Evolution Indicator of Wireless Communications from 1G to 5G

Key parameter	1G	2G	3G	4G	5G
Period	1980–1990	1990–2000	2000–2010	2010–2020	2020–2030
Technology	Analogue voice	GSM	CDMA-2000	Wi-Fi, WiMAX, LTE	5G NR, IPv6, LAN, WAN, PAN
Multiplexing	FDMA	TDMA, CDMA	CDMA	CDMA, OFDM	OFDM, BDMA
Data rate	2.4–14.4 kb/s	14.4–64 kb/s	3.1–14.7 Mb/s	100 Mb/s-1 Gb/s	1 Gb/s and above
Bandwidth	150 kHz	5–20 MHz	25 MHz	100 MHz	1–2 GHz
Architecture	SISO	SISO	SISO	MIMO	Massive-MIMO
Main Network	PSTN	PSTN	Packet	Internet	Internet
Features	Voice	Voice, SMS	Voice, data	Video	VoIP, ultra-HD
Band type	Narrow	Narrow	Broad	Ultra-broad	Ultrawide
Highlight	Mobility	Digitization	Internet	Real-time streaming	Extra-high rate
Switching	Circuit	Circuit, packet	Packet	All packet	All packet
Handoff	Horizontal	Horizontal	Horizontal	Horizontal, vertical	Horizontal, vertical

- ***1G and 2G.*** The initial mobile network generation, known as 1G, made its debut in the 1980s, with a primary focus on vocal sound facilities. These systems delivered a data rate of 2.4 kbps. Conversely, 1G was constructed using analogue broadcast, resulting in several contests, including limited capability, unpredictable service quality, and safety vulnerabilities [7]. In response to these challenges, the second generation of mobile communication networks, or 2G, emerged in the 1990s. 2G systems remained grounded on digital modulation skills and introduced encoded statistics facilities, for example, Short Message Services (SMS), together with conventional speech communications [8]. These 2G networks adhered to the Global System for Mobile Communications (GSM) standard and offered a data rate of 64 kbps. Both 1G and 2G mobile communication networks were integrated with the Public Switched Telephone Network (PSTN), which encompasses a mix of various telecommunication systems, comprising telephone lines, swapping centres, fibre-optical systems, mobile nets, and satellite television networks.
- ***3G.*** In the early 2000s, the third generation of mobile communication networks, 3G, was familiarized to address the growing demand for a wide range of data services, including video calls and internet browsing. To meet these demands, 3G networks employed a variety of technologies, such as Wideband code-division multiple access (WCDMA), CDMA2000, unpaired spectrum utilizing time-division synchronous code-division multiple access (TD-SCDMA) with Time Division Duplex (TDD), and Worldwide

Interoperability for Microwave Access (WiMAX) [9]. These 3G networks were capable of delivering data rates of up to 2 Mb/s [10], achieved through the use of High-Speed Packet Access (HSPA) machinery.

- ***4G.*** In 2009, the fourth part of generation mobile communication, commonly referred to as 4G, emerged, showcasing the Long-Term Evolution (LTE) system. LTE networks exhibited versatility by supporting both paired and unpaired spectrums, enabling operation in both Time Division Duplex (TDD) and Frequency Division Duplex (FDD) modes. These LTE networks incorporated several technologies, including orthogonal frequency–division multiplexing (OFDM), coordinated multiple transmission/reception (CoMP), and multiple-Input and multiple-output (MIMO) techniques. These advancements were strategically designed to facilitate broader transmission bandwidth, achieve higher data rates, and provide more extensive cell phone broadband networks.

  The LTE cell phone communication scheme received further enhancements with the introduction of LTE-Advanced networks in 2011. LTE-Advanced networks expanded their operation to unrestricted range. A 4G LTE network equipped with a 2 × 2 MIMO (multiple-input, multiple-output) system could deliver an extreme data proportion of up to 150 Mb/s. In contrast, LTE-Advanced networks, utilizing a 4 × 4 MIMO system over a 100 MHz accumulated bandwidth, could achieve an impressive extreme data proportion of up to 1 Gb/s [11].
- ***5G.*** 5G networks are rapidly becoming a commercial reality, designed to offer a diverse range of capabilities. Unlike their predecessors, 5G networks can not only accommodate wide area network (WAN) bands but also cater to radios for wireless local area networks (LAN) and personal area networks (PAN). This adaptability makes 5G highly suitable for consolidating spectral resources, facilitating high-definition (HD) video streaming, and supporting a variety of data-intensive applications. In fact, 5G has the potential to deliver faster throughput compared to the LTE-Advanced (LTE-A) system, with a maximum data rate reaching around 20 Gb/s [12].

  Nonetheless, the proliferation of smart devices and applications, driven by the advancement of the Internet of Everything (IoE) networks, presents a new challenge. IoE networks encompass a broad array of applications, including independent and airborne automobiles, healthcare services, smart applications, and other time-sensitive uses. These IoE networks demand pervasive detecting and calculating abilities which might surpass the capacities of 5G nets. Additionally, the data rates reinforced through 5G networks are not likely to withstand the ever-growing explosion of data traffic. As a result, ongoing research aims to improvise the current wireless organizations to encounter the evolving demands of IoE claims and amenities.
- ***6G.*** The future of 6G wireless communication systems holds several key objectives and advancements:

  1. **High data rate and spectral efficiency.** 6G aims to achieve high data rates and spectral efficiency, particularly by operating in higher frequency bands. This will enable faster and more efficient data transmission.

2. **Energy efficiency.** 6G networks are estimated to be ten times extra energy-efficient than their 5G predecessors, contributing to greener and more sustainable communication technologies.
3. **Connectivity and coverage.** 6G will focus on increasing connectivity and providing full coverage. This includes support for satellite communication to bridge connectivity gaps.
4. **Security, secrecy, and privacy.** Ensuring robust security, secrecy, and privacy measures will be a priority in 6G networks to protect sensitive data and communications.
5. **Ultra-high reliability and low latency.** 6G will target ultra-high reliability and low-latency communication, enabling real-time and mission-critical applications.
6. **Support for high mobility.** 6G networks are estimated to support tremendously high mobility, accommodating scenarios with speeds of up to 1,000 km/hr, such as fast-moving vehicles and high-speed transport.
7. **Integration of artificial intelligence (AI) and machine learning (ML).** AI and ML will be integral to 6G, allowing for intelligent network management, optimization, and enhancing user experiences.
8. **Tactile internet communication.** 6G will aim to fully support tactile internet, enabling applications that require response times of less than 1 millisecond. This capability is crucial for applications like remote surgery and haptic communication.

In summary, the vision for 6G wireless communication networks is to deliver groundbreaking advancements in data rates, energy efficiency, coverage, security, reliability, mobility, and intelligence, positioning it as a transformative technology for the next era of connectivity and communication.

2.3 OVERVIEW OF WIRELESS CHANNEL

The outcomes of every communication organization are fundamentally estimated by the media through which information is transmitted. This media, whether it is an optical fibre, a computer's hard disk drive, or a wireless connection, is commonly known as the *communication network*. Communication channels can be divided into dual primary assemblies depending on the nature of the connection between the transmitter and the receiver.

1. **Wired channels.** Wired channels have a physical, solid connection between the transmitter and the receiver. This connection could be in the form of cables, optical fibres, or other physical pathways. Wired channels are typically more reliable and predictable compared to wireless channels.
2. **Wireless channels.** Wireless channels lack a physical, solid connection between the transmitter and the receiver. Information is transmitted through the air using electromagnetic waves. Wireless channels are characterized by their unreliable behaviour in comparison to wired channels. One key characteristic of wireless channels is the potential for rapid and unpredictable changes in their state.

The unpredictable behaviour of wireless channels, including fading, interference, and signal propagation variations, makes communication over such channels challenging. The state of a wireless channel can change rapidly, causing fluctuations in signal quality and reliability. Additionally, wireless networks can be additionally renowned by the specific propagation surroundings in which they operate. Various propagation environments include:

- **Urban.** Channels in densely populated urban areas, characterized by buildings and structures that can affect signal propagation.
- **Suburban.** Channels in suburban areas with a mix of residential and commercial zones, which have different signal propagation characteristics.
- **Indoor.** Channels within buildings, where signal propagation is influenced by walls, ceilings, and floors.
- **Underwater.** Channels for underwater communication, which poses unique challenges due to water's properties.
- **Orbital.** Channels in space, such as satellite communication, which have their own distinct characteristics.

Understanding the specific characteristics and challenges of each type of channel is essential for designing effective and reliable communication systems in various environments. Point-by-point explanation of the most commonly referenced types of channels, providing clarity on different aspects related to wireless channels in digital communication systems, is shown in what follows [13]:

- ***Propagation channel.*** The propagation channel in a communication system is the segment that exists in the middle of the receiver and transmitter antennas. This channel is primarily predisposed by the factors that affect the propagation of EM waves. It is typically assumed to be linear and reciprocal in nature, meaning, that its characteristics remain constant under different conditions. Understanding the properties and performance of the propagation channel is essential for designing communication systems that can effectively transmit and receive information. Factors such as attenuation signal fading, interference, and signal-to-noise ratio show a critical role in calculating the quality and reliability of communication over the channel. Engineers and researchers use various techniques to mitigate the effects of the propagation channel, ensuring that signals can be reliably and accurately transmitted between communication devices.
- ***Radio channel.*** This entity comprises the propagation channel, along with both the transmitting and receiving antennas. The channel exhibits linearity and reciprocity, provided the antennas remain linear, two-way, and passive. In this context, the signal is primarily exaggerated by attenuation, but the extent of attenuation within the propagation channel can vary, as it can be adjusted by the antennas in use. While the signal transmitted within this entity remains the same as in the propagation channel, it may be scaled due to the influence of the antennas.
- ***Modulation channel.*** This component encompasses the radio channel, along with all system elements, such as amplifiers and various phases of radio

frequency circuits. These components extend from the output of the modulator on the transmitter side to the input of the demodulator on the reception side. The transfer characteristics of the elements between modulator and antennas, or demodulator and antennas, play a pivotal role in determining the linearity of the system. Notably, the inclusion of amplifiers, a system component, disrupts reciprocity, since amplifiers are non-reciprocal. Consequently, the channel itself becomes non-reciprocal. The amplification process applied to the received signal at this stage introduces cumulative effects that can degrade the signal quality. Noise and interference are notable examples of these effects. While these factors may already exist in the radio channel, the addition of noise from electric circuits at this stage makes it challenging to provide a comprehensive characterization of the cumulative additive effects. The signal within this context comprises baseband symbols modulated atop the carrier frequency.

- ***Digital channel.*** This entity comprises the modulation channel, along with the modulator and demodulator components. It plays a critical role in defining bit error patterns and serves as the bridge connecting the digital baseband signal at the transmitter to the digital signal at the receiver. Notably, the channel lacks both linearity and reciprocity. At this channel level, no additional effects are introduced; instead, the impaired signal is interpreted as a bit sequence. In cases where the signal has been significantly corrupted, the interpreted bit sequence deviates from the original intended bit sequence for communication. The inputs to this channel are in the form of bits, which could also be organized into packets. These bits are subsequently structured and converted into analogue representations known as symbols. These symbols originate from the baseband signal. The analogue signal is then processed through a modulator, which further modifies it.

2.4 WIRELESS POWER TRANSFER (WPT) TECHNOLOGIES

Wireless power transfer (WPT) solutions can be categorized into different technological approaches, as described in the sketch presented in Figure 2.3 [14].

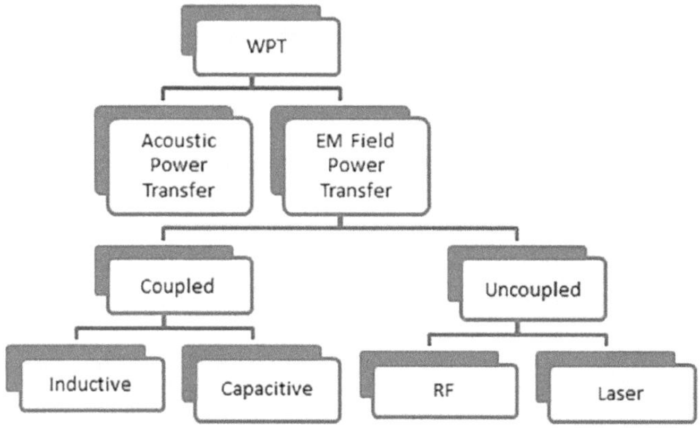

FIGURE 2.3 Classification of wireless power transfer systems.

TABLE 2.2
Characteristics of Various WPT Technologies

WPT tech.	Power transfer	Range	Frequency	Efficiency	Biological impact
Inductive, IPT	W to MW	cm	kHz to MHz	High	Minor
Capacitive, CPT	W to KW	mm/cm	kHz to MHz	High	Minor
Laser, LPT	W/KW	m/km	> THz	Medium	Medium/significant
Radio frequency, RFPT	mW/KW	m/km	MHz to GHz	Low	Medium/significant
Acoustic, APT	mW/KW	m/cm	kHz to MHz	Medium	Significant/medium

These approaches encompass electromagnetic (EM) and power transfer using acoustic waves (acoustic power transfer, APT). Within these categories, there are coupled and uncoupled systems that leverage various principles to transmit power wirelessly. Here's an overview of these approaches:

1. **Coupled systems**
    - **Inductive power transfer (IPT).** In IPT, power is transferred through magnetic fields. This method typically involves coils of wire on both the transmitter and receiver sides, and power is transferred through magnetic induction.
    - **Capacitive power transfer (CPT).** CPT involves the transfer of power through electric fields. It relies on the principles of capacitors, where electric potential differences between electrodes allow power to be transferred wirelessly.
2. **Uncoupled systems**
    - **Radio frequency power transfer (RFPT).** RFPT utilizes radio frequency waves to transfer power. It can operate over various frequency ranges and is commonly used for wireless charging of devices like smartphones and electric vehicles.
    - **Laser power transfer (LPT).** LPT involves the transmission of power through focused laser beams. This method is often used in applications where precise and directed energy transfer is required, such as in space-based solar power systems.

Each of these wireless energy transfer approaches has its own advantages and limitations. The choice of technology depends on the specific requirements and constraints of the application. Here is a summary of the advantages and disadvantages of each technique, which can be further detailed in Table 2.2 [14].

2.4.1 Electromagnetic (EM) Coupled Technologies

In inductive and capacitive wireless power systems, the central characteristic is the proximity of the transmitter (TX) and receiver (RX) components, which enables

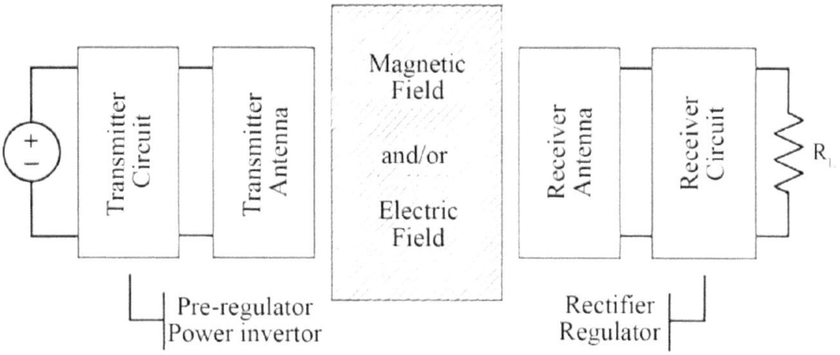

FIGURE 2.4 Arrangement of coupled WPT system.

highly efficient power transfer. The element that connects transmitter and receiver, influenced by various design factors, holds a crucial role in determining the overall link efficiency. The structure of these interconnected wireless power transfer (WPT) technologies is illustrated in Figure 2.4. To elucidate the process, initially, a power source working in tandem through a transmitter part circuit generates a sine wave that has been enhanced with adjustable frequency and amplitude. Essential building blocks like a pre-regulator, power source inverter, and waveform generator are indispensable for this function. Second, electric or magnetic fields are employed to establish the connection between the transmitter and reception antennas. Finally, the receiver circuit includes an AC/DC converter, smoothing capacitors, and if necessary, a switched-mode power supply (SMPS).

- ***Power supply.*** Due to the inability of WPT organizations to directly draw power from the AC mains, an initial conversion to DC power becomes essential. Several methods exist for implementing an AC/DC converter. In mains powered systems, a fly back converter is a common choice to attain a secure and steady DC voltage. Alternatively, the transmitter can also be powered by a battery.
- ***Transmitter circuit.*** In specific configurations, a pre-regulator is positioned between the power source and the inverter to fine-tune the amplitude of the sine wave supplied to the antenna. Pre-regulators are often employed in switched-mode power supplies (SMPSs), including step-up, step-down, and buck-boost conversions [15]. These pre-regulators exhibit minimal losses, rendering them highly efficient.
- ***Compensation network and antenna.*** To attain optimal link efficiency, it is essential to implement an appropriate circuit. Link efficiency, in this context, is quantified as ratio of energy obtained at the reception antenna to power emitted from the transmitter antenna. Achieving this balance often involves the use of a matching network, which comprises both parallel and series connections of capacitive and inductive impedances. In the literature, terms like 'primary tank' and 'secondary tank' are employed to describe the combined configuration of the compensation network and the antenna [16].

- ***Receiver circuit.*** A rectifier, in conjunction with flattening capacitors, converts the alternating current voltage generated by the antenna and compensating circuitry into a stable direct current output. Understanding the behaviour of the rectifier circuit is crucial for determining the equivalent AC-type load, particularly in strongly coupled structures, as it can significantly enhance efficiency. Incorporating a feedback link to measure and report actual subordinate currents and voltages gives the transmitter the ability to fine-tune the frequency and amplitude, optimizing overall connection efficiency. Notably, in applications like battery charging, both the DC and AC loads often vary. Rectification circuits such as half-wave and full-wave (with or without voltage doubling) rectifiers are often-used diode circuits for AC–DC conversion [16].
- ***DC/DC regulator.*** Here are numerous rationales for implementing a controller to the end of the receiver. Firstly, it is essential for maintaining a constant voltage (CV) to source components like an MCU. Furthermore, a steady current (CC) is necessary to drive elements such as LEDs. Lastly, a grouping of CC and CV is employed for charging specific devices, such as lithium batteries, adhering to a predefined CC–CV charge profile. To achieve these goals, a DC/DC regulator is typically employed, which can encompass both a low-dropout regulator (LDO) and a switched-mode power supply (SMPS). In general, SMPSs are recognized for their superior energy efficiency.

2.4.2 Inductive Coupling

Magnetic linkages represent DC/DC converters that rely on a transformer with loose coupling. In a typical system, there is a magnetic connection between the primary and secondary coils, both operated by AC current. The degree of coupling between both the coils in core-fewer transformers is determined by the connection factor, denoted as 'km'. According to Faraday's law of induction [16], the changeable magnetic flux within the secondary-side coil generates the force of having electromotive nature. Efficiency is notably influenced via resistive losses in the coils and the interconnecting factor among them. The joint link, which hinges on the mutual-type inductance (M) in henrys and the self-inductance of the transmitting and receiving-side coils (L1 and L2) in henrys, can be calculated using equation (1). The subscript 'm' signifies that this operation is based on the philosophies of AC nature magnetic fields.

$$km = \frac{M}{\sqrt{L1.L2}} \tag{2.1}$$

2.4.3 Electrodynamic Coupling

An electromagnetically coupled wireless power transfer (WPT) system enables the transfer of generated energy from the transmitting coil to an electro-mechanical receiver by utilizing magnetic fields of minimal amplitude and frequency. The effective range for this energy transfer typically extends from a few millimetres to centimetres. In this set-up, a transmitter produces a rotational force to the permanent-type magnet positioned to the receiver side. The magnitude of this force relies on the separation and alignment of the components of receiver and transmitter. This motorized form of energy is subsequently transformed into electrical type of energy by means of various transducers, namely, electrodynamics, electrostatic, and piezoelectric transducers. It is important to note that the quantity of energy transferred is typically in the range of a small number, namely, hundred microwatts. In contrast to inductively attached WPT organizations, this approach exhibits relatively lower efficiency, typically falling within the range of 4.1% to 12% [17]. This is primarily attributed to the system's operation at a low frequency, necessitating the use of coils through a considerable amount of windings.

2.4.4 Capacitive Coupling

This approach relies on the interaction of electrical grounds among the plates to facilitate transfer of energy. As seen in Figure 2.5, a standard system comprises a pair of capacitors and four metal plates. When an alternating voltage is applied to this system, it produces an electrical field. On the receiving end, this electrical field produces the current. In sufficiently high value of frequencies, all these capacitors are effectively auctioned as conductors. This scheme offers several advantages, including lower costs for both, that is, the receiver and the transmitter; concentrated susceptibility to proficiency loss in the existence of nearby metallic substances; and a more compact form factor compared to inductive power transfer (IPT). The separation among both the plates can vary as little as millimetres to tens of centimetres.

However, it is worth noting that the maximum field strength, leading to electric breakdown, limits the power density gap for capacitive power transfer (CPT) in the air to a level approximately 400 times lower than that of IPT. When attempting wireless power transfer over longer distances, such as 15 cm, various challenging techniques are required. These might involve the use of large voltages, larger plates, higher transaction frequencies, and increased electrical fields. It is essential to be aware that elevated electric fields can also raise safety concerns for the surrounding environment [18].

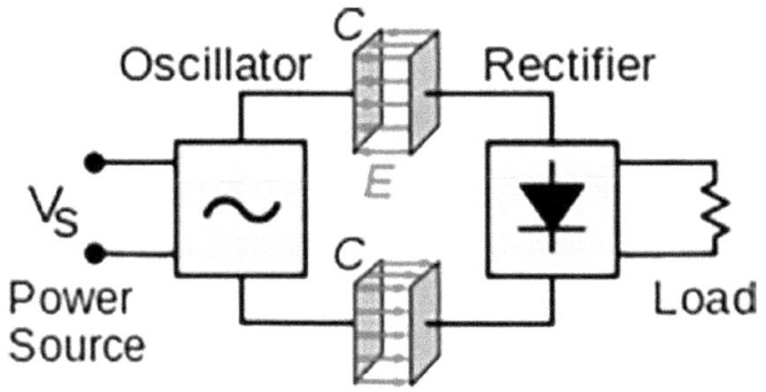

FIGURE 2.5 Capacitive coupled WPT system.

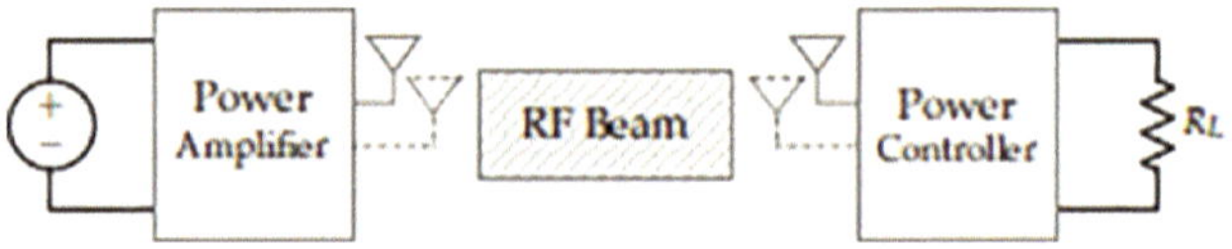

FIGURE 2.6 RFPT system.

2.4.5 Electromagnetic (EM) Uncoupled Technologies

The following section covers a brief description of WPT technologies that are uncoupled. Radio frequency and laser transmission of power are specifically addressed. In contrast to paired WPT, the receiver and the transmitters are present far away from each other. In contrast to the non-radioactive coupled systems covered in earlier sections, these organizations are similarly known as radioactive WPT.

- ***RF power transfer.*** Due to the rising usage of wireless, low-power strategies, RF power transfer (shown in Figure 2.6) has recently attracted a lot of attention. Because radio waves are omnipresent, power can be collected through a broad region, in the unreachable areas. Radio frequency energy has the smallest power concentration in comparison with other renewable energy collecting techniques like solar energy, thermal energy, and resonance energy [19]. RF energy power density ranges between 0.2 $nWcm^{-2}$ and 1 Wcm^{-2} [20]. In practise, this outcome is in the μW range in the reception side.
- ***Laser power transfer.*** Committed organizations for photosensitive transfer of power, known as optical wireless power transmission (OWPT), can be set up in the same way that RF systems are. The domestic model can be viewed as an OWPT scheme, where it is evident that the obtained power from a tiny photovoltaic (PV) system is insufficient because of light source's diversified and diffusive nature. Because laser light is concentrated in a resolute tiny beam of high power density, replacing the light source with a laser opens up new possibilities. As a result, the laser beam may be obtained at great distances utilizing a little surface [21]. This configuration is known as LPT, and it can give a few mWs to several kWs of power to a device.

2.5 SIMULTANEOUS WIRELESS INFORMATION AND POWER TRANSFER (SWIPT)

In current years, there has been a growing challenge in mobile communication systems concerning the limitations in battery capacity and charging speed. To address this issue, energy harvesting (EH) has emerged as an encouraging method to substantially increase the operational lifespan of battery-dependent devices. Between the various EH equipment, one particularly noteworthy approach is SWIPT, which seeks to harness a portion of the energy supported by the wireless communication

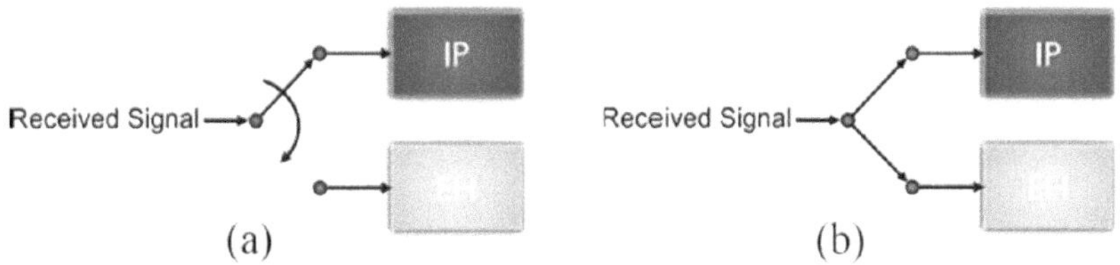

FIGURE 2.7 (a) Time switching and (b) power splitting.

signals. SWIPT has found successful applications, notably in energy-constrained relay devices primarily powered by harvested energy, referred to as EH relays. This technology has garnered significant attention from both academic and industry sectors.

Differing from energy cooperation approaches, SWIPT entails the transmission of both energy and statistics above the similar frequency group. Notably, a study in [22] expanded the examination of essential trade-offs in SWIPT to include frequency-selective networks. In the existing works, SWIPT arrangements could be categorized into two main groups. In the first group, it is presumed that the receivers possess the capability to excerpt evidence and simultaneously crop energy from the received signals [22]. However, it is important to note that current practical circuitry for energy harvesting from wireless signals is not yet equipped to unswervingly process the statistics embedded in these signals [23]. This has prompted the exploration of the additional group, in which the receivers employ either time switching (TS), as illustrated in Figure 2.7a, or power splitting (PS), as depicted in Figure 2.7b, to coordinate the processes of information processing (IP) and energy harvesting (EH). Here, a portion of the received signal's period or authority is assigned for evidence processing, although the residual portion is reserved for energy harvesting.

2.5.1 Cooperative Communication and Relay Network

Supportive communicating has occurred as a strategy to expand the coverage in wireless communication networks. This innovative approach not only extends the reach of wireless connections but also holds the potential to significantly enhance energy productivity by reducing the distances among transmitting (Tx) and receiving (Rx) devices within every connection. Within the realm of cooperative relaying networks, there has been a substantial body of research focused on energy-efficient resource allocation and scheduling schemes, with the aim of striking the right balance between spectrum efficiency (SE) and energy efficiency (EE), as documented in [24]. In wireless networks featuring multiple relay options, relay selection (RS) serves as a real-world clarification for harnessing the benefits of cooperative diversity. This has garnered significant interest within the academic community [25–27].

Much of the existing research has concentrated towards RS problem below the postulation, which relays operates on orthogonal networks and possesses self-governing authority restrictions. In such scenarios, it becomes evident that choosing a single relay is the optimal choice in relations to range efficiency. This is because the relay through the maximum endwise signal-to-interference-and-noise ratio (SINR) can attain the maximum data amount by utilizing the whole network bandwidth completely. However, the idea of selection of the relay has been prolonged to encompass scenarios where numerous relays collaborate within a shared bandwidth, as detailed in [26, 27]. In such cases, the traditional approach of single relay selection (SRS) is no longer the best strategy. Choosing numerous relays becomes advantageous, as it leads to an enhanced signal-to-noise ratio (SNR) at the receiver side without the need to divide the network bandwidth, as is the case with orthogonal network distribution.

2.5.2 Application of SWIPT in Relay Network

Energy harvesting (EH) relaying is an amalgamation of simultaneous wireless information and power transfer (SWIPT) and relay technology. In classic EH relay systems, the relays are the sole components equipped with SWIPT capabilities. To elucidate the concept of EH relaying, consider the following straightforward example: In a three-node half-duplex (HD) relay system comprising a source, a relay, and a target, the source diffuses signs to the relay during the initial half of the broadcast sequence. The relay, in turn, segregates the received signals, either by time allocation or power division, into two distinct portions – one for information transmission, and the other for energy harvesting. In the latter half of the transmission cycle, the relay relays the acquired statistics to the terminus by means of the collected energy.

This approach stands in stark contrast to traditional relays, which rely on their individual energy resources to simplify communication among other nodes. In the case of EH relays, their operational reliance on batteries is notably diminished or even entirely eliminated, as they primarily or exclusively harness energy from the signals they receive.

2.6 GRAND CHALLENGES AND OPPORTUNITIES FOR ADVANCED WIRELESS COMMUNICATION

Wireless communications have played a transformative role in our daily lives, initially connecting people without the need for physical tethers and subsequently linking individuals to the vast expanse of the internet. This has led to the establishment of a fully interconnected society. In the past couple of decades, the pervasive availability of mobile broadband (MBB) admittance through the internet has becomes the major subject in wireless communications. Freshly, this theme has transitioned into a tangible authenticity with the overview of progressive communication techniques such as multiple-input, multiple-output (MIMO), multi-carrier broadcast, channel-adaptive transmission, and more. These scientific advancements have laid a robust groundwork for forthcoming improvements in MBB amenities [28].

However, the rising demand for exceptionally high-data-rate solutions, along with the rise of sophisticated Internet of Things (IoT) and Industry 4.0 uses, presents new practical hurdles to wireless communications. For a varied array of linked devices, modern Internet of Things apps require ultra-reliable, low-latency, and highly environmentally friendly connectivity. Precision localization and multidimensional sensing are required in future human-centred solutions. The seamless combination of computers, interaction, and control with the use of artificial intelligence and machine learning is critical for achieving Industry 4.0 goals [29].

The future of wireless communication systems must excellently sustenance a worldwide and pervasive cyber-physical substructure catering to a wide range of presentations, each demanding innovative system structures, range admittance strategies, and source distribution resolutions. This must be done while considering energy efficiency, security, and privacy concerns [6]. To successfully distribute ultra-high data charges, immense connection, and comprehensive reportage while obliging vastly dissimilar excellence of provision necessities, vastly inventive technical explanations are indispensable for tackling the various challenges confronting wireless communications. The forthcoming sections outline several of the significant challenges inherent to future wireless systems.

2.6.1 Challenge 1: Safety, Confidentiality, and Isolation

Wireless communication systems are increasingly tasked with the transmission of trustworthy statistics. However, the inherent distribution nature of wireless transmissions poses a significant vulnerability to eavesdropping, putting the transmitted information at risk. To effectively support highly sensitive applications requiring stringent secrecy measures, there is a pressing need for novel transmission technologies aimed at not only safeguarding but also enhancing information security. One promising avenue for enhancing the security of wireless transmissions involves delving into the intricacies of wireless propagation, a domain known as physical layer security technology [30]. The key challenge lies in how to harness the full potential of this exploration while preserving the privacy of authorized users. The wireless community remains deeply committed to the ongoing pursuit of innovative transmission solutions that bolster both secrecy and privacy [31].

2.6.2 Challenge 2: Capitals and Range Consumption

The range available for wireless communications is progressively dwindling, prompting the investigation of fresh spectrum bands, such as millimetre wave and terahertz (THz). These higher-frequency bands offer more extensive bandwidth; however, their coverage area is typically more limited due to elevated propagation loss. To address this limitation, directional beamforming transmission technology has emerged as the prevailing mitigation strategy. The key challenge lies in the efficient adaptation of beamforming directions to the evolving environment, particularly in complex multiuser scenarios [32].

2.6.3 Challenge 3: Infrastructure

In response to the escalating demand for wireless services, an alternative approach to enhancing resource utilization in wireless systems involves the adoption of innovative system structures. The deployment of a significant range of collective cell-free system organization offers a solution to circumvent the limitations associated with the stationary source distribution of outmoded cellular arrangements [33]. Leveraging a combined method of signal processing empowered by cloud computing approach enables numerous base stations to work together in serving numerous operators by circulated antenna groups. This collaborative effort enhances spectrum utilization efficiency and allows for dynamic responsiveness to changes in service requirements and propagation conditions. However, the challenge is in realizing the full potential of performance improvements while maintaining acceptable system complexity and implementation costs. This encompasses factors such as instantaneous channel assessment, bringing together equally in the frequency and time domains, and coordinated dispensation at interconnected stations [34]. Substantial research endeavours are currently underway to strike the right balance between performance, complexity, and cost.

2.6.4 Challenge 4: Integration of Wireless Statistics and Power Transferral

In anticipation of the substantial energy demands posed through the enormous quantity of associated nodes in forthcoming wireless systems, multiple approaches have been put forth to establish self-sustaining communication structures. One of these strategies involves the incorporation of wireless information and power transfer (WIPT), which offers a practical means of replenishing power for wireless expedients. This approach holds considerable promise in addressing the power constraints of wireless networks, particularly in light of the imminent challenges posed by the subsequent group of IoT devices, which will demand even additional command because of their extensive computational requirements [35].

It is important to note that the broadcast range of wireless energy transmission is inherently inadequate. The supreme energy productivity achievable in the long distance typically does not surpass 50%. Consequently, there is a pressing need for further research in the long-distance zone to enhance the directionality and energy broadcasting efficiency, especially across various broadcasting technologies.

2.6.5 Challenge 5: Wireless Access Methods

Upcoming peers of wireless communication organizations must encounter a comprehensive set of demanding requirements, including high spectral efficiency, minimal latency, extensive device connectivity, rapid data rates, exceptional reliability, equitable user treatment, robust quantity, provision for miscellaneous quality-of-service (QoS) standards, power efficiency, and a substantial lessening in costs [36]. To address these heterogeneous demands, non-orthogonal multiple access (NOMA) has emerged as a compelling clarification. NOMA leverages non-orthogonal designs to sustenance additional operators than the accessible period, frequency, or code-domain means would traditionally allow. This approach enhances spectral efficiency and facilitates extensive connectivity within a limited number of resource blocks.

Additional encouraging wireless entrance method is rate-splitting, which serves as a robust non-orthogonal broadcast policy with efficient intervention organization [37]. In this approach, a portion of interference is decoded, while the remainder is treated as noise, in contrast to the two predictable intervention modification systems that either entirely give interference as clatter or completely decode it. The key challenges associated with these non-orthogonal approaches primarily revolve around reducing implementation complexity, integrating them with developing presentations such as intellectual reproducing planes and unmanned aerial vehicles (UAVs) and ensuring safety provisions to brand them practical and applicable in real-world systems.

2.6.6 Challenge 6: Modulation and Coding

In the context of future wireless networks, the development of novel coding and modulation schemes plays a pivotal role in translating the anticipated theoretical gains into practical realities. Specifically, advanced channel coding and modulation techniques, in conjunction with integrated space–air, ground, and underwater networks, can be instrumental in realizing comprehensive connectivity, accessible everywhere and at all times. Furthermore, to attain significantly improved spectral efficiency in upcoming wireless networks, a strategic approach involves the adoption of superior modulation scheme and channel coding arrangements, or the introduction of novel modulation means.

Moreover, the jointly designed modulation and channel schemes hold promise in minimizing information loss that can occur when they are constructed independently, where detection and decoding processes are performed separately. In this context, the comprehensive design of channel coding, intonation, recognition, and interpreting – targeting high statistics rates, small expectancy, high dependability, and wider bandwidths – warrants deep examination to guarantee effective enactment in the subsequent generation of the wireless networks [38].

2.6.7 Challenge 7: Funds and Interface Organization

Future wireless networks are poised to introduce new resource allocation and interference management strategies to accommodate the growing array of emerging applications. Within this context, machine learning–based techniques have garnered substantial attention for their capacity to enhance system performance while simultaneously reducing computational overhead. Real-time adaptability further positions ML-based strategies as a promising solution for addressing the demanding requirements of upcoming applications and services [39]. In such scenarios, the system will be intricately contrived with circulated machine learning methods integrated into various stations, and their rapid adaptation to evolving network conditions will serve as a crucial performance metric.

2.7 FUTURE OPPORTUNITIES IN WIRELESS COMMUNICATION

Machine learning and artificial intelligence methods are poised to assume pivotal roles in shaping the landscape of future 6G networks. Artificial intelligence, in essence, imparts human-like behaviours to machines, enabling them to make informed decisions and accomplish specific objectives. Machine learning, as a subset of artificial intelligence, empowers supercomputer managers to perceive, acquire information, think, and brand bright judgements created by collected data. Classically, machine learning (ML) acquires crucial insights from datasets or through communications with their environments [40].

The integration of machine learning and artificial intelligence methods is anticipated to permeate the optimization, planning, and scheme of forthcoming 6G wireless communication systems. This encompassing transformation aims to support a myriad of critical services, including healthcare, business engineering, virtual schooling, virtual shopping, virtual betting, and consumer amenities [41]. Moreover, these techniques will form the bedrock of future 6G wireless communication systems, endowing them with capabilities for intelligent resource management, policymaking, adept approximation and finding, smooth routing, computerized conservation, network processes, and management [42].

To unlock the full potential of machine learning and artificial intelligence in wireless communication systems, their seamless integration into future 6G networks is essential. This is especially crucial for meeting the stringent quality of service requirements inherent to the strategy of wireless communication systems. A nuanced indulgence of the ultimate routine bounds when applying artificial intelligence and machine learning is indispensable.

Conventional artificial intelligence and machine learning algorithms may face challenges when implementing them in portable or wearable devices, primarily because of their difficulty and consumption of energy. Consequently, the development of efficient algorithms suited for small devices becomes imperative. Striking a balance between prediction accuracy and training complexity is another consideration. Moreover, addressing practical encounters connected to the prolonged meeting times of machine learning processes necessitates additional research to identify well-organized algorithms with swifter convergence.

REFERENCES

[1] D. Tse and P. Viswanath, Fundamentals of Wireless Communication; Cambridge University Press; Cambridge, 2005.

[2] IMT Traffic Estimates for the Years 2020 to 2030, M series, Mobile, radio determination, amateur and related satellite services document ITU 0–2370, 2015.

[3] Cisco, "Cisco annual internet report (20182023)," White Paper, 2020. Available online at: www.cisco.com/c/en/us/solutions/collateral/executiveperspectives/annual-internet-report/white-paperc11-741490.html.

[4] M. Alsabah, M.A. Naser, B.M. Mahmmod, S.H. Abdulhussain, M.R. Eissa, A. Al-Baidhani, N.K. Noordin, S.M. Sait, K.A. Al-Utaibi, and F. Hashim, "6G wireless communications networks: a comprehensive survey", IEEE Access, vol. 9, pp. 148191–148243, 2021. doi:10.1109/ACCESS.2021.3124812.

[5] T. Huang, W. Yang, J. Wu, J. Ma, X. Zhang, and D. Zhang, "A survey on green 6G network: architecture and technologies," IEEE Access, vol. 7, pp. 175758–175768, 2019.

[6] S. Dang, O. Amin, B. Shihada, and M.-S. Alouini, "What should 6G be?" Nature Electron., vol. 3, no. 1, pp. 20–29, 2020. doi:10.1038/s41928-019-0355-6

[7] A. Gupta and E.R.K. Jha, "A survey of 5G network: architecture and emerging technologies," IEEE Access, vol. 3, pp. 1206–1232, 2015.

[8] K. David and H. Berndt, "6G vision and requirements: is there any need for beyond 5G?" IEEE Veh. Technol. Mag., vol. 13, no. 3, pp. 72–80, 2018.

[9] S. Chen, J. Zhao, and Y. Peng, "The development of TD-SCDMA 3G to TD-LTE-advanced 4G from 1998 to 2013," IEEE Wireless Commun., vol. 21, no. 6, pp. 167–176, 2014.
[10] P. Sharma, "Evolution of mobile wireless communication networks-1G to 5G as well as future prospective of next generation communication network," Int. J. Comput. Sci. Mobile Comput., vol. 2, no. 8, pp. 47–53, 2013.
[11] I.F. Akyildiz, D.M. Gutierrez-Estevez, R. Balakrishnan, and E. Chavarria-Reyes, "LTE-advanced and the evolution to beyond 4G (B4G) systems," Phys. Commun., vol. 10, pp. 31–60, 2014.
[12] C.-X. Wang, F. Haider, X. Gao, X.-H. You, Y. Yang, D. Yuan, H.M. Aggoune, H. Haas, S. Fletcher, and E. Hepsaydir, "Cellular architecture and key technologies for 5G wireless communication networks," IEEE Commun. Mag., vol. 52, no. 2, pp. 122–130, 2014.
[13] R. Steele and L. Hanzo (Eds.), Mobile Radio Communications; John Wiley & Sons Ltd: New York, 2000.
[14] J. Van Mulders, D. Delabie, C. Lecluyse, C. Buyle, G. Callebaut, L. Van der Perre, and L. De Strycker, "Wireless power transfer: systems, circuits, standards, and use cases," Sensors, vol. 22, no. 15, p. 5573, 2022. doi:10.3390/s22155573.
[15] Wireless Power Consortium, Qi Specification Power Delivery; V1.3.; Wireless Power Consortium: Piscataway, NJ, 2021.
[16] K. Schuylenbergh and R. Puers (Eds.), Inductive Powering; Analog Circuits and Signal Processing; Springer: Dordrecht, 2009.
[17] V.R. Challa, J.O. Mur-Miranda, and D.P. Arnold, "Wireless power transmission to an electromechanical receiver using low-frequency magnetic fields," Smart Mater. Struct., vol. 21, p. 115017, 2012.
[18] C. Lecluyse, B. Minnaert, and M. Kleemann, "A review of the current state of technology of capacitive wireless power transfer," Energies, vol. 14, p. 5862, 2021.
[19] H. Sun, M. Yin, W. Wei, J. Li, H. Wang, and X. Jin, "MEMS based energy harvesting for the internet of things: a survey. Microsyst. Technol., vol. 24, pp. 2853–2869, 2018.
[20] M. Cansiz, D. Altinel, and G.K. Kurt, "Efficiency in RF energy harvesting systems: a comprehensive review," Energy, vol. 174, pp. 292–309, 2019.
[21] D.H. Nguyen, C. Qin, T. Matsushima, and C. Adachi, "Thing-to-thing optical wireless power transfer based on metal halide perovskite transceivers," arXiv 2020, arXiv:abs/2009.06163.
[22] P. Grover and A. Sahai, "Shannon meets Tesla: wireless information and power transfer," in Proc. Intl. Symp. Information Theory, Austin, TX, pp. 2363–2367, June 13–18, 2010.
[23] X. Zhou, R. Zhang, and C.K. Ho, "Wireless information and power transfer: architecture design and rate-energy tradeoff," IEEE Trans. Commun., vol. 61, no. 11, pp. 4754–4767, 2013.
[24] K. Singh, A. Gupta, T. Ratnarajah, and M.-L. Ku, "A general approach toward green resource allocation in relay-assisted multiuser communication networks," IEEE Trans. Wireless Commun., vol. 1276, no. c, p. 1, 2017.
[25] H. Cui, M. Ma, L. Song, and B. Jiao, "Relay selection for two-way full duplex relay networks with amplify-and-forward protocol," IEEE Trans. Wireless Commun., vol. 13, no. 7, pp. 3768–3777, 2014.
[26] L. Blanco and M. Najar, "Subset relay selection in wireless cooperative networks using sparsity-inducing norms," in Intl. Symp. Wireless Communications Systems, Barcelona, pp. 501–505, August 26–29, 2014.
[27] Y. Jing and H. Jafarkhani, "Single and multiple relay selection schemes and their achievable diversity orders," IEEE Trans. Wireless Commun., vol. 8, no. 3, pp. 1414–1423, 2009.
[28] S. Chen, Y.-C. Liang, S. Sun, S. Kang, W. Chen, and M. Peng, "Vision, requirements,

and technology trends of 6G—how to tackle challenges of system coverage, capacity, user data-rate and movement speed," arXiv 2020, arXiv:2002.04929.
[29] M. Wollschlaeger, T. Sauter, and J. Jasperneite, "The future of industrial communication: automation networks in the era of the internet of things and industry 4.0," IEEE Indus. Electron. Mag. vol. 11, pp. 17–27, 2017. doi:10.1109/MIE.2017.2649104.
[30] Y.-S. Shiu, S.Y. Chang, H.-C. Wu, S. C.-H. Huang, and H.-H. Chen, "Physical layer security in wireless networks: a tutorial," IEEE Wireless Commun., vol. 18, pp. 66–74, 2011. doi:10.1109/MWC.2011.575 1298
[31] B. Nour, K. Sharif, F. Li, and Y. Wang, "Security and privacy challenges in information-centric wireless internet of things networks," IEEE Secur. Privacy, vol. 18, pp. 35–45, 2020. doi:10.1109/MSEC.2019.2925337.
[32] Z. Zhang, Y. Xiao, Z. Ma, M. Xiao, Z. Ding, X. Lei, G. K. Karagiannidis, and P. Fan, "6G wireless networks: vision, requirements, architecture, and key technologies," IEEE Veh. Technol. Mag., vol. 14, pp. 28–41, 2019. doi:10.1007/978-3-030-01150-5.
[33] S. Buzzi, C. D'Andrea, A. Zappone, and C. D'Elia, "User-centric 5G cellular networks: resource allocation and comparison with the cell-free massive MIMO approach," IEEE Trans. Wireless Commun., vol. 19, pp. 1250–1264, 2020. doi:10.1109/TWC.2019.2952117.
[34] Y. Jin, J. Zhang, S. Jin, and B. Ai, "Channel estimation for cell-free mmWave massive MIMO through deep learning," IEEE Trans. Veh. Technol., vol. 68, pp. 10325–10329, 2019. doi:10.1109/TVT.2019.2937543.
[35] X. Lu, P. Wang, D. Niyato, D.I. Kim, and Z. Han, "Wireless networks with RF energy harvesting: a contemporary survey," IEEE Commun. Surveys Tut., vol. 17, pp. 757–789, 2015. doi:10.1109/COMST.2014.2368999.
[36] K. Yang, N. Yang, N. Ye, M. Jia, Z. Gao, and R. Fan, "Non-orthogonal multiple access: achieving sustainable future radio access," IEEE Commmun. Mag., vol. 57, pp. 116–121, 2019. doi:10.1109/MCOM.2018.1800179.
[37] B. Clerckx, Y. Mao, R. Schober, and H.V. Poor, "Rate-splitting unifying SDMA, OMA, NOMA, and multicasting in MISO broadcast channel: a simple two-user rate analysis," IEEE Wireless Commun. Lett., vol. 9, pp. 349–353, 2020. doi:10.1109/LWC.2019.2954518.
[38] M. Latva-aho and K. Keppännen, "Key drivers and research challenges for 6G ubiquitous wireless intelligence (White Paper)," in 6G Flagship, 2019.
[39] F. Tarik, M. R. A. Khandaker, K.-K. Wong, M. Imran, M. Bennis, and M. Debbah, "A speculative study on 6G," arXiv 2019, arXiv:1902.06700.
[40] K. Sharma and X. Wang, "Toward massive machine type communications in ultra-dense cellular IoT networks: current issues and machine learning-assisted solutions," IEEE Commun. Surveys Tuts., vol. 22, no. 1, pp. 426–471,2019.
[41] E. Peltonen, M. Bennis, M. Capobianco, M. Debbah, A. Ding, F. Gil-Castiñeira, M. Jurmu, T. Karvonen, M. Kelanti, A. Kliks, T. Leppänen, L. Lovén, T. Mikkonen, A. Rao, S. Samarakoon, K. Seppänen, P. Sroka, S. Tarkoma, and T. Yang, "6G white paper on edge intelligence," arXiv 2020, arXiv:2004.14850.
[42] M. Elsayed and M. Erol-Kantarci, "AI-enabled future wireless networks: challenges, opportunities, and open issues," IEEE Veh. Technol. Mag., vol. 14, no. 3, pp. 70–77, 2019.

3 Millimetre Wave for Next-Generation Networks

Opportunities, Challenges, and Possible Solutions

Mohd Javed Khan, Pooja Gaur, Saif Ahmad, and Indrasen Singh

3.1 INTRODUCTION

High-frequency radio waves in the millimetre band of the frequency spectrum are used in millimetre wave (mmWave) technology; these frequencies typically range from 30 GHz to 300 GHz [1]. It utilizes a specific, very short wavelength region of the radio frequency spectrum between 24 GHz and 100 GHz in an effort to enhance the amount of bandwidth available. The range of frequencies in the mmWave spectrum that can be used for communication is referred to as the mmWave bandwidth. The precise frequency range and regulatory factors in many nations determine the mmWave communication bandwidth. In comparison to lower-frequency bands that have typically been employed for wireless communication, mmWave frequencies often offer a substantially greater bandwidth. The Federal Communications Commission (FCC), for instance, in the United States has designated frequency ranges in the mmWave spectrum for 5G wireless connectivity. These frequency bands span the ranges of 24–28 GHz, 37–40 GHz, and 64–71 GHz [2]. These bands each offer many gigahertz (GHz) of bandwidth, enabling high data rates and expanded wireless network capacity. Consider the 28 GHz spectrum, which is frequently used for 5G deployments, to get a sense of the bandwidth that is accessible in mmWave. Depending on the particular regulatory circumstances and how the technology is used, it delivers up to 1 GHz or more of bandwidth. In order to handle applications like streaming ultra-high-quality video, virtual reality, and augmented reality, a broad bandwidth is required. The effective bandwidth that may be used in mmWave communication relies on a number of variables, including channel conditions, interference, and the design and implementation of the communication system [3]. Additionally, shorter wavelengths at higher mmWave frequencies can make signal propagation through obstacles like buildings and other obstructions difficult.

Telecommunications, short-range radar, and airport security scanners are just a few applications for millimetre wave technology [4, 5]. Applications like the wireless

DOI: 10.1201/9781032625034-3

transmission of ultra-high-definition video over short distances and communications from tiny, low-power Internet of Things devices are perfect for it [6]. Multiple-gigabit wireless systems, wireless metropolitan area networks, 5G cellular telecommunications, and wireless PAN all utilize millimetre wave technology [7]. In addition, it has promise for a number of applications, including wireless sensing, imaging, and communications. Millimetre wave technologies will be crucial to achieving the anticipated network.

3.1.1 How Do 5G and Millimetre Wave Technology Interact?

MmWave technology is one of the essential technologies that enable 5G networks to achieve faster speeds, lower latency, and larger capacity [8]. The usage of millimetre wave technology is not required by all 5G networks, though. Additionally, low band (below 1 GHz) and sub-6 GHz (between 1 GHz and 6 GHz) are frequency bands that are included in the 5G standard [9]. The coverage and penetration of these bands are greater than those of mmWave, albeit at the expense of reduced data rates and bandwidth. 5G networks can use a variety of frequency bands to offer users a variety of services and experiences, depending on the availability and demand. In densely populated areas or hotspots, millimetre wave, for instance, can be used for high-speed downloads or streaming, while sub-6GHz, for larger coverage or rural areas, can be employed [10]. Voice calls and Internet of Things devices that do not need high data rates can be used with low band.

Users must have appropriate devices that support the particular frequency bands utilized by their carriers in order to connect to millimetre wave 5G networks [11]. For its millimetre wave 5G network, Telstra, for instance, employs the 26 GHz spectrum in Australia [12]. Additionally, users must be in close proximity to a millimetre wave base station or small cell, which are typically mounted on structures or other street furniture. Users who move around or come across obstructions may notice fluctuations or dips in their connection, because mmWave signals can be easily blocked by walls, trees, or even human bodies.

3.1.2 How Do 6G and mmWave Technology Interact?

The sixth-generation (or 6G) mobile system standard is currently under development and is anticipated to be available by the year 2030 [13]. For different applications, including remote-controlled factories, self-driving cars, and smart wearable, the 6G network will likely incorporate mmWave technology as well as terahertz radiation (300 to 3,000 GHz) [14]. These technologies will enable ultra-fast speeds of up to 1 terabit per second, ultra-low latency, and vast connections.

Following are some explanations of how mmWave technology works with 6G networks:

1. mmWave technology can offer mobile devices high-speed wireless access in crowded urban settings when the conventional microwave and lower frequency bands are too narrow or overloaded. Innovative uses for mmWave technology include holographic communication, virtual reality, and augmented reality, which call for immersive, high-resolution experiences.

2. mmWave technology is also applicable for fronthaul and backhaul connectivity between base stations and remote radio heads, as well as between base stations and core networks. This can make installing fibre-optic lines less expensive and less complicated, especially in difficult-to-reach places. In order to increase the spectral efficiency and coverage of wireless networks, mmWave technology can also support beamforming and massive MIMO techniques.
3. To achieve faster data rates and lower latency than the 5G network, the 6G network will use both mmWave technology and terahertz radiation. Since terahertz radiation has a shorter wavelength than mmWaves, it can provide a wider bandwidth and better resolution. The propagation loss, air absorption, and molecular scattering of terahertz radiation are all higher than for mmWaves. Because of this, the 6G network will need to employ cutting-edge methods like orbital angular momentum multiplexing, intelligent reflecting surfaces, reconfigurable intelligent surfaces, and quantum communication to address these difficulties.
4. In order to offer flexible and tailored services for various users and circumstances, the 6G network will also make use of software-defined networking (SDN) and network slicing technologies. Centralized management and dynamic configuration of network resources are made possible by SDN's ability to separate the control plane from the data plane of the network. Multiple virtual networks that can accommodate various quality of service (QoS) criteria and performance metrics can be created via network slicing on top of a physical network. In contrast, a network slice for eMBB can offer high data rates and ample bandwidth for leisure applications like video streaming and online gaming, while a network slice for URLLC can offer low latency and high reliability for critical applications like remote surgery and autonomous driving.

The sixth generation of wireless technology, or 6G, is anticipated to deliver data rates of several hundred gigabits per second to several terabits per second with low latency and great dependability for a variety of applications, including extended reality and 3D gaming. Nearly the entire surface of the Earth as well as nearby space will also be covered by 6G. As it can enable ultra-high-speed wireless broadband access, massive connectivity, and low latency for a variety of scenarios, including indoor wireless local area network (WLAN) or wireless personal area network (WPAN), cellular access, vehicle-to-everything (V2X), high-speed train (HST), unmanned aerial vehicle (UAV), and the new space-air-ground-sea communication scenarios, mmWave technology will play a significant role in 6G wireless communications. However, mmWave technology also faces some technical and implementation challenges for 6G communications, such as high-precision component fabrication and integration, careful design and optimization of the system architecture, signal processing, modulation schemes, and protocols, specialized measurement techniques and equipment for antennas, accurate and precise characterization of dielectric materials, and robustness and reliability of antennas against a variety of environmental factors.

3.2 OPPORTUNITIES IN MMWAVE

This section will provide a thorough overview of the opportunities in mmWave technology.

3.2.1 Communications and Mobile Devices

High-speed data transmission and the development of wireless networks beyond 5G can both be made possible by millimetre wave technology. In dynamic scenarios, such as indoor WLAN or WPAN, cellular access, vehicle-to-everything (V2X), high-speed rail (HST), unmanned aerial vehicle (UAV), and the emerging space-air-ground-sea communication scenarios, millimetre wave technology can also assist mobility. Additionally, millimetre wave technology can take the place of the current fibre-optic transmission links that link mobile base stations [15]. Millimetre waves help tech businesses test and invest in WLAN infrastructure. There is a rising need for more effective ways to transport and retain multimedia content as consumer demand for higher-quality content soars. Millimetre wave transmission of HD multimedia data is one possible approach. Ultra-high definition (UHD) video can be wirelessly sent to HDTVs using millimetre waves [16]. Devices for high-definition video transmission from digital set-top boxes to HD game stations and other sources can be equipped with tiny transmission modules.

3.2.2 Consumer and Business

Millimetre wave technology is applicable to applications such as wireless power transfer, wireless charging, gesture recognition, virtual reality, augmented reality, and smart homes. In order to take advantage of new mmWave technology potential, millimetre wave technology can also enable new programmable multifunctional surfaces with ad hoc communication protocols. A smart mirror that displays personalized information and communicates with people through gestures, for instance, can be made using millimetre waves. Intense VR/AR experiences that need low-latency, high-resolution data transmission can also be made with millimetre waves [17].

3.2.3 Healthcare

Millimetre wave technology is useful for medical imaging, diagnosis, therapy, and monitoring. Vital indicators, including blood pressure, breathing, and heart rate, may all be non-invasively and contactlessly sensed using millimetre wave technology. As an illustration, ionizing radiation can be avoided by using millimetre waves to produce high-resolution photographs of the inside of organs and tissues. Tumours, infections, and other anomalies in the body can also be discovered using millimetre waves. The condition of patients' health can also be remotely and wirelessly monitored using millimetre waves [18].

3.2.4 Industrial Automation

Industrial automation, process control, quality assurance, and security are all possible uses for millimetre wave technology [19]. High-resolution radar imaging and sensing for avoiding collisions, detecting objects, and locating them are also possible using millimetre wave technology. Millimetre waves, for instance, can be used to map out industrial settings in three dimensions and monitor the movement of people and equipment. Without endangering the products or materials, millimetre waves can also be utilized to examine their quality. Detecting intruders and unauthorized access with millimetre waves can also help industrial facilities become more secure.

3.2.5 Automotive and Transportation

Autonomous driving, intelligent transportation systems (ITS), and vehicle-to-infrastructure (V2I) communication are all possible with millimetre wave technologies in the automotive and transportation sectors [20]. Vehicle placement and navigation with a high level of accuracy can also be accomplished with millimetre wave technology. Radar sensors that can identify barriers, people, and other vehicles on the road, for instance, can be made using millimetre waves. The coordination of traffic flow and avoidance of crashes can also be accomplished by using millimetre waves to communicate with infrastructure and other vehicles. As a source of precise location data for navigation systems, millimetre waves can also be used.

3.2.6 Military and Defence

Military and defence applications for millimetre wave technology include surveillance, reconnaissance, target acquisition, tracking, identification, and jamming. Additionally, for military activities, millimetre wave technology might offer covert and secure communication. Millimetre waves, for instance, can be utilized to build radar systems that can track and identify enemy movements. Weapons with exceptional precision that may disable or obliterate adversary targets can likewise be made using millimetre waves. A communication system that is impervious to jamming and eavesdropping can be developed using millimetre waves [21].

3.2.7 Satellite Communication

Millimetre wave technology can be used for satellite communication, Earth observation, remote sensing, and space exploration [22]. For satellite-to-ground and inter-satellite communication, millimetre wave technology can also offer high-capacity links. Millimetre waves, for instance, can be utilized to build satellite networks that can offer worldwide coverage and rapid data delivery. Sensors that can track the ecology and resources of the Earth can also be made using millimetre waves. Instruments for exploring space and other planets can be built using millimetre waves as well.

3.2.8 Non-Invasive and Contactless Vital Sign Sensing

The viability of employing millimetre waves to detect human vital signs like heart rate, blood pressure, and respiration has been established by researchers [23]. Without touching the body or fastening any apparatus to it, these sensors can deliver precise and dependable data. These sensors can be applied to applications for monitoring and wellness in the medical field. For instance, scientists have created a millimetre wave sensor that can assess the human body's heart rate and blood pressure by spotting the minute movements of the chest wall brought on by the heartbeat and blood flow. The sensor uses a frequency-modulated continuous wave (FMCW) radar method to measure the difference in distance between the sensor and the chest wall while operating at 60 GHz. The sensor can measure blood pressure with an accuracy of 2 mmHg and a heart rate precision of 0.3 beats per minute (bpm). Another illustration is a millimetre wave sensor that can gauge a person's respiratory rate by observing the movement of their chest during breathing. The sensor uses continuous wave (CW) radar technology to measure the phase difference between the broadcast and received signals at a frequency of 94 GHz. The sensor can accurately measure breathing at 0.1 breaths per minute (brpm). Patients' health states can be remotely and wirelessly monitored by these devices.

3.2.9 Radar Sensing and Imaging with High Resolution

Millimetre wave radar systems have enhanced performance and functionality for a variety of uses, including industrial automation, quality control, security, collision avoidance, object detection, and localization. These radar systems can use cutting-edge signal processing and machine learning techniques to give high-resolution images and measurements of the surroundings and objects. As an illustration, scientists have created a millimetre wave radar system that can provide 3D images of industrial settings and monitor the motion of people and machinery [24]. The system uses a multiple-input, multiple-output (MIMO) radar method to create a sizable virtual array of antennas and operates at 77 GHz. The system is capable of 4.5 cm of range resolution, 1.8° of azimuth resolution, and 2.7° of elevation resolution. Deep learning techniques can be used by the system to categorize and segment the scene's objects [25]. A millimetre wave radar system is another illustration that can examine the quality of goods and materials without causing harm to them. Synthetic aperture radar (SAR) technology is used by the system, which works at 60 GHz, to produce high-resolution photographs of the targets. The system is capable of 0.5 mm range resolution and 1 mm cross-range resolution [26]. Machine learning techniques can be used by the system to find flaws and anomalies in the objects. These radar systems can improve the productivity and security of commercial operations.

3.2.10 mmWave Terahertz Communications

This is the investigation of the terahertz (THz) frequency band for wireless communications, which ranges from 300 GHz to 10 THz. THz communications may be able to provide even greater data rates and reduced latency than mmWave, in addition to novel spectroscopic, security, and biological application areas [27]. High propagation loss, molecule absorption, noise, and technology constraints are further issues with THz communications.

3.2.11 Secure and Stealthy Communication for Military Operations

In order to improve the security and covertness of millimetre wave communication networks for military operations, researchers have created new techniques and technologies. These techniques include frequency hopping, beam steering, encryption, authentication, jamming resistance, and interference reduction. These techniques can offer dependable and secure communication lines for jamming, tracking, identification, target acquisition, surveillance, and reconnaissance. For instance, researchers have created a millimetre wave communication system that may use physical layer security methods to encrypt and authenticate the delivered data. The system generates a random key that is shared by the transmitter and receiver at 28 GHz using a chaotic oscillator [28]. In order to route the signal to the target recipient and thwart eavesdroppers, the system can also use beamforming techniques. Another illustration is a millimetre wave communication system that uses adaptive approaches to withstand interference and jamming. The device uses cognitive radio technology to sense the spectrum environment and choose the best available channel while operating at 60 GHz. In order to avoid jamming, the system can also use frequency hopping techniques to shift the signal's frequency at random. These communication systems are capable of assisting secure and clandestine military operations.

3.2.12 High-Capacity Connectivity for Satellite-to-Ground and Inter-Satellite Communication

Millimetre waves may be used to offer high-capacity communications for satellite-to-ground and inter-satellite communication, according to research [29]. Global coverage and fast data transfer for satellite communication, Earth observation, remote sensing, and space exploration applications are made possible by these links. As an illustration, scientists have shown that a millimetre wave link between two low Earth orbit (LEO) satellites is capable of transmitting data at a rate of 10 Gbps. The link uses a quadrature phase shift keying (QPSK) modulation technique and operates between 71 and 76 GHz. Beam tracking methods can also be used by the link to keep the satellites aligned. Another illustration is a 2 Gbps millimetre wave link between a ground station and a satellite in geostationary Earth orbit (GEO). The link uses an orthogonal frequency-division multiplexing (OFDM) modulation technique and operates between 81 and 86 GHz. To counteract atmospheric attenuation, the link can potentially use adaptive coding and modulation techniques. For satellite applications, these networks can offer high-capacity data transmission.

3.3 CHALLENGES IN MMWAVE TECHNOLOGY

High-speed wireless communications and sensing applications in numerous fields are made possible by millimetre wave (mmWave) technology. The performance and dependability of mmWave systems are, however, constrained by a number of technological issues. These challenges include:

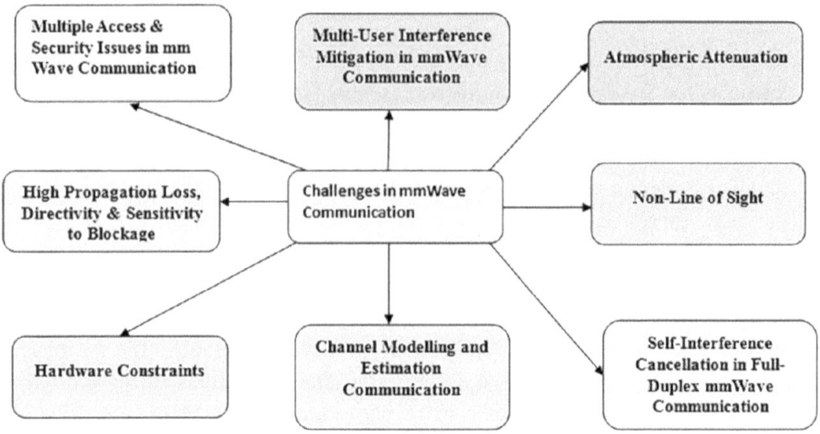

FIGURE 3.1 Challenges in mmWave technology.

3.3.1 High Propagation Loss, Directivity, and Sensitivity to Blockage

Due to this issue, mmWave signals have a smaller range and are more susceptible to fading and shadowing than lower-frequency signals. As a result, mmWave transmission can become less reliable and of worse quality [30]. The transmission power must be focused, and the link quality must be improved, for millimetre wave systems to successfully address this issue. This calls for quick beam tracking and precise channel estimates, both of which are difficult to achieve in dynamic and complicated situations. Because users and devices move around, there are obstructions and reflectors, the weather is changing, etc., the channel state information, for instance, may change quickly. As a result, adaptive and reliable algorithms must be created for millimetre wave systems so that they can estimate and track channel state information in real time and modify the beam direction as necessary [31].

3.3.2 Multiuser Interference Mitigation in mmWave–NOMA

In order to overcome this obstacle, mmWave systems must be able to handle higher inter-user interference, particularly in non-orthogonal multiple access (NOMA) settings. Multiple users can share the same spectrum resources with NOMA by employing various power levels or coding. As a result, mmWave systems may operate more efficiently and fairly for all users. Though NOMA also increases inter-user interference, particularly in mmWave regions where users' beams may overlap, this interference is more prevalent. Thus, to lessen interference and improve performance, mmWave–NOMA systems must develop efficient beamforming, power allocation, and user pairing algorithms [32]. For instance, when developing these strategies for mmWave–NOMA systems, it is necessary to take into account the channel characteristics, user preferences, demands for quality of service, etc. [33].

3.3.3 Self-Interference Cancellation in Full-Duplex mmWave Communications

This difficulty means that millimetre wave systems must reduce the potent self-interference signal that seeps from the transmitter to the receiver of the same device while employing full-duplex communication [34]. The simultaneous transmission and receipt of data on the same frequency band is made possible by the full-duplex communication technique. As a result, mmWave systems' network efficiency can be increased, and connection throughput can be doubled. Full-duplex communication, however, is also susceptible to self-interference, or the undesired signal that leaks from the transmitter to the receiver of the same device. Advanced self-interference cancellation techniques must be developed for the system in order to accomplish full-duplex millimetre wave communication. These approaches must be able to suppress the strong self-interference signal without compromising the targeted signal. In order to achieve strong self-interference suppression, for instance, mmWave systems must use passive or active cancellation methods, such as antenna separation, analogue or digital cancellation, etc. [35].

3.3.4 Multiple Access and Security Issues in mmWave Communications

This difficulty means that mmWave communication is susceptible to a number of security risks, including eavesdropping, jamming, spoofing, etc., and also faces a number of access issues, including how to divide up the available spectrum among various users and devices, how to coordinate the beam alignment and handover processes, how to control interference and congestion in dense networks, etc. In order to guarantee the security and dependability of millimetre wave communication, these problems necessitate cutting-edge solutions. For instance, millimetre wave systems must employ physical layer security approaches to safeguard the confidentiality and integrity of data transmission [36]. These techniques include artificial noise production, secure beamforming, secret key generation, etc. To further enable flexible and scalable resource allocation among various users and devices, mmWave systems must use effective multiple access schemes like hybrid division multiple access (HDMA), orthogonal frequency-division multiple access (OFDMA), code-division multiple access (CDMA), etc. [37].

3.3.5 Channel Modelling and Estimation

The *channel* is the medium used to convey a signal from a transmitter to a receiver. The communication system's performance and signal quality may be impacted. In light of this, it is crucial to model and estimate the channel well. mmWave channels, however, differ from ordinary microwave channels in that they exhibit sparse dispersion, a high bandwidth, and an angular spread. There are fewer reflectors or scatterers in the surroundings when there is sparse scattering, which reduces the possibility of multipath effects. Wider frequency ranges can be used by mmWave systems to carry more data due to their large bandwidth. Due to their narrower beams and ability to cover smaller angular regions, mmWave transmissions exhibit angular spread. Because of these factors, modelling and estimation of mmWave channels are more difficult than those of microwave channels. The great dimensionality, complexity, and variability of mmWave channels make it difficult to describe and estimate channels for mmWave systems [38].

3.3.6 Hardware Constraints

To process the wideband signals, mmWave systems need analogue-to-digital converters (ADCs) and digital-to-analogue converters (DACs) that are fast and have excellent resolution [39]. However, it is expensive, difficult to integrate these components with other components on a single chip, and they consume a lot of power. The solution to this problem is for mmWave systems to embrace novel architectures and designs that minimize the hardware requirements and enhance system performance and efficiency. Due to the large-scale antenna arrays and the high bandwidth, millimetre wave systems have a far more complex transceiver architecture than sub-6 GHz systems. As a result, millimetre wave systems must cope with a greater number of RF analogue components and data conversion procedures, which might result in more hardware issues and processing lags. In order to overcome this difficulty, millimetre wave systems must include unique system architectures that can strike a compromise between performance and complexity, like hybrid analogue–digital beamforming, lens antenna arrays, or spatial modulation.

3.3.7 Non-Line-of-Sight (NLOS) Communication

The term 'non-line-of-sight' (NLOS) refers to a circumstance in which a structure, such as a wall, tree, building, or person, blocks the direct line of sight between the transmitter and the receiver. In this instance, in order for the mmWave signal to reach the receiver, it must rely on reflections, diffractions, or scattering from the environment. But this also brings forth a number of challenges and constraints for mmWave systems, such as:

1. **Higher path loss.** Due to the numerous interactions with the barriers, NLOS paths typically exhibit more attenuation than LOS paths. At the receiver, this weakens the signal and degrades its quality. The mmWave signal can also be totally blocked or absorbed by specific substances, such as metal or concrete, which prevents NLOS communication [40].
2. **Lower reliability.** NLOS pathways are more vulnerable to interference and noise from other sources, such as other mmWave devices or ambient radiation. In turn, this causes the receiver's bit error rate (BER) and signal-to-noise ratio (SNR) to suffer. In addition, because they are dependent on the shifting surroundings and mobility of the transmitter and receiver, NLOS paths are more dynamic and unpredictable than LOS paths [41].
3. **Greater complexity.** To locate and track the NLOS pathways during NLOS communication, more complicated approaches and algorithms are needed. The angle of arrival (AOA) and angle of departure (AOD) of the reflection route, as well as the target's distance and direction, are a few examples

of measurements that must be made. Sub-Nyquist sampling, mean-shift clustering, passive reflectors, and other sophisticated signal processing and optimization techniques are used in this.

3.3.8 Atmospheric Attenuation

Attenuation in the atmosphere is the weakening of a signal brought on by the electromagnetic waves that are absorbed and scattered by air molecules and other airborne particles. One of the main problems that restrict the transmission range and dependability of millimetre wave systems, especially in humid and wet environments, is atmospheric attenuation [42].

For mmWave frequencies, the following are the primary sources of atmospheric attenuation:

1. **Oxygen absorption.** Oxygen molecules' large resonance peak, which occurs at about 60 GHz, results in a significant attenuation of mmWave transmissions in this frequency range. At sea level, the attenuation can approach 15 dB/km, which indicates that the signal power is reduced by more than 95% after travelling 1 km [43].
2. **Water vapour absorption.** The resonance peaks of water vapour molecules in the millimetre wave spectrum are at 22 GHz, 183 GHz, and 325 GHz, respectively. The amount of moisture in the air and its temperature affect how much attenuation water vapour causes. Depending on the area and the time of year, the attenuation might range from a few dB/km to tens of dB/km [44].
3. **Rain scattering.** mmWave transmissions can be scattered by raindrops in a variety of ways, which lowers the antenna's ability to receive them. Rainfall rate and drop size distribution both affect how much attenuation occurs. When it is raining heavily, the attenuation might reach hundreds of decibels per kilometre [45].

3.4 SOLUTIONS FOR THE CHALLENGES IN MMWAVE TECHNOLOGY

In this section, we examine various cutting-edge approaches to overcoming these difficulties and improving the functionality of mmWave systems. With its high data speeds, low latency, and widespread connection, millimetre wave technology is a crucial enabler for 5G and future wireless networks. However, there are numerous technical difficulties that mmWave communications must overcome. These difficulties include the following.

3.4.1 Solution for High Propagation Loss, Directivity, and Sensitivity to Blockage

The following are some potential solutions to this problem:

1. Increasing the signal strength by using **reconfigurable intelligent surfaces (RISs)** to reflect and guide millimetre wave transmissions in the required

directions. RISs are man-made objects that are capable of manipulating electromagnetic waves in a predictable way [46].
2. Extending the coverage and range of mmWave transmissions and overcoming the blocking effects by using **relays or repeaters**. The mmWave signals can be received, amplified, and retransmitted to the desired locations by relays or repeaters, which are intermediate devices [47].
3. Combining signals from various millimetre wave lines or from various frequency bands using **multi-connectivity**. By utilizing the variety and redundancy of many paths or resources, multi-connectivity is a technique that can increase the reliability and resilience of millimetre wave communication.

3.4.2 Solution for Multiuser Interference Mitigation in mmWave–NOMA

The following are some potential solutions to this problem:

1. Creating the best beam patterns for each user while minimizing inter-user interference using **hybrid beamforming**. In order to achieve great spectrum efficiency and minimal hardware complexity in millimetre wave systems, hybrid beamforming is a technique that blends analogue and digital beamforming [32].
2. Enabling the users to work together and cancel the interference by using **cooperative NOMA**. Through the use of a relay or direct link, users can share signals or information using the cooperative NOMA approach, which also enables successive interference cancellation (SIC) at the receivers [48].
3. Making use of **cognitive radio NOMA** to take advantage of available spectrum in millimetre wave bands while avoiding interference. In order to share spectrum resources with other unlicensed users, cognitive radio NOMA enables users to detect and access spectrum resources that are not being used by licenced users [49].

3.4.3 Solution for Self-Interference Cancellation in Full-Duplex mmWave Communications

The following are some potential solutions to this problem:

1. Increasing the space between the transmitter and receiver antennas of a single device in order to lessen self-interference leakage. Utilizing the path loss and directivity of millimetre wave signals, spatial separation is a technique that can increase the isolation and attenuation of the self-interference signal.
2. At the RF stage, using a copy of the transmitted signal, use analogue cancellation to remove the self-interference signal from the received signal. Using passive elements like couplers, phase shifters, attenuators, etc., analogue

cancellation is a technique that can produce high cancellation performance and low latency.

3. Making use of advanced signal processing algorithms, such as digital cancellation, to eliminate the remaining self-interference signal from the received signal at the baseband stage. By utilizing adaptive filters, neural networks, machine learning, and other techniques, digital cancellation is a technique that can further enhance cancellation performance and accuracy.

3.4.4 Solution for Multiple Access and Security Issues in mmWave Communications

For this problem, a few potential solutions include:

1. Taking advantage of the properties of millimetre wave signals, such as the creation of artificial noise, secure beamforming, secret key generation, etc., to ensure the secrecy and integrity of the data transfer. Physical layer security approaches are ways to make the physical layer more secretive and unpredictable, improving the security of millimetre wave transmission [50].
2. Utilizing efficient multiple access schemes to make use of the various millimetre wave signal dimensions will allow for flexible and scalable resource distribution among various users and devices. For example, code-division multiple access (CDMA), orthogonal frequency-division multiple access (OFDMA), and high-speed data multiple access (HDMA). By enabling several users and devices to use the same spectrum resources, multiple access schemes are techniques that can enhance the effectiveness and performance of millimetre wave communication.

3.4.5 Solution for Channel Modelling and Estimation

Here are some ideas for how to overcome this obstacle:

1. **Compressed sensing techniques.** A technique for compressed sensing is one that can recover a sparse signal from a few observations by taking advantage of its sparsity structure. Compressed sensing allows mmWave systems to maintain high accuracy and resolution while reducing channel estimation's complexity and overhead.
2. **Models for statistical and ray-tracing channels.** In order to capture the mmWave channel characteristics, mmWave systems can use statistical and ray-tracing channel models to get over the issues associated with channel modelling and estimation [51]. The channel parameters, such as path loss, delay spread, angle spread, etc., are described by statistical channel models using probability distributions. Geometric optics are used in ray-tracing channel models to replicate the pathways taken by mmWave signals as they travel through space, accounting for environmental factors such as reflection, diffraction, and scattering.

3.4.6 Solution for Hardware Constraints

Some potential solutions regarding these challenges include:

1. **Reducing the quantity of ADCs and DACs.** Reducing the quantity of ADCs and DACs in the system is one technique to lower the hardware requirements. An example would be to use a shared ADC or DAC for numerous antenna elements rather than one ADC or DAC for each antenna element, which, in a large array, could number in the hundreds or thousands. Lens antennas, subarray processing, and hybrid beamforming are some examples of techniques that can be used to accomplish this [51].
2. **Using lower-resolution ADCs and DACs in the system.** This is an additional method of reducing the hardware requirements. For instance, 1-bit or 2-bit ADCs or DACs can be used in place of 8-bit or 10-bit ADCs or DACs, which can take a lot of power and space but are more effective and compact. Techniques like quantization-aware signal processing, noise shaping, or oversampling can be used to achieve this [51].
3. **Narrowing the bandwidth of ADCs and DACs.** Using ADCs and DACs with a smaller bandwidth within the system is a third method of reducing the hardware requirements. ADCs or DACs that can handle only a small portion of the spectrum, such as a few GHz or less, can be used in place of those that can handle the entire mmWave spectrum, which can range from tens to hundreds of GHz. Frequency-division multiplexing, channelization, and sub-Nyquist sampling are a few examples of techniques that can be used to accomplish this [52].

3.4.7 Solution for Non-Line-of-Sight (NLOS) Communication

For resolving this issue, some potential solutions include:

1. **Using passive reflectors.** Redirecting the signal towards the receiver with passive metallic reflectors of various sizes and shapes is one method to improve the coverage and quality of NLOS mmWave networks. To provide the mmWave signal with alternate pathways, passive reflectors, which are simple to deploy and require no processing or power, can be placed in advantageous positions. To reflect the signal at various gains and angles, one can use, for instance, flat metallic sheets, parabolic dishes, or corner reflectors. The reflectors themselves must be carefully crafted and positioned, and the transmitter and reflector must be precisely aligned and tracked in order for this approach to work [53].
2. **Using active relays.** Using active relays [54], which can amplify and forward or decode and forward the signal from the transmitter to the receiver, is one more technique to increase the performance and dependability of NLOS mmWave networks. Beamforming, interference cancellation, and network coding are just a few of the extra features that active relays can offer, but they also need power and processing. The usage of multi-hop

relays, which can establish numerous intermediate links between the transmitter and the receiver, or full-duplex relays, which can transmit and receive simultaneously on the same frequency band, are two examples. The system's cost and complexity go up as a result of this method, and it also poses new problems, including synchronization, security, and the need to choose the right relays.

3. **Using mmWave imaging.** Using mmWave imaging [55] to create a high-resolution 3D image of the LOS and NLOS objects in the environment is a third method for enabling and optimizing NLOS mmWave communication. In order to record the size, location, shape, and material characteristics of the barriers and reflectors in the picture, mmWave imaging can take advantage of the high bandwidth and directionality of mmWave signals. In order to create a 3D map of the environment, for instance, one can use synthetic aperture radar (SAR) or inverse synthetic aperture radar (ISAR) techniques. The optimum NLOS pathways for mmWave communication can then be identified, located, and tracked using this data. This strategy, however, also calls for very advanced signal processing and optimization techniques, as well as substantial computational and storage resources.
4. **Lens antenna arrays.** This method focuses the mmWave signals in various directions by utilizing a lens-shaped antenna element rather than a lot of antenna elements. Due to the lens antenna's ability to provide high gain and directivity with a single RF chain, the transceiver architecture may be simplified, and the impact of hardware flaws is lessened. The lens antenna may handle several beams simultaneously using various frequency bands or polarization states, which can improve the system's spectral efficiency and durability [56].

3.4.8 Solution for Atmospheric Attenuation

A few solutions for dealing with problems include:

1. **Frequency selection.** Reducing atmospheric attenuation by picking an appropriate frequency band for millimetre wave transmission is very effective [57]. Several frequency bands, including those at 28 GHz, 38 GHz, and 73 GHz that have relatively low attenuation, are appropriate for millimetre wave communication systems. However, some frequency bands with extremely high attenuation, such as 60 GHz, 118 GHz, and 183 GHz, are not appropriate for long-distance transmission.
2. **Antenna designs.** The signal quality and coverage of mmWave systems can be enhanced by designing an antenna with a high gain and low loss. Through the use of a narrow beam to direct the broadcast power in the desired direction and an increase in the received power at the antenna, antenna gain can make up for route loss and atmospheric attenuation. The reflection, scattering, and ohmic losses that antenna loss causes in the antenna structure can reduce the signal quality and efficiency. Consequently, optimizing the antenna design can boost the effectiveness and dependability of millimetre wave systems [58].

3. **Diversity approaches.** By using diversity techniques, mmWave systems' resistance to air attenuation can be strengthened. The wireless channel's spatial, temporal, or frequency variety can be taken advantage of by diversity approaches to prevent fading and interference. The likelihood of signal blockage or degradation by atmospheric obscurants can be decreased, for instance, by using numerous antennas or various pathways to broadcast or receive the same signal using spatial diversity. The impact of time-varying channel conditions can be lessened by temporal diversity, which can use various time slots or codes to send or receive the same signal. The effect of frequency-selective fading or interference can be lessened through frequency diversity, which allows the same signal to be transmitted or received over a variety of frequency bands or subcarriers [59].

3.5 THE FUTURE OF MMWAVE TECHNOLOGY

With the goal of delivering massive machine-type communication (mMTC), increased mobile broadband, and ultra-reliable and low-latency communication (URLLC), millimetre wave technology is developing as a crucial enabler for the next-generation 5G networks. MmWave technology can enable these many applications by offering high-capacity backhaul and access lines, massive MIMO antennas, beamforming and beam steering algorithms, network slicing, and virtualization.

The future potential of millimetre wave technology is highly interesting, according to certain recent study articles and reports. The following are a few potential uses and trends:

1. **Joint communication and sensing.** This idea enables a single system to carry out data transmission and environmental sensing using millimetre wave signals. A millimetre wave system, for instance, has the ability to transmit and receive data as well as track the movement and presence of nearby objects and people. New applications, including gesture recognition, indoor localization, object identification, and vital sign monitoring, can be made possible by this. Waveform design, beamforming, interference control, and resource allocation are some of the issues and solutions for simultaneous communication and sensing [60].
2. **Satellite communication.** This type of wireless communication uses satellites as relay stations to offer worldwide coverage, broadband connectivity, and affordable access for rural areas. By offering high-throughput connectivity between satellites and ground stations as well as between satellites in low Earth orbit (LEO), medium Earth orbit (MEO), and geostationary Earth orbit (GEO), mmWave technology can improve the performance and efficiency of satellite communication systems [22]. Atmospheric attenuation, rain fading, the Doppler effect, synchronization, and security are some problems with satellite communication and its solutions.
3. **Internet of Things (IoT).** This is a network of actual objects, sensors, actuators, and software programmes that can talk to each other and share information via the internet. Due to its huge bandwidth, low power requirements, and compact

form factor, mmWave technology can enable the massive interconnectedness and diversity of IoT devices. Device heterogeneity, network scalability, energy efficiency, data privacy, and reliability are a few issues with the IoT and its solutions [61].

The size of the global millimetre wave technology market was evaluated at USD 2.75 billion in 2022 and is projected to increase at a compound annual growth rate (CAGR) of 39.5% from 2023 to 2030. The main drivers of this growth are the rising demand for high-speed data services, the expanding use of mmWave technology across a variety of industries, including MDA (military, defence, and aerospace), automotive, healthcare, telecommunication, and consumer electronics, as well as the ongoing research and development efforts in mmWave technology.

3.6 CONCLUSION

A state-of-the-art invention, millimetre wave technology opens up new possibilities for sensing, imaging, and security, as well as faster and more dependable wireless communication. Prior to being widely adopted, it must first clear a number of technological and legal obstacles. The significant atmospheric interference and attenuation of millimetre wave signals, which lowers the transmission's range and quality, is one of the key problems. The limited coverage and accessibility of the service are another issue due to the limited transmission and reflection of millimetre wave signals through walls and other obstructions. Thirdly, the system's capacity and compatibility are limited by the limited number and fragmented nature of the spectrum bands that are available. Beamforming, massive MIMO, hybrid network designs, adaptive modulation and coding, and spectrum sharing are a few of the methods that researchers and engineers have suggested to address these problems. Beamforming is a method that strengthens the signal and minimizes interference by directing and concentrating the signal in a particular direction using many antennas. Massive MIMO is a technique that boosts the system's capacity and effectiveness by using a lot of antennas to serve several customers at once. Millimetre wave technology is used in hybrid network architectures along with other wireless technologies like sub-6 GHz or Wi-Fi to enable flexible and seamless communication. The performance and robustness of the system are optimized by the use of adaptive modulation and coding techniques, which modify the error correction code and modulation scheme in response to channel conditions. By allowing different operators or technologies to use the same spectrum band, spectrum sharing techniques help maximize spectrum use and coexistence. By enhancing millimetre wave systems' functionality, effectiveness, and adaptability, these solutions also strive to make them compatible with other wireless technologies and allow them to coexist peacefully. By overcoming these obstacles, millimetre wave technology will be able to fully realize the potential of 5G and beyond and benefit a variety of markets and sectors, including healthcare, transportation, entertainment, education, and public safety. In order to aid future study and development in this developing sector, we hope that this chapter will serve as a useful reference and guide.

REFERENCES

[1] Pi, Z., & Khan, F. (2011). An introduction to millimeter-wave mobile broadband systems. *IEEE Communications Magazine*, *49*(6), 101–107.
[2] Tseliou, G., Adelantado, F., & Verikoukis, C. (2019). A base station agnostic network slicing framework for 5G. *IEEE Network*, *33*(4), 82–88.
[3] Sung, J. (2020). *Compressed-sensing based channel state information acquisition in mmWave hybrid beamforming communication systems* (Doctoral dissertation). The University of Texas at Austin.
[4] Menzel, W. (2010, September). Millimeter-wave radar for civil applications. In *The 7th European Radar Conference* (pp. 89–92). IEEE.
[5] Robertson, D. A., Macfarlane, D. G., Cassidy, S. L., Bryllert, T., Gandini, E., & Llombart, N. (2016, March). Submillimeter wave 3D imaging radar for security applications. In *IET Colloquium on Millimeter-Wave and Terahertz Engineering & Technology 2016* (pp. 1–5). IET.
[6] Shafique, K., Khawaja, B. A., Sabir, F., Qazi, S., & Mustaqim, M. (2020). Internet of things (IoT) for next-generation smart systems: A review of current challenges, future trends and prospects for emerging 5G-IoT scenarios. *IEEE Access*, *8*, 23022–23040.
[7] Rangan, S., Rappaport, T. S., & Erkip, E. (2014). Millimeter-wave cellular wireless networks: Potentials and challenges. *Proceedings of the IEEE*, *102*(3), 366–385.
[8] Al-Falahy, N., & Alani, O. Y. (2017). Technologies for 5G networks: Challenges and opportunities. *IT Professional*, *19*(1), 12–20.
[9] Ishteyaq, I., & Muzaffar, K. (2022). Multiple input multiple output (MIMO) and fifth generation (5G): An indispensable technology for sub-6 GHz and millimeter wave future generation mobile terminal applications. *International Journal of Microwave and Wireless Technologies*, *14*(7), 932–948.
[10] Zhang, J., Tang, P., Tian, L., Hu, Z., Wang, T., & Wang, H. (2017). 6–100 GHz research progress and challenges from a channel perspective for fifth generation (5G) and future wireless communication. *Science China Information Sciences*, *60*, 1–18.
[11] Al-Falahy, N., & Alani, O. Y. (2019). Millimeter wave frequency band as a candidate spectrum for 5G network architecture: A survey. *Physical Communication*, *32*, 120–144.
[12] Patzold, M. (2018). 5G readiness on the horizon [mobile radio]. *IEEE Vehicular Technology Magazine*, *13*(1), 6–13.
[13] Alraih, S., Shayea, I., Behjati, M., Nordin, R., Abdullah, N. F., Abu-Samah, A., & Nandi, D. (2022). Revolution or evolution? Technical requirements and considerations towards 6G mobile communications. *Sensors*, *22*(3), 762.
[14] Rikkinen, K., Kyosti, P., Leinonen, M. E., Berg, M., & Parssinen, A. (2020). THz radio communication: Link budget analysis toward 6G. *IEEE Communications Magazine*, *58*(11), 22–27.
[15] Karjalainen, J., Nekovee, M., Benn, H., Kim, W., Park, J., & Sungsoo, H. (2014, June). Challenges and opportunities of mm-wave communication in 5G networks. In *2014 9th International Conference on Cognitive Radio Oriented Wireless Networks and Communications (CROWNCOM)* (pp. 372–376). IEEE.
[16] Kleine-Ostmann, T., & Nagatsuma, T. (2011). A review on terahertz communications research. *Journal of Infrared, Millimeter, and Terahertz Waves*, *32*, 143–171.
[17] Yang, T. H., Kim, J. R., Jin, H., Gil, H., Koo, J. H., & Kim, H. J. (2021). Recent advances and opportunities of active materials for haptic technologies in virtual and augmented reality. *Advanced Functional Materials*, *31*(39), 2008831.
[18] Jamshidi, M., Lalbakhsh, A., Talla, J., Peroutka, Z., Hadjilooei, F., Lalbakhsh, P., . . . & Mohyuddin, W. (2020). Artificial intelligence and COVID-19: Deep learning approaches for diagnosis and treatment. *IEEE Access*, *8*, 109581–109595.

[19] Aslam, M. M., Du, L., Zhang, X., Chen, Y., Ahmed, Z., & Qureshi, B. (2021). Sixth generation (6G) cognitive radio network (CRN) application, requirements, security issues, and key challenges. *Wireless Communications and Mobile Computing, 2021*, 1–18.

[20] Busari, S. A., Khan, M. A., Huq, K. M. S., Mumtaz, S., & Rodriguez, J. (2019). Millimeter-wave massive MIMO for cellular vehicle-to-infrastructure communication. *IET Intelligent Transport Systems*, *13*(6), 983–990.

[21] Cotton, S. L., Scanlon, W. G., & Madahar, B. K. (2009). Millimeter-wave soldier-to-soldier communications for covert battlefield operations. *IEEE Communications Magazine*, *47*(10), 72–81.

[22] Giordani, M., & Zorzi, M. (2020, February). Satellite communication at millimeter waves: A key enabler of the 6G era. In *2020 International Conference on Computing, Networking and Communications (ICNC)* (pp. 383–388). IEEE.

[23] Iyer, S., Zhao, L., Mohan, M. P., Jimeno, J., Siyal, M. Y., Alphones, A., & Karim, M. F. (2022). mm-wave radar-based vital signs monitoring and arrhythmia detection using machine learning. *Sensors*, *22*(9), 3106.

[24] Čopič Pucihar, K., Attygalle, N. T., Kljun, M., Sandor, C., & Leiva, L. A. (2022). Solids on Soli: Millimeter-wave radar sensing through materials. *Proceedings of the ACM on Human-Computer Interaction*, *6*(156), 1–19.

[25] Gilpin, R., Bray, J., & Hefnawi, M. (2022). MIMO Radar testbed based on USRP N320/321 software-defined radios. *e-Prime-Advances in Electrical Engineering, Electronics and Energy*, *2*, 100082.

[26] Cooper, K. B., Dengler, R. J., Chattopadhyay, G., Schlecht, E., Gill, J., Skalare, A., . . . & Siegel, P. H. (2008). A high-resolution imaging radar at 580 GHz. *IEEE Microwave and Wireless Components Letters*, *18*(1), 64–66.

[27] Tripathi, S., Sabu, N. V., Gupta, A. K., & Dhillon, H. S. (2021). Millimeter-wave and terahertz spectrum for 6G wireless. In *6G Mobile Wireless Networks* (pp. 83–121). Springer International Publishing.

[28] Onuki, K., Cho, K., Horio, Y., & Miyano, T. (2022). Secret-key exchange through synchronization of randomized chaotic oscillators aided by logistic hash function. *IEEE Transactions on Circuits and Systems I: Regular Papers*, *69*(4), 1655–1667.

[29] Tachikawa, W., Suetsuna, A., Wang, M., Yoshii, K., & Shimamoto, S. (2023, March). Applying PDMA for ground-HAPS uplink network with hybrid FSO/RF communication. In *2023 IEEE Wireless Communications and Networking Conference (WCNC)* (pp. 1–6). IEEE.

[30] Niu, Y., Li, Y., Jin, D., Su, L., & Vasilakos, A. V. (2015). A survey of millimeter wave communications (mmWave) for 5G: Opportunities and challenges. *Wireless Networks*, *21*, 2657–2676.

[31] Zhang, J., Huang, Y., Wang, J., You, X., & Masouros, C. (2020). Intelligent interactive beam training for millimeter wave communications. *IEEE Transactions on Wireless Communications*, *20*(3), 2034–2048.

[32] Zhan, J., & Dong, X. (2021). Interference cancellation aided hybrid beamforming for mmwave multi-user massive MIMO systems. *IEEE Transactions on Vehicular Technology*, *70*(3), 2322–2336.

[33] Li, J., Niu, Y., Wu, H., Ai, B., Chen, S., Feng, Z., . . . & Wang, N. (2022). Mobility support for millimeter wave communications: Opportunities and challenges. *IEEE Communications Surveys & Tutorials*, *24*(3), 1816–1842.

[34] Roberts, I. P., Jain, H. B., & Vishwanath, S. (2020, June). Equipping millimeter-wave full-duplex with analog self-interference cancellation. In *2020 IEEE International Conference on Communications Workshops (ICC Workshops)* (pp. 1–6). IEEE.

[35] Morgenstern, C., Chiriyath, A. R., Dutta, A., Herschfelt, A., Rong, Y., Molnar, A. C., . . . & Bliss, D. W. (2022, March). Analog self-interference mitigation for IBFD, joint

radar-communications in vehicular applications. In *2022 IEEE Radar Conference (RadarConf22)* (pp. 1–6). IEEE.

[36] Tang, J., Wen, H., Zeng, K., Liao, R. F., Pan, F., & Hu, L. (2019). Light-weight physical layer enhanced security schemes for 5G wireless networks. *IEEE Network*, *33*(5), 126–133.

[37] Park, C., & Rappaport, T. S. (2007). Short-range wireless communications for next-generation networks: UWB, 60 GHz millimeter-wave WPAN, and ZigBee. *IEEE Wireless Communications*, *14*(4), 70–78.

[38] He, H., Wen, C. K., Jin, S., & Li, G. Y. (2018). Deep learning-based channel estimation for beamspace mmWave massive MIMO systems. *IEEE Wireless Communications Letters*, *7*(5), 852–855.

[39] Dutta, S., Barati, C. N., Ramirez, D., Dhananjay, A., Buckwalter, J. F., & Rangan, S. (2019). A case for digital beamforming at mmWave. *IEEE Transactions on Wireless Communications*, *19*(2), 756–770.

[40] MacCartney, G. R., Zhang, J., Nie, S., & Rappaport, T. S. (2013, December). Path loss models for 5G millimeter wave propagation channels in urban microcells. In *2013 IEEE Global Communications Conference (GLOBECOM)* (pp. 3948–3953). IEEE.

[41] Ford, R., Zhang, M., Mezzavilla, M., Dutta, S., Rangan, S., & Zorzi, M. (2017). Achieving ultra-low latency in 5G millimeter wave cellular networks. *IEEE Communications Magazine*, *55*(3), 196–203.

[42] Frey, T. L. (1999). The effects of the atmosphere and weather on the performance of a mm-wave communication link. *Applied Microwave and Wireless*, *11*, 76–81.

[43] Marcus, M., & Pattan, B. (2005). Millimeter wave propagation: Spectrum management implications. *IEEE Microwave Magazine*, *6*(2), 54–62.

[44] Dias-Lalcaca, P., Packham, N. J. C., & Gebbie, H. A. (1984). The effect of ultraviolet radiation on water vapour absorption between 5 and 50cm^{-1}. *Infrared Physics*, *24*(5), 437–441.

[45] Zhensen, W., Yi, Y., & Lihong, C. (1992). Monte Carlo simulation for millimeter wave propagation and scattering in rain medium. *International Journal of Infrared and Millimeter Waves*, *13*, 981–994.

[46] Ntontin, K., Boulogeorgos, A. A. A., Selimis, D. G., Lazarakis, F. I., Alexiou, A., & Chatzinotas, S. (2021). Reconfigurable intelligent surface optimal placement in millimeter-wave networks. *IEEE Open Journal of the Communications Society*, *2*, 704–718.

[47] Moerman, A., Van Kerrebrouck, J., Caytan, O., de Paula, I. L., Bogaert, L., Torfs, G., . . . & Lemey, S. (2022). Beyond 5G without obstacles: mmWave-over-fiber distributed antenna systems. *IEEE Communications Magazine*, *60*(1), 27–33.

[48] Zhu, L., Xiao, Z., Xia, X. G., & Wu, D. O. (2019). Millimeter-wave communications with non-orthogonal multiple access for B5G/6G. *IEEE Access*, *7*, 116123–116132.

[49] Tlebaldiyeva, L., Nauryzbayev, G., Arzykulov, S., & Eltawil, A. M. (2021). Performance of NOMA-based mmWave D2D networks under practical system conditions. *IEEE Access*, *9*, 160958–160974.

[50] Wu, Y., Khisti, A., Xiao, C., Caire, G., Wong, K. K., & Gao, X. (2018). A survey of physical layer security techniques for 5G wireless networks and challenges ahead. *IEEE Journal on Selected Areas in Communications*, *36*(4), 679–695.

[51] Fuschini, F., Zoli, M., Vitucci, E. M., Barbiroli, M., & Degli-Esposti, V. (2019). A study on millimeter-wave multiuser directional beamforming based on measurements and ray tracing simulations. *IEEE Transactions on Antennas and Propagation*, *67*(4), 2633–2644.

[52] Uwaechia, A. N., & Mahyuddin, N. M. (2020). A comprehensive survey on millimeter wave communications for fifth-generation wireless networks: Feasibility and challenges. *IEEE Access*, *8*, 62367–62414.

[53] Khawaja, W., Ozdemir, O., Yapici, Y., Guvenc, I., & Kakishima, Y. (2018, May). Coverage enhancement for mm wave communications using passive reflectors. In *2018 11th Global Symposium on Millimeter Waves (GSMM)* (pp. 1–6). IEEE.

[54] Biswas, S., Vuppala, S., Xue, J., & Ratnarajah, T. (2015). On the performance of relay aided millimeter wave networks. *IEEE Journal of Selected Topics in Signal Processing*, *10*(3), 576–588.

[55] Aladsani, M., Alkhateeb, A., & Trichopoulos, G. C. (2019, May). Leveraging mmWave imaging and communications for simultaneous localization and mapping. In *ICASSP 2019–2019 IEEE International Conference on Acoustics, Speech and Signal Processing (ICASSP)* (pp. 4539–4543). IEEE.

[56] Zeng, Y., & Zhang, R. (2016). Millimeter wave MIMO with lens antenna array: A new path division multiplexing paradigm. *IEEE Transactions on Communications*, *64*(4), 1557–1571.

[57] Liebe, H. J. (1985). An updated model for millimeter wave propagation in moist air. *Radio Science*, *20*(5), 1069–1089.

[58] Elfergani, I., Hussaini, A. S., Abdalla, A. M., Rodriguez, J., & Abd-Alhameed, R. (2019). Millimeter wave antenna design for 5G applications. In *Optical and Wireless Convergence for 5G Networks* (pp. 139–156). IEEE.

[59] Ghosh, A., Thomas, T. A., Cudak, M. C., Ratasuk, R., Moorut, P., Vook, F. W., . . . & Nie, S. (2014). Millimeter-wave enhanced local area systems: A high-data-rate approach for future wireless networks. *IEEE Journal on Selected Areas in Communications*, *32*(6), 1152–1163.

[60] Alloulah, M., & Huang, H. (2019). Future millimeter-wave indoor systems: A blueprint for joint communication and sensing. *Computer*, *52*(7), 16–24.

[61] Hussain, N., & Kim, N. (2022). Integrated microwave and mm-wave MIMO antenna module with 360 pattern diversity for 5G internet of things. *IEEE Internet of Things Journal*, *9*(24), 24777–24789.

4 Cognitive Radio
A Key Enabling Technology for 5G and Beyond Communication

Vishwas Srivastava and Binod Prasad

4.1 INTRODUCTION

Due to the advancement of technology, many smart devices have evolved nowadays to improve the ease of life of human beings, such as smart healthcare, autonomous machines, Industry 4.0, etc. These applications are comprised of a large number of nodes, and such proliferation of the uses of these applications has resulted in billions of nodes. Further, these smart devices have limited performance due to their size, transmission power, and bandwidth. One of the important resources in a given network is the frequency band (channel), over which the signal is transmitted and is limited in nature. Fifth-generation (5G) and beyond communication technology evolved to support such a large number of applications with a given quality of services. Demand by such a large number of nodes resulted in spectrum scarcity. Cognitive radio (CR) can play one of the important roles to mitigate the spectrum scarcity issue in 5G and beyond communication technology. CR allows dynamic spectrum access by a secondary user (unlicensed user) as long as it does not hamper the services of the primary user (licensed user). Therefore, to use a given band for secondary user (SU) transmission, detection of PU activity or presence in the given band is one of the important tasks. The performance of CR is affected due to uncertainty in the channel, such as fading and shadowing. The impact of uncertainty in the channel on sensing performance is generally measured in terms of probability of false alarm (P_f) and probability of missed detection (P_m). Different detection techniques used in the literature are energy detection, matched filter-based detection, cyclostationary process, and wavelet-based methods. In literature, cooperative spectrum sensing (CSS) has been proposed as an efficient approach to improve the sensing performance of CR networks. The sensing performance of cooperative spectrum sensing can further be improved by using machine learning (ML) and other novel technologies, such as intelligent reflecting surface (IRS). This chapter focuses on the role of machine learning (ML) to improve the spectrum sensing performance of a CR network. Further, the direction for how the IRS can assist in improving the detection performance is also highlighted. The following section gives a brief introduction to spectrum sensing, using various approaches such as cooperative, non-cooperative, and machine learning.

DOI: 10.1201/9781032625034-4

4.2 ROLE OF SPECTRUM SENSING IN COGNITIVE RADIO

Spectrum sensing (SS) is the first and most important task of the CR, where an SU finds the vacant spectrum by determining the presence of the PU. If PU is present, it means the spectrum is in use and not available for sharing; otherwise, the spectrum is available and can be used by SUs for their data transmission [1, 2]. In the literature, various methods have been suggested to perform SS [3]; some common methods are matched filter detection, cyclostationary feature detection, and energy detection. In terms of spectrum sensing, each of these techniques has certain advantages and disadvantages. For instance, a matched filter works by matching the received signal with a matched filter or predefined template corresponding to the expected signal waveform [4, 5]. It works well in low signal-to-noise ratio (SNR); however, it does require prior knowledge of the operating environment. On the other hand, cyclostationary feature detection provides an advanced form of spectrum sensing that uses the periodic nature of communication signals [6, 7]. It can be adapted to work with different types of signals and does not require any prior knowledge of the operating environment; however, its computational complexity is high, and thus, it might not be the best option for systems that have limited computing power. Compared to this, the energy detection method is very simple to implement and does not require any prior knowledge of the environment [3, 8, 9]. It compares the received signal energy with the predefined threshold to identify the vacant spectrum. Though it does not perform well in low SNR, its simplicity makes it popular among researchers. Due to these limitations, efficient SS is still a challenging task in CR and has received a lot of research interest. Various approaches to perform SS as mentioned in the literature [2, 10, 11] can be categorized as follows:

- Non-cooperative SS
- Cooperative SS
- Machine learning–based SS (ML-SS)

4.2.1 Non-Cooperative SS (NCSS)

In this section, the spectrum sensing performance by non-cooperative SU user – that is, it does share the sensing information with other SU – to determine the activity of PU is discussed. We considered a system model as depicted in Figure 4.1 with a single SU to determine the presence of PU by comparing the received signal energy with a predefined threshold which can be extended for a multiple-SU non-cooperative scenario, where each user will follow a similar sensing strategy, as discussed later.

Following [11] and [12], the received signal $r(t)$ at SU for a given time instant t can be given as

$$\begin{aligned} r(t) &= w(t) && \text{PU is absent } H_0 \\ r(t) &= h.s(t) + w(t) && \text{PU is present } H_1 \end{aligned} \tag{4.1}$$

where $w(t)$ is the additive white Gaussian noise (AWGN), $s(t)$ is the transmitted signal by PU, and h is the channel coefficient between PU and SU. The SU listens

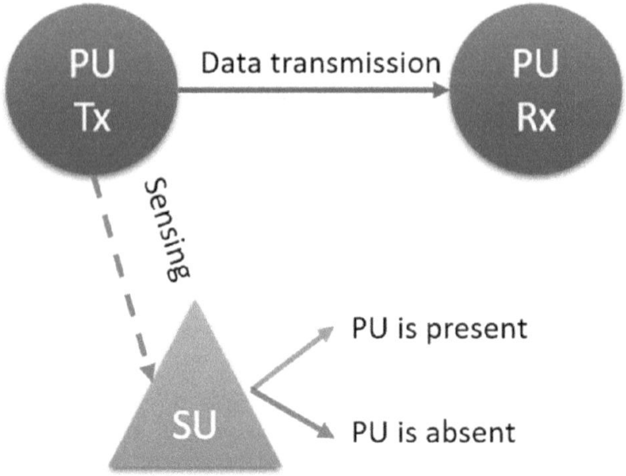

FIGURE 4.1 Non-cooperative SS.

to the licensed user band to determine the activity of PU and collects a large number of samples (P); the normalized estimated energy of the same can be given as [12]

$$E_{est} = \frac{1}{\sigma^2} \sum_{p=1}^{P} r(t)^2 \tag{4.2}$$

where σ^2 is the variance of AWGN. To determine the presence of PU, estimated energy is compared with the threshold as

$$\begin{aligned} SU_d &= 1, \text{if } E_{est} \geq \gamma_{th} \\ SU_d &= 0, \text{if } E_{est} < \gamma_{th} \end{aligned} \tag{4.3}$$

where *1* or *0* denotes the presence or absence of PU, respectively; SU_d is the decision given by SU; E_{est} is the estimated energy at SU; and γ_{th} is the predetermined threshold. Due to environmental issues such as fading coefficient, shadowing, or any uncertain parameter, SU may wrongly detect the presence of PU and lead to sensing errors, like the probability of false alarm (P_f) or probability of misdetection (P_m), which is defined as [12]:

$$\begin{aligned} P_f &= Pr(E_{est} \geq \gamma_{th} \mid \text{PU is absent}) \\ P_m &= Pr\left(E_{est} \langle \gamma_{th} \mid \text{PU is present}\right) \end{aligned} \tag{4.4}$$

Similarly, the probability of detection (P_d) is defined as

$$P_d = Pr\left(E_{est} \geq \gamma_{th} \mid \text{PU is present}\right) \tag{4.5}$$

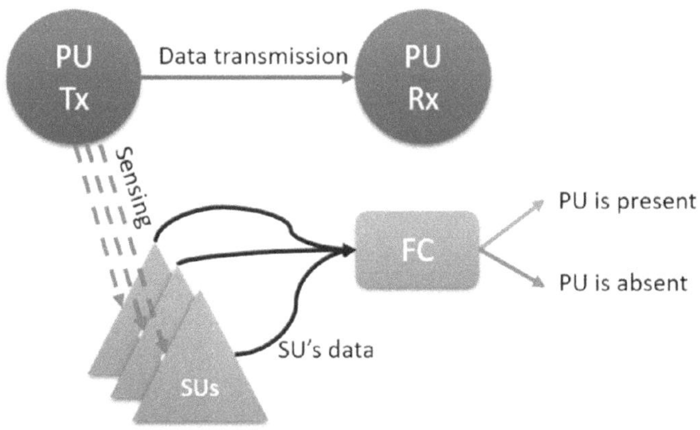

FIGURE 4.2 Cooperative SS (CSS).

where P_f denotes the presence of PU while it is absent, P_m denotes the absence of PU while it is actually present, and P_d indicates that PU is present when it is in fact present.

4.2.2 Cooperative SS (CSS)

To improve the sensing performance and mitigate the impact of fading, cooperative spectrum sensing (CSS) has emerged as a potential approach in the literature. The sensing error, that is, P_f and P_m, under CSS, is found to be less as compared to single user detection. Figure 4.2 shows a cooperative spectrum sensing scenario where multiple SUs work together to correctly identify the presence of PU. In this approach, multiple SUs send their information to the fusion centre (FC). The CSS approach can be of two types, namely, decision fusion and data fusion [13].

4.2.2.1 Decision Fusion

In *decision fusion*, each SU sends its individual binary decision about PU availability to the FC. Then FC may select different combining schemes, such as OR, AND, majority rule, or any other novel scheme, to combine the decision received from each SU and make the final decision about the PU availability. A brief discussion about each of the combining schemes is given as follows.

OR. In this method, each SU involved in the SS process independently determines if a primary user (PU) signal is there or not. Using the logical OR operator, the fusion centre must combine these distinct decisions to arrive at its conclusion. In other words, the fusion centre will assume that a PU is present if any SU detects a PU signal.

$$\begin{aligned} FC_d &= 1, \text{ If } \left(SU_d^1 \vee SU_d^2 \vee \ldots SU_d^M = 1\right) \\ FC_d &= 0, \text{ otherwise} \end{aligned} \tag{4.6}$$

where $SU_d^1, SU_d^2, \ldots. SU_d^M$ are the PU availability decisions made by each SU at FC and FC_d denotes the decision given by FC.

AND. To use this method, the FC must get independent confirmation from each of the participating SUs that a PU signal is present. The FC will conclude that there is no PU if any SU is unable to identify the signal.

$$FC_d = 1, \text{If } \left(SU_d^1 \wedge SU_d^2 \wedge \ldots. SU_d^M = 1\right)$$
$$FC_d = 0, \text{ otherwise} \tag{4.7}$$

Majority rule. Under this scheme, the determination of a PU is established by the agreement of k SUs concurrently verifying the PU's presence.

$$FC_d = 1, \text{ If} \left(\sum_{i=1}^{m} SU_d^i = 1\right)$$
$$FC_d = 0, \text{ otherwise} \tag{4.8}$$

4.2.2.2 Data Fusion

In contrast to decision fusion, each SU sends its individual estimated sensing energy to the FC in the *data fusion* approach. This approach uses square law combining (SLC), maximum ratio combining (MRC), etc.

SLC. In this method, estimated energy from all the SUs is summed together and then compared with the threshold.

$$FC_d = 1, \text{If} \left(\sum_{i=1}^{m} E_{est}^i \geq \gamma_{th}\right)$$
$$FC_d = 0, \text{ otherwise} \tag{4.9}$$

MRC. MRC is an advanced CSS method. It improves detection performance by elegantly combining the estimated signal energy from multiple SUs with particular weights.

$$FC_d = 1, \text{ If} \left(\sum_{i=1}^{m} w_i E_{est}^i \geq \gamma_{th}\right)$$
$$FC_d = 0, \text{ otherwise} \tag{4.10}$$

where E_{est}^i is the received estimated signal energy at i^{th} SU and w_i is the average SNR over M SUs $(i=1,2, \ldots M)$: $w_i = \frac{\gamma_i}{\sum \gamma}$.

4.2.3 Machine Learning–Aided SS (ML-SS)

The age we live in now is the age of machine learning (ML) or artificial intelligence (AI). With significant advancements in computing resources, including processing power and storage capacity, ML helped us in solving complex problems. In the context of spectrum sensing, it refers to a classification task, and using several supervised and unsupervised ML methods helps us achieve higher accuracy in detecting the presence of PU [11, 12, 14–17].

4.2.3.1 Supervised Learning

In supervised learning, the algorithm learns to categorize signals as PU or non-PU through training on labelled data. In this labelled dataset, input features consist of signal energy values $\mathbf{E} \epsilon \mathbb{R}^{N \times M}$ (where $\boldsymbol{E}$ denotes all the estimated energy E_{est}^{i}), while the output vector consists of corresponding channel status $\boldsymbol{v} \epsilon (0,1)^{N \times M}$ for all SU. The variables N and M represent the size of the training set and the number of SU, respectively. The FC was subsequently trained using this labelled dataset. FC extracts the pattern and learns the mapping between the input and the output throughout the training phase, treating $\mathbf{E}$ as an independent feature and $\boldsymbol{v}$ as a dependent feature. After training (as shown in Figure 4.3), the model is capable of making correct predictions when new unlabelled data is given. The most common techniques in this category are logistic regression, support vector machines (SVM), k-nearest neighbours (KNN), artificial neural networks (ANN), etc.

4.2.3.1.1 Logistic Regression

The logistic regression model produces a value between 0 and 1 based on its prediction of an event's probability. The predicted outcome is 'no' if this number is below a given threshold and 'yes' otherwise [18]. For hypothesis H_i, the *logit* value can be expressed as

$$p(\mathrm{H}_i | \mathbf{E}) = \frac{1}{1 + e^{-E.w^T}} \tag{4.11}$$

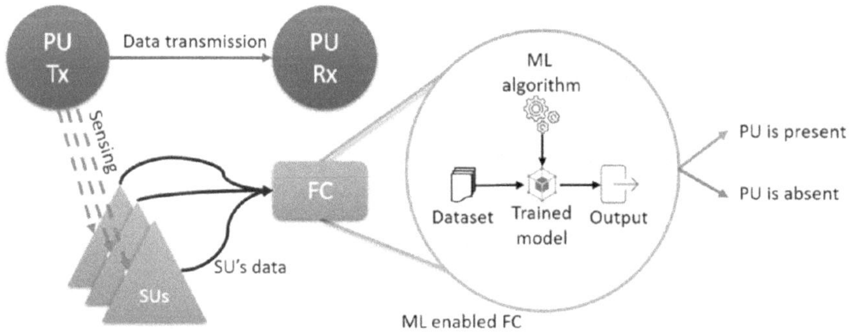

FIGURE 4.3 ML-aided spectrum sensing.

where $p(\mathrm{H}_i|\mathbf{E})$ specifies the probability of presence/absence of PU given sensed input energy $\mathbf{E}=[E_1, E_2, \ldots E_M]$, $\mathbf{w}$ is the weight vector consisting $[w_1, w_2, \ldots w_M]$, and i indicates 0/1.

4.2.3.1.2 *Support Vector Machine (SVM)*

Using weight $\mathbf{w}$ and bias w_0, the SVM finds the hyperplane $f(\mathbf{E})$ that separates the data into two categories or classes. The data points that support this hyperplane are referred to as the support vectors. In some cases, the data cannot be separated by a line or plane in lower dimensions, which means that it is not linearly separable. In such cases, the data is projected to higher dimensions, where the classes can be separated via a plane. Depending on the choice of kernel ϕ, the separation margin between the classes might be linear, sigmoid, or polynomial; thus, the $f(\mathbf{E})$ can be expressed as [12]:

$$f(\boldsymbol{E}) = \mathbf{w}^T \phi(\boldsymbol{E}) + w_0 \tag{4.12}$$

We need to determine the weight $\mathbf{w}$ and bias w_0 in such a way that

$$f(\mathbf{E}) > 0 : \mathrm{H}_1$$

$$f(\mathbf{E}) \leq 0 : \mathrm{H}_0$$

4.2.3.1.3 *K-Nearest Neighbours (KNN)*

KNN classifies data into labels based on proximity. The approach uses a majority vote of k-nearest data points to label the data for the classification problem, and the votes denote the classes $\epsilon(\mathrm{H}_1, \mathrm{H}_0)$ and proximity among data points is calculated by Euclidian distance [19].

4.2.3.1.4 *Naive Bayes*

The naive Bayes approach predicts the labels by applying the Bayes theorem. One presumption of this approach is that there is no correlation between the features; thus, the naive Bayes probability can be written as [20]:

$$p(\mathrm{H}_i|\mathbf{E}) = \frac{p(\mathbf{E}|\mathrm{H}_i)\, p(\mathrm{H}_i)}{p(\mathbf{E})} \tag{4.13}$$

4.2.3.1.5 *Multi-Layer Perceptron (MLP)*

Since the data travels forward direction in MLP, it is said to be a feed-forward neural network. It has three layers: input layer (initial layer), hidden layer (intermediate layer), and the output layer (final layer). The input layer gets the data in, the hidden layer processes the data, and the output layer gives the decision. Backpropagation helps the neurons in the layers learn the patterns. It adjusts the weights and biases in an attempt to reduce the cost function and improve accuracy. Weight adjustment is done with a backward pass following each forward pass. It computes and distributes the loss gradient layer by layer backward. Output at the forward phase can be expressed as [12]

$$
\begin{aligned}
o_n^1 &= \sigma\left(\sum_{i=1}^{M} w_{in}^1 \mathbf{E}_i\right) \\
o_n^2 &= \sigma\left(\sum_{i=1}^{m} w_{in}^2 o_n^1\right) \\
&\vdots \\
o_n^l &= \sigma\left(\sum_{i=1}^{m} w_{in}^l o_n^{l-1}\right)
\end{aligned}
\tag{4.14}
$$

where o_n^l denotes n^{th} neuron's output at l^{th} layer with weight $\sigma(.)$. $\sigma(.)$ denotes the sigmoid function [11].

Since spectrum sensing is a classification problem, the final L^{th} layer contains just one neuron. Thus, the likelihood of hypothesis H_1 can be represented as follows:

$$
\begin{aligned}
p(H_1|\mathbf{E}) &= o^L \\
p(H_0|\mathbf{E}) &= 1 - o^L
\end{aligned}
\tag{4.15}
$$

4.2.3.1.6 *Unsupervised Learning*

On the other hand, it involves clustering and anomaly detection. It does not require labelled data. Alternatively, it groups data into clusters or finds outliers. This can be useful, where any deviation from the normal is considered a PU signal. The most common techniques in this category are k-means clustering, Gaussian mixture models (GMM), etc.

4.3 INTELLIGENT REFLECTING SURFACE (IRS)–ASSISTED COMMUNICATION

In the preceding section, ML techniques in the context of CR network to improve sensing performance is discussed. Reconfigurable intelligent surfaces (RIS), often referred to as metasurfaces or intelligent reflecting surfaces (IRS), emerged as one of the potential techniques that can be used to improve sensing performance. In this section, the role of IRS in further improving the sensing performance is discussed. We present here a direction to the reader for improving spectrum sensing performance using soft computing techniques. IRS presents a viable way to address the problems caused by signal fading in cognitive radio systems and spectrum sensing. The passive components that make up an IRS are capable of modifying the amplitude and phase of incoming electromagnetic waves. IRS's adaptive capabilities help it mitigate the fading problem by reconfiguring the radio propagation environment [21, 22].

Figure 4.4 illustrates that the main goal of incorporating IRS into cognitive radio is to improve the signal-to-noise ratio (SNR) at the secondary user (SU). The more elements there are in the IRS, the better the SNR becomes. We considered grey wolf optimization techniques to optimize the phase of IRS elements. The results

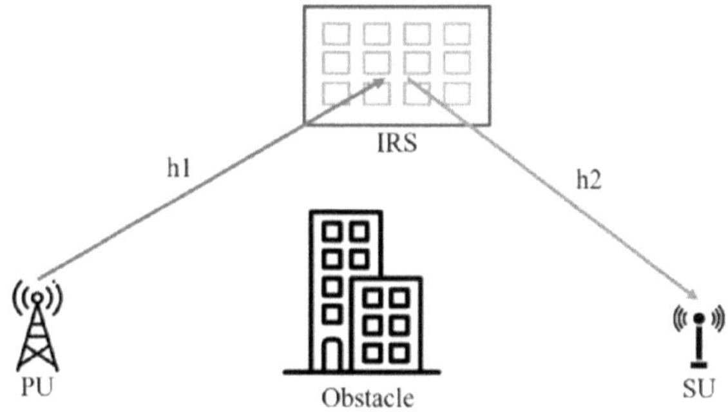

FIGURE 4.4 A basic system model for IRS-assisted SS.

shown here compare the SNR obtained by configuring IRS elements randomly with the SNR obtained from configuring IRS elements utilizing the grey wolf optimizer approach [23]. Grey wolf optimizer is an effective algorithm motivated by the social behaviours of grey wolves in the wild. This optimizer uses cooperative hunting techniques to solve challenging mathematical and computational problems, much like real wolves do when they hunt and solve problems as a pack.

For IRS-assisted spectrum sensing, a system model is shown in Figure 4.4, having one PU–SU pair and one IRS. IRS is situated in between PU and SU at a considerable height. Channels between PU, IRS, and SU are considered to be Rayleigh fading. There is no direct channel between PU and SU, while SU is getting signals from PU with the help of IRS only. The kth sample of received signal y_k at SU can be expressed as [22]

$$\begin{aligned} y_k &= n_k && \text{PU is absent } \mathrm{H}_0 \\ y_k &= \sqrt{P}\left(\mathbf{h}1\phi\,\mathbf{h}2\right)s_k + n_k && \text{PU is present } \mathrm{H}_1 \end{aligned} \tag{4.16}$$

where n_k is the AWGN noise, P is the transmit power of PU, $\mathbf{h}1$ denotes the channel between PU and IRS, $\mathbf{h}2$ denotes the channel between IRS and SU. ϕ represents the phase shift coefficient of the IRS (where $\phi = \text{diag}\left[\phi_1, \phi_2, \ldots \phi_m\right]$ and $\phi_m = e^{j\theta_m}$), and s_k is the transmitted signal of PU. The motive of this IRS-assisted spectrum sensing is to maximize the received SNR at SU by tuning the reflecting coefficient; therefore, a nature-inspired maximization algorithm is used to maximize the SNR.

4.4 NUMERICAL SIMULATION AND RESULT DISCUSSION

This section provides the result analysis conducted by two experiments, namely, ML-based and IRS-based spectrum sensing. Experiment 1 for ML-based spectrum sensing is carried out using a synthetic dataset produced by 20,000 realizations of Monte Carlo simulations. The parameters considered for this experiment are given in Table 4.1.

Figure 4.5 shows the performance of non-cooperative, cooperative, and ML-based SS. Here, different PU–SU distances must be taken into account to account for propagation losses. As a result, as distance rises, the received signal power attenuates,

TABLE 4.1
Default Parameters

Parameters	Value
Number of PU and SU	1 and 3
$P\,(H_1)$	0.5
Channel bandwidth	5 MHz
Transmit power	0.1 mW
Total sample	50
Fading channels	Rayleigh
Size of training dataset	1,000
PU to all SU distances	500, 800, 1,000, in metres

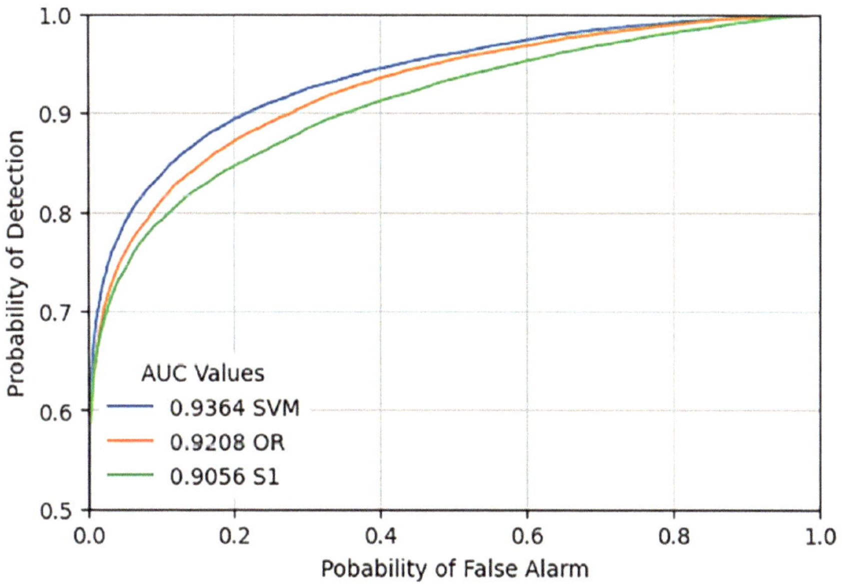

FIGURE 4.5 ROC curve in Rayleigh fading.

affecting the SU's ability to detect PU occupancy. As discussed previously, cooperative efforts among multiple SUs in sensing can enhance overall detection accuracy; however, an excessive number of SUs may result in higher calculation overhead without significant gain in performance.

We created receiver operating characteristic (ROC) curves to assess performance for ML-assisted SS and conventional CSS techniques, as well as the effectiveness of distinct SUs (as demonstrated in N-CSS). In the Rayleigh fading scenario, the area under the curve (AUC) values are shown in Figure 4.5, where you can observe the efficiency of different spectrum sensing techniques. Notably, the method that achieves the best AUC value is SVM and performs better than all the other methods.

We performed an experiment where we calculated the AUC for various ML techniques taking into consideration variable numbers of SUs in order to study the effect of SU count on spectrum sensing. The relationship between spectrum sensing performance and the number of SUs is illustrated in Figure 4.6. The AUC value rises with the number of SUs, suggesting that the detection of PU activity can be enhanced by having more sensing data. However, in the specific simulation instance we investigated, we found that at least four SUs are needed to get better performance. This indicates that adding more SUs does not improve performance; instead, it raises calculation overhead.

For evaluating the spectrum sensing performance in the second experiment with IRS, the parameters used are given in Table 4.2.

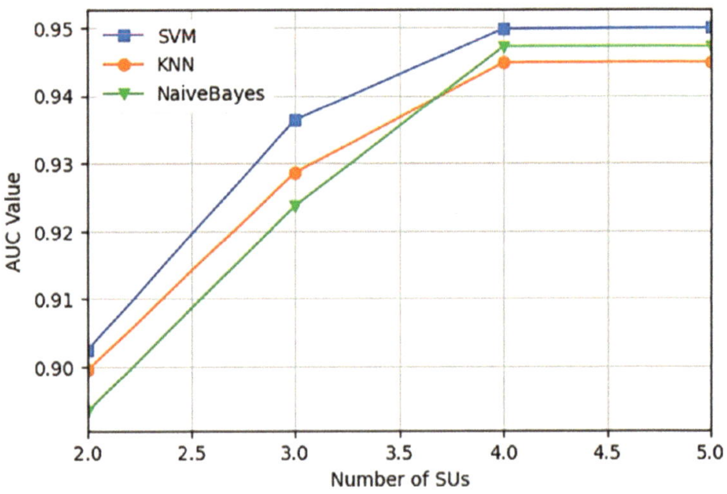

FIGURE 4.6 Number of SUs vs. AUC values.

TABLE 4.2
Default Parameters for IRS-Assisted Spectrum Sensing

Parameters	Values
Number of PU and SU	1
PU active probability	$P(H_1) = 0.5$
SNR	−10 dB
Number of IRS elements	10–40
Reference path loss	50 dB
Distance (PU–IRS, IRS–SU, and PU–SU)	In metre (80, 80, and 150)
Fading channels	Rayleigh

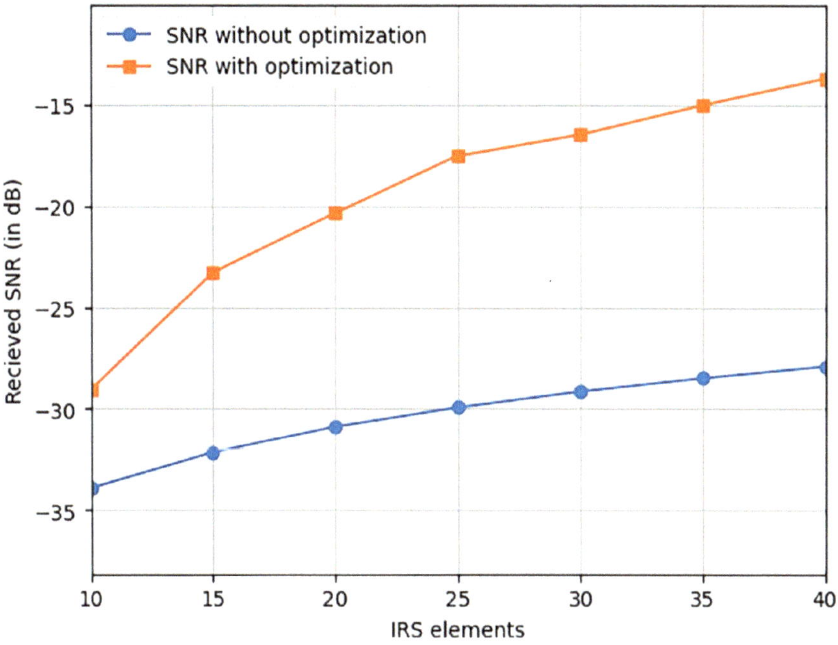

FIGURE 4.7 SNR (in dB) vs. number of IRS elements.

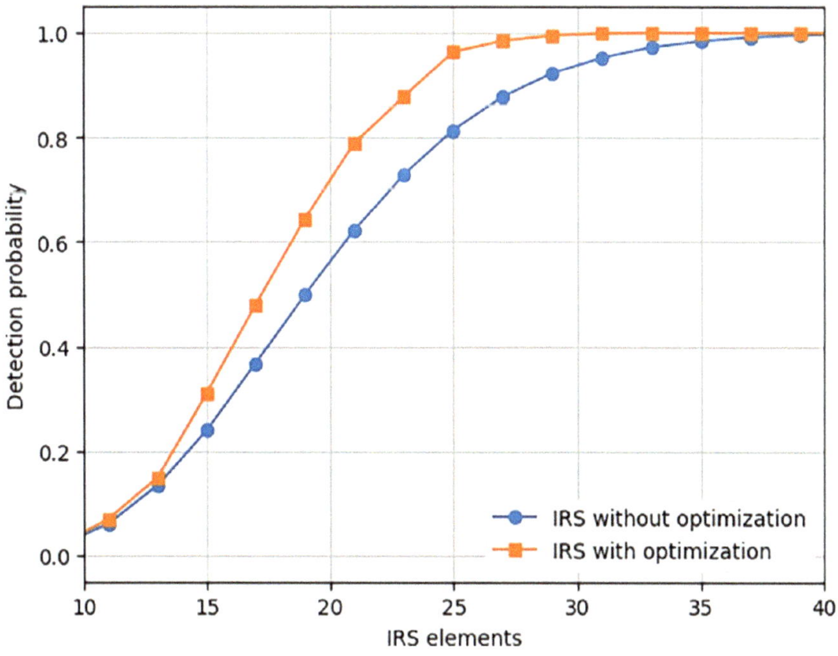

FIGURE 4.8 Detection probability (P_d) vs. number of IRS elements.

Results shown in Figure 4.7 demonstrate the performance of spectrum sensing using IRS with and without optimization. It is visible from the figure that, as the number of IRS elements increases, the SNR and probability of detection increase; however, after performing optimization, there is significant improvement in the SNR and probability of detection. Figure 4.8 shows the performance in terms of detection probability $\left(P_d\right)$ with optimization and without optimization. As the number of IRS elements increases, the detection probability also increases.

4.5 CONCLUSION

This chapter provides a brief introduction to cognitive radio (CR) and how CR helps us improve spectrum efficiency by opportunistically sharing the licensed spectrum with the unlicensed users. This chapter mainly focuses on the spectrum sensing part of the CR, where various techniques, such as classical, ML-based, and IRS-based spectrum sensing, are provided.

To demonstrate the performance, two experiments of spectrum sensing using ML and IRS are conducted. In the first experiment, we compare classical with ML-based spectrum sensing models in CRN. The generated graphical data indicates that the ML model performs better than the classical spectrum sensing technique and demonstrates that, using energy samples from different SUs, the evaluated machine learning models accurately predict channel state. In the second experiment, the performance of IRS-assisted spectrum sensing is evaluated in the case of optimization and without optimization. Results demonstrate that spectrum sensing accuracy is significantly improved with optimization.

4.6 ACKNOWLEDGEMENTS

This work is supported by the Science and Engineering Research Board (SERB-DST), Govt. of India, under grant no. EEQ/2021/000942.

REFERENCES

[1] Y.-C. Liang, Y. Zeng, E. C. Y. Peh, and A. T. Hoang, "Sensing-throughput tradeoff for cognitive radio networks," *IEEE Transactions on Wireless Communications*, vol. 7, no. 4, pp. 1326–1337, 2008.

[2] G. Ganesan and Y. Li, "Cooperative spectrum sensing in cognitive radio, part ii: Multiuser networks," *IEEE Transactions on Wireless Communications*, vol. 6, no. 6, pp. 2214–2222, 2007.

[3] Y. Arjoune and N. Kaabouch, "A comprehensive survey on spectrum sensing in cognitive radio networks: Recent advances, new challenges, and future research directions," *Sensors*, vol. 19, no. 1, p. 126, 2019.

[4] F. Salahdine, H. El Ghazi, N. Kaabouch, and W. F. Fihri, "Matched filter detection with dynamic threshold for cognitive radio networks," in *2015 international conference on wireless networks and mobile communications (WINCOM)*. IEEE, pp. 1–6, 2015.

[5] X. Zhang, R. Chai, and F. Gao, "Matched filter based spectrum sensing and power level detection for cognitive radio network," in *2014 IEEE global conference on signal and information processing (Global-SIP)*. IEEE, pp. 1267–1270, 2014.

[6] P. S. Yawada and A. J. Wei, "Cyclostationary detection based on non- cooperative spectrum sensing in cognitive radio network," in *2016 IEEE international conference on cyber technology in automation, control, and intelligent systems (CYBER)*. IEEE, pp. 184–187, 2016.
[7] I. Ilyas, S. Paul, A. Rahman, and R. K. Kundu, "Comparative evaluation of cyclostationary detection based cognitive spectrum sensing," in *2016 IEEE 7th annual ubiquitous computing, electronics & mobile communication conference (UEMCON)*. IEEE, pp. 1–7, 2016.
[8] A. Ranjan and B. Singh, "Design and analysis of spectrum sensing in cognitive radio based on energy detection," in *2016 international conference on signal and information processing (IConSIP)*. IEEE, pp. 1–5, 2016.
[9] D. M. M. Plata and A. G. A. Reatiga, "Evaluation of energy detection for spectrum sensing based on the dynamic selection of detection-threshold," *Procedia Engineering*, vol. 35, pp. 135–143, 2012.
[10] T. Yucek and H. Arslan, "A survey of spectrum sensing algorithms for cognitive radio applications," *IEEE Communications Surveys & Tutorials*, vol. 11, no. 1, pp. 116–130, 2009.
[11] C. H. A. Tavares, J. C. Marinello, M. L. Proenca Jr, and T. Abrao, "Machine learning-based models for spectrum sensing in cooperative radio networks," *IET Communications*, vol. 14, no. 18, pp. 3102–3109, 2020.
[12] R. S. Jabbal, V. Srivastava, and B. Prasad, "Machine learning aided spectrum sensing," *2023 international conference on electrical, electronics, communication and computers (ELEXCOM)*, Roorkee, pp. 1–6, 2023.
[13] D. Teguig, B. Scheers, and V. Le Nir, "Data fusion schemes for cooperative spectrum sensing in cognitive radio networks," *2012 Military communications and information systems conference (MCC)*. IEEE, pp. 1–7, 2012.
[14] C. Clancy, J. Hecker, E. Stuntebeck, and T. O'Shea, "Applications of machine learning to cognitive radio networks," *IEEE Wireless Communications*, vol. 14, no. 4, pp. 47–52, 2007.
[15] D. Wang and Z. Yang, "A novel spectrum sensing scheme combined with machine learning," in *2016 9th international congress on image and signal processing, biomedical engineering and informatics (CISP-BMEI)*. IEEE, pp. 1293–1297, 2016.
[16] K. M. Thilina, K. W. Choi, N. Saquib, and E. Hossain, "Machine learning techniques for cooperative spectrum sensing in cognitive radio networks," *IEEE Journal on Selected Areas in Communications*, vol. 31, no. 11, pp. 2209–2221, 2013.
[17] F. Azmat, Y. Chen, and N. Stocks, "Analysis of spectrum occupancy using machine learning algorithms," *IEEE Transactions on Vehicular Technology*, vol. 65, no. 9, pp. 6853–6860, 2015.
[18] J. S. Cramer, "The origins of logistic regression," in *Tinbergen Institute Working Paper No. 2002-119/4*, 2002.
[19] R. O. Duda and P. E. Hart, Pattern classification, John Wiley & Sons, 2006.
[20] K. Bouallegue and M. Crussiere, "Detection of primary user assisted by machine learning over multipath channels," in *2022 IEEE wireless communications and networking conference (WCNC)*. IEEE, pp. 824–829, 2022.
[21] Q. Wu, S. Zhang, B. Zheng, C. You, and R. Zhang, "Intelligent reflecting surface-aided wireless communications: A tutorial," *IEEE Transactions on Communications*, vol. 69, no. 5, pp. 3313–3351, 2021.
[22] A. Nasser, H. A. H. Hassan, A. Mansour, K.-C. Yao, and L. Nuaymi, "Intelligent reflecting surfaces and spectrum sensing for cognitive radio networks," *IEEE Transactions on Cognitive Communications and Networking*, vol. 8, no. 3, pp. 1497–1511, 2022.
[23] S. Mirjalili, S. M. Mirjalili, and A. Lewis, "Grey wolf optimizer," *Advances in Engineering Software*, vol. 69, pp. 46–61, 2014.

5 Device-to-Device Communication

Cooperative Device-to-Device Communication

Kankanala Kavitha and Suseela Vappangi

5.1 INTRODUCTION AND LITERATURE SURVEY

5.1.1 Introduction

Cooperative device-to-device (C-D2D) communication necessitates efficient spectrum, power, and network resource allocation strategies. In addition, mechanisms for the dynamic formation and management of cooperative clusters, taking into account mobility and fluctuating network conditions, are required to maintain connectivity and ensure reliable communication. Additionally, security and privacy aspects of C-D2D communication require consideration. It is essential to ensure secure and trusted communication within ad hoc cooperative networks to prevent unauthorized access and protect sensitive data. This study calls for additional research on C-D2D communication in order to investigate novel designs, methods, and architectures that address the challenges. The prospective benefits of C-D2D communication extend to multiple domains, such as public safety, the Internet of Things (IoT), and smart cities, where dependable and efficient communication is crucial.

5.1.2 Overview of Device-to-Device (D2D) Communication

The probabilities of device-to-device (D2D) communication are extremely reliant on D2D connection fidelity. The lack of assistance from a base station (BS) in the form of efficient data recovery and signal processing capabilities will inevitably result in communications through D2D links being vulnerable to factors such as fading and ambient noise. As a consequence, the low transmission rate that would be achieved as a result of this vulnerability would reduce the incentive for users and network providers to implement D2D communication. The performance of cooperative communication (CC) [1] in dealing with overfading to achieve maximum capacity for the channel and accuracy in a cost-effective way has been established, and this makes it an appropriate choice for D2D communication. The basic idea is for a relay node to send the signal it gets from a transmitter to its own receiver. This way, diversity gain is achieved by the same signal moving along different paths from direct and relay transmissions.

 DOI: 10.1201/9781032625034-5

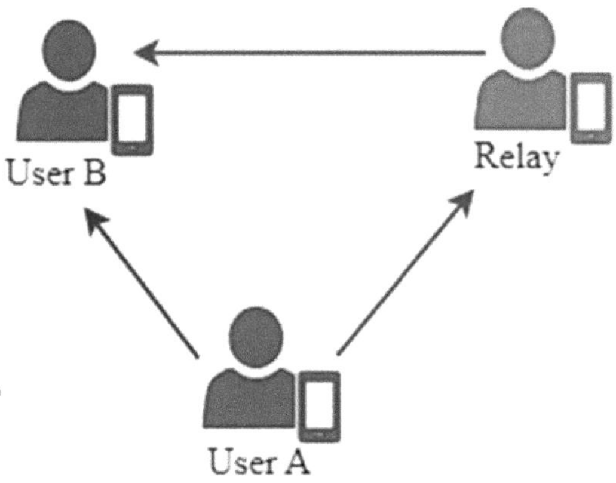

FIGURE 5.1 Cooperative D2D communication modelled as a network of three nodes.

The three-node paradigm of CC is depicted in Figure 5.1. In this configuration, user A sends data to user B with the assistance of a relay node. This configuration is quite frequent. Every single transmission is carried out one frame at a time. In conventional direct transmission (DT), user A sends data to user B for the duration of the frame. Each frame is partitioned into two time periods when CC is applied. User A sends its signal to user B over the first time period. Due to the broadcast nature of wireless communication, the relay node can also receive this transmission. During the second time interval, the relay transmits the signal to user B after first amplifying or decoding the signal that it has received. After that, in order to reconstruct the initial signal, user B merges the signals that came in in two different time intervals when the signal's attenuation was varying. When D2D communication is facilitated by CC, commonly referred to as cooperative D2D communication, other challenges arise for the optimal allocation of resources in wireless networks. These new challenges include relay selection, channel allocations, and transmission rescheduling.

- First, because there could be more than one D2D link operating in a cellular network, there is no way to give each one its own channel. In the meanwhile, cognitive radio is used. CC must be taken into account when allotting bands to communication links in a network with limited channels. This includes D2D links, cellular uplinks, and downlinks.
- To ensure a certain level of quality of service (QoS), multiple links for communication using a single channel must be meticulously arranged. This issue turns into a more complex one while CC is considered, as the use of a relay node could raise the link transmission rate; however, it could also boost the range of interference for a link, resulting in a decrease in spatial multiplexing.

5.1.3 Evolution of D2D Communication

First-generation (1G) wireless networking evolved to fifth-generation (5G) wireless communication. In 2020, 5G was introduced to satisfy the needs of users and accommodate the escalating volume of data traffic. Investigators were compelled to employ 5G technology due to the exponential growth in high-data-rate multimedia applications, such as video streaming, live conferences, and three-dimensional (3D) gaming.

5.1.4 Benefits of D2D Communication

D2D communication underpinning cellular networks (CNs) offers multiple benefits to both network operators and mobile consumers, as detailed in [2]:

- **User perspective.** D2D communication systems benefit from the close proximity of mobile devices as compared to traditional communication, that is, via BS, in terms of data rate, latency, and energy efficiency.
- **Network perspective.** Relay communication, higher spectrum efficiency, increased coverage of the network, reduced BS traffic load, and local management of short-range transmissions are only some of the benefits of D2D communication that can be realized through its usage of CN bandwidth.

5.1.5 Practical Scenario of D2D Communication

Let us imagine a world where a wide variety of equipment are interconnected and exchanging data across wireless networks. Some portable electronics are linked via cellular uplink and downlink connections established by BS, while others are linked directly between themselves. The provided scenario makes use of three distinct links to bring attention to the following: (1) the cellular uplink (shown with cyan colour); (2) the interference links (shown with dotted lines) between the cellular user (CLU) and the D2D receiver (DRx) and the D2D transmitter (DTx) and the BS; and (3) the D2D link (shown with black colour). There are three distinct types of D2D links depicted in [3], as follows:

- **Cellular links.** The DTx and DRx communicate via BS within this approach. DTx transfers information to the BS, while DRx retrieves information via the BS. In this case, the BS functions as a relay between the DTx and DRx and helps D2D users (DUs) achieve high throughput.
- **Dedicated links.** To facilitate communication, the BS in this case reserves towards a specific portion of the radio spectrum to be utilized by the DTx and DRx. This enhances the DCS's spectrum and energy efficiency while mitigating DU–CLU interference.
- **Reuse links.** By reusing the CLU's resources in an underlay configuration, DTx and DRx are able to communicate messages directly with one another.

5.1.6 Classifications of D2D Communication

Depending on the channel of exchange, D2D interactions can be broken down into three distinct categories as [4, 5]:

- **Peer-to-peer or direct link D2D communication.** Direct link D2D communication is very similar to the point-to-point transmission, which is

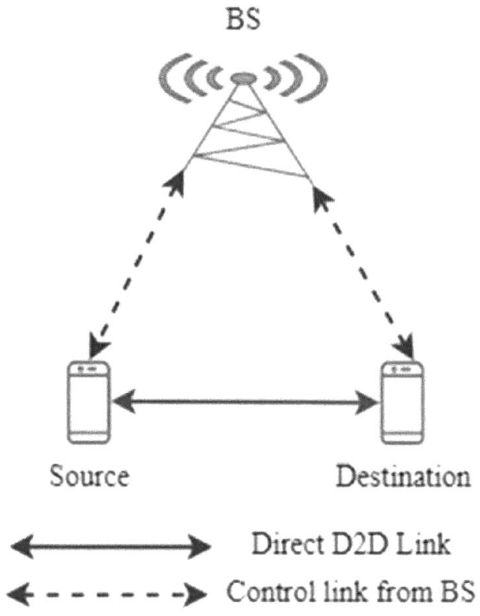

FIGURE 5.2 Peer-to-peer or direct link D2D communication.

depicted in Figure 5.2. No intermediary nodes are needed for direct communication between the sender and the recipient.

- **Cooperative communication-aided D2D network.** It is a method of relaying data across networks to increase their territory. A user wishes to connect with some other user, but the latter is outside of coverage (or has bad channel constraints), as shown in Figure 5.3. By serving as a relay, intermediate users can increase the network's reach and mitigate the impact of channel fading, as in [6, 7].
- **Multi-hop D2D communication.** In the first scenario, numerous devices work together to aid in D2D communication by taking turns routing data to the destination. In [8–10], researchers investigate multi-hop D2D communications for nearby user equipment (UEs) to facilitate superior content uploading services in cellular networks. The primary goal is to speed up the uploading process for all UEs while still letting them submit content of the desired quality and size.
- **Cluster-based D2D communications.** In a data-centric context, a group of DUs sharing a common file request can organize into a cluster and employ multicast to send the file more efficiently and with less delay. The suggested novel cluster-based D2D design is shown in [11–13], which utilizes cluster-based round robin scheduling to equitably divide radio resources across cluster nodes. The goal is to maintain a low level of resource use to allow for the effective sharing of information resources by more traditional mobile communication services.

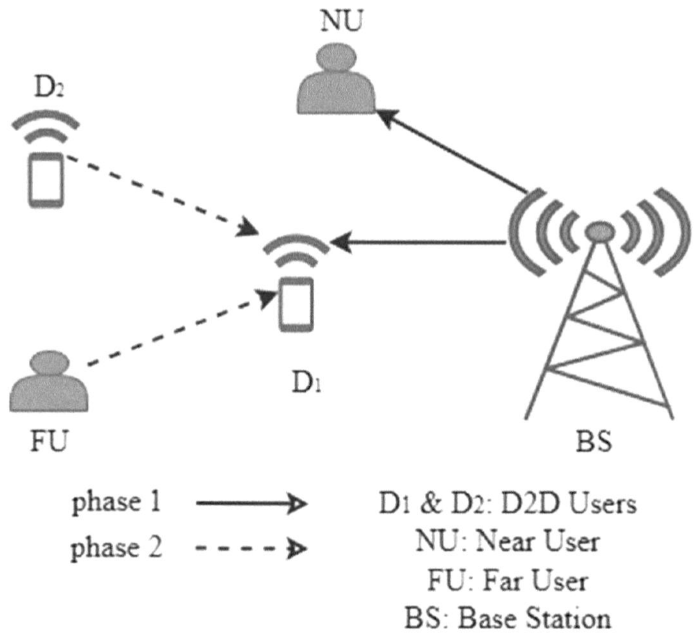

FIGURE 5.3 Cooperative communication-aided D2D network.

5.1.7 Applications of D2D/C-D2D Communication

- Vehicular communication
- Publish safety applications
- Distributed sensing in IoT applications
- Cooperative/collaborative mobile gaming
- Emergency applications
- Energy-efficient communications

5.1.8 Literature Review

In this part, the contributions of the several researchers to the fields of D2D and cooperative D2D communications have been addressed. It starts by discussing the latest developments in cellular network offloading technology, then go on to the work already being done on D2D communication. Following this, we introduce our work in the related field of cooperative D2D communication, utilizing cutting-edge approaches and frameworks.

5.1.8.1 D2D Communications

In [14], the authors proposed a D2D-aided cooperative NOMA method that serves two similar gain (SG) nearby users and a user at a distance in two cluster time slots using the overheard signal. During the same cluster, two nearby users can communicate using D2D during two time intervals. With mathematical composition, the suggested scheme's delay-tolerant capacity, outage probability, and delay-limited capacity are examined. According to the D2D-enhanced uplink NOMA method suggested in [15], it is conceivable for the same group of cellular users to engage in both NOMA-based cellular transmission and D2D communications. Independent Rayleigh fading channels have been used to analyze the performance of D2D uplink NOMA in terms of its ergodic rate (ER) and ergodic sum rate (ESR). Theoretical and simulation outcomes both show that D2D uplink NOMA is more effective than OMA.

In [16], D2D communications can make advantage of existing cellular infrastructure or the unlicensed spectrum already in use by Wi-Fi. However, ensuring the QoS for users on both of them and performing NOMA user clustering appropriately remain important open questions. To achieve this, they explore a joint resource allocation issue consisting of user clustering, power control, and D2D mode selection, with the aim of optimizing the network performance. Offloading cellular traffic to Wi-Fi or other networks has been the subject of extensive study [17, 18] in recent years for the purposes of saving bandwidth or energy. In [19], we see a quantitative analysis of the growth in mobile data traffic and a proposed strategy for unloading that traffic. Three different IP traffic offloading solutions are presented and compared in [20], all of which aim to operate at the internet layer and simply implement the standard TCP/IP protocol suite as defined by the Internet Engineering Task Force (IETF2) without requiring any knowledge of access technology. The authors of [17] provide an empirical investigation of the efficacy of third-generation (3G) mobile data offloading via Wi-Fi. Their modelling of a full day's worth of information demonstrates that Wi-Fi already offloads about 65% of all mobile data consumption and saves 55% of battery life by not sacrificing transmission speed. The authors of [18] proposed two approaches for the delay-tolerant offloading of large ideas endorsed by social media on 3G networks. They discovered that both solutions are effective at offloading a substantial amount of traffic, which has an encouraging effect on battery life.

Because of the positive effects of transferring data traffic at BSs in cellular networks, D2D communication has become a popular research topic currently. In their early work, [21] suggested a D2D wireless system that functions as an overlay network to an International Mobiles Telecommunication–Advanced (IMT-Advanced) cellular network for promoting regional peer-to-peer communication. To identify generalized scenarios, D2D communications were thoroughly studied from a variety of perspectives, including interference management, power control, spectrum sharing, etc. For instance, the authors of [22] recommend an effective and practical approach to find the extent of interference between cellular and D2D users in a certain area at the BS. This method then uses the multiuser diversity that comes with the cellular network to lower interference. The investigators of [23] have devised a distributed dynamic spectrum procedure whereby ad hoc DUs gain opportunistic access to cellular user's actively employed spectrum.

The authors of [24] investigated the throughput optimization problem over shared resources while satisfying prioritized mobile communication constraints using an identical network paradigm. A novel interference management scheme is proposed in [25] to enhance the efficiency of D2D communication. They calculate the closed-form solution to the outage probability (OP) and provided a selection strategy that minimizes the OP of the proposed scheme. The authors of [26] recommend a semi-distributed two-stage resource management model for D2D communication. At first phase, the BS allocates resource blocks directly to BS-to-user device (B2D) and D2D connections. During the second phase, the BS timeline communications use the resource blocks assigned to B2D links, whereas the main user device for every D2D link uses the resource blocks assigned to D2D links to perform distributed link adaptation processes. A two-tier 5G cellular network with a macrocell tier (i.e. base-station-to-device communications) and a device tier (i.e. device-to-device communications) is described in [27]. In [28], the authors employ the best possible approaches for the system utilization of resources and selecting a mode acquired under the realistic user and mobility scenarios to examine the basic complications of how D2D communication enhances the structure of cellular systems and what is the possible consequence of D2D communication. Particularly, the authors accomplish the mathematical upper bound on frameworks' information-downloading performance through coming up with a max-rate optimization challenge that optimizes the rate of data downloading from all the cellular BSs to the media downloading devices via all possible methodologies of telecommunication.

5.1.8.2 Cooperative D2D Communications (C-D2D)

In [6], the authors have proposed and investigated a novel cooperative NOMA-based hybrid cellular and bidirectional D2D transmission scheme. In BD2D-assisted cooperative NOMA, two distant users could simultaneously send their signals to the BS through a relay and then exchange D2D signals. For each of the information sources, analytical expressions are generated for the ergodic capacity (EC), ergodic sum capacity (ESC), and OP via perfect and imperfect successive interference cancellation (SIC). The authors in [7] explore an investigation at the problem of resource allocation in the context of a downlink cooperative communication system. When cellular users (CUs) and the BS are unable to connect directly with one another, D2D user pairs take part in the cooperative communication to finish forwarding information required by CU and themselves through NOMA. In order to compensate for fading, diversity is provided using relay-assisted C-D2D communication. It simulates the effects of multi-input, multi-output (MIMO) diversity [29] by having multiple inputs and multiple outputs. In this scenario, the relay nodes volunteer to help transmit data across the sending UE and the receiving UE [30]. To get many separate faded signals at the destination UE without employing the MIMO antenna array, employing relay nodes across the source UE and the destination UE is recommended [30]. Improvements in low latency, high data rate, expanded coverage area, lower energy consumption, and dependable communication are all provided by cooperative communication [31, 32].

Figure 5.4 illustrates a typical model of a cooperative D2D communication system in which a pair of relays helps the D2D UE communicate. In this model of the system, the source UE (UEs) hopes to provide data to the destination UE (UEd).

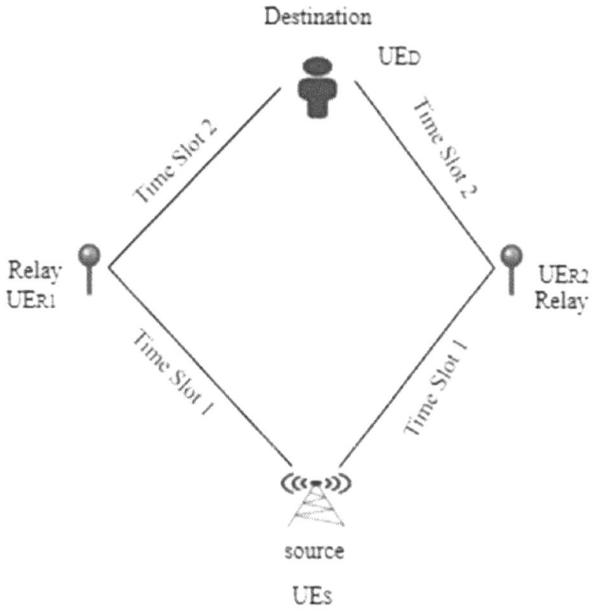

FIGURE 5.4 A two-hop relay-assisted cooperative D2D communication system model.

However, the distance between UEs and UEd is greater than the reliable transmission range of UEs. Therefore, UEs cannot send the signal to other UEs directly. As a result, relays aid UEs in transmitting signals. In what follows, we present two time slots in which the information signal is transmitted [33, 34].

Time slot 1. During the first time slot, the UEs send out a signal that is picked up by both UE_{R1} and UE_{R2}.

Time slot 2. In the second time slot, the relays apply the relaying policy to the incoming information signal. The data signal is then sent to UE_d for further processing. The coherent receiver was able to carry out maximal ratio combining (MRC) and maximum likelihood (ML) detection at the close of time slot 2. In this case, the channel state information (CSI) knowledge of the ML receiver is guaranteed.

5.1.9 Chapter Outline

The rest of the chapter is organized as follows. We present an envisioned demonstration on D2D/C-D2D communication networks with expanded provisions in Section 5.2. In Section 5.3, a summary of security/privacy aspects of cooperative D2D communications is provided, along with future research directions. Finally, we will conclude our research work in Section 5.4.

5.2 PROPOSED DEVICE-TO-DEVICE (D2D) COMMUNICATION MODEL

D2D communication can be broken down into two distinct categories, in-band D2D and out-of-band D2D, based on how each makes use of the available spectrum. The first method involves the use of a shared spectrum band [23] between the cellular network and D2D communication. Underlay D2D communication and overlay D2D communication are two subsets of in-band D2D communication. Underlay D2D communication improves spectrum efficiency by having D2D and cellular users share the same frequencies. Cellular performance can be enhanced by offering high spectral efficiency (SE) [22].

To counteract these issues [35, 36], LTE-A versions 12 and 13 have recently integrated D2D communication [37]. D2D is also expected to be a core technology of 5G, with the goal of enhancing SE through the use of neighbourhood knowledge [38, 39]. In the meantime, D2D communication offers cellular users high-capacity gains with little limitations on network upgrades and servicing. The evolved node B (eNB)/BS is not required for data transmission between two nearby cellular users in a general D2D architecture. However, the BS is still responsible for controlling or signalling amongst users [27, 40]. Underlay and overlay D2D communication [41] are two popular D2D communication models for wireless networks. Underlay D2D communication [42] is similar to underlay cognitive radio communication in that it allows cellular and D2D users to share the licensed cellular spectrum while keeping the interference threshold constant. However, D2D users can sometimes receive less attention in comparison to cognitive radio; this is because they are still part of the same cellular system. If the quality of service (QoS) for cellular users is not affected, the BS will provide overlay D2D communication rather than underlay D2D communication [43]. While this resolves the problem of cellular and D2D links interfering with one another, it leads to poor utilization of the bandwidth. While the C-D2D paradigm has not been substantially examined in the literature, overlay and underlay D2D architectures have. To improve spatial performance, the C-D2D systems divide a point-to-point D2D channel into numerous time slots amid various objects. In [44], a method for allocating cellular network resources finely for single-channel C-D2D communication is proposed. According to [44], a relay can work with mobile clients in two ways: either one frame at a time or one slot at a time. In addition, interference range of every node is limited to prevent accidental communication through neighbouring sources. The outcomes demonstrate that having access to a wide number of channels improves performance in terms of the minimum transmission rate. The performance improvement for neighbouring mobile devices is the only target of the strategy presented in [44]. There were no suggestions offered on how to better facilitate geographical diversity through the combination of cellular and DUs.

More recently developed C-D2D frameworks have focused on offloading cellular traffic via the deployment of a single UE relay or a group of UE relays [45]. As of yet, no comprehensive analytical model has been explored to determine the optimal DU selection strategy and the maximum amount of DUs that can be facilitated by a CU whereas still meeting the QoS requirements of all concurrent users. Lack of D2D prototype test beds to demonstrate protocol performance is another challenge in the field of D2D communication. Therefore, only simulations can be used to judge how well existing D2D frameworks perform. Unfortunately, the actual wireless environment is difficult to simulate with most modelling methods. As a result, the veracity of analytical work on D2D communication is compromised. Regardless of the investigator's efforts to enhance cellular communications with cooperative cognitive radios (CCR) and C-D2D communication, there is currently little published work on this topic. Based on what has been said thus far, it is clear that the CCR and C-D2D will pave the way for the next generation of cellular networks, most likely 5G, by significantly improving network performance.

5.2.1 Underlay D2D Communication

Through the usage of underlay D2D communication, licensed cellular spectrum can be used by both cellular and D2D users at the same time without exceeding an interference threshold. In Figure 5.5, we see the architecture of a D2D communication layer as an underlay. Since the D2D user and the cellular user are competing for the same uplink resources, there will be interference between the two types of transmission. A CU and a D2D Tx share the same time/frequency resource block when sending data to the base station and the end user, respectively. Therefore, the cellular user causes interference for the BS through D2D Tx, while the BS causes interference for D2D Rx. Therefore, in an underlay D2D system, managing interference from CU to DU and vice versa is of paramount importance. Interference management for

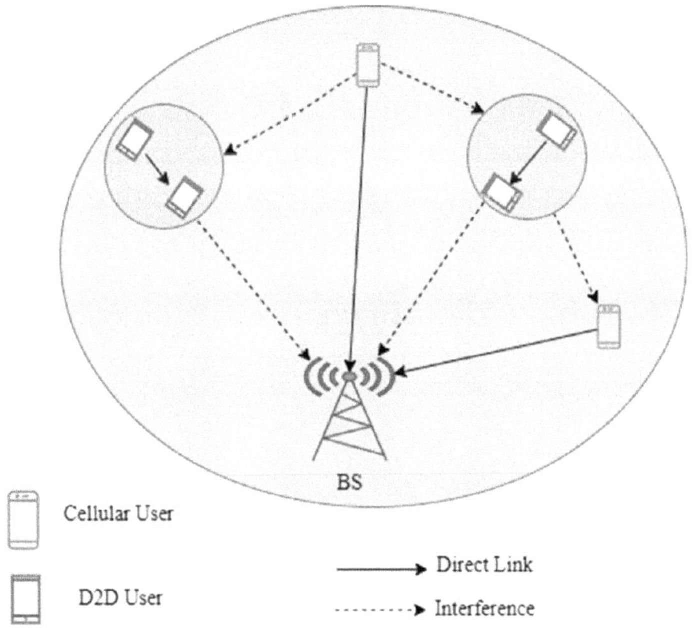

FIGURE 5.5 Underlay D2D communication.

LTE-Advanced (LTE-A) underlying D2D frameworks was the subject of research in [35]. Optimizing power consumption while sharing resources between a D2D user and a cellular user is the focus of [24]. In [46], the conditions for resource sharing in an underlay D2D cellular system, where distance is a factor, are explored. To reduce D2D interference, the authors have developed an analytical method to determine the optimal distance between the CU and D2D receiver.

5.2.2 Overlay D2D Communication

Overlay D2D communication, as opposed to underlay direct-to-device communication, involves the BS allocating specific time slots or spectrum to the D2D channel if the QoS of the CU is not impaired [43]. It gets rid of the disturbance that CUs/DUs experience with each other. Nevertheless, it is possible that it does not make efficient use of the spectrum resources that are available. In Figure 5.6, we see a schematic representation of an overlay D2D communication framework. A D2D user is able to share the cellular uplink resources in an uninterrupted manner in this scenario. To be more specific, if the required QoS for the cellular uplink communication can be met by the portion of the time/frequency resource blocks that is available, then the base station will allocate D subcarriers (or T1 time) to the CU for base station transmissions, and the leftover N-D subcarriers (or T-T1 time) can be used for D2D communication. Due to the fact that cellular and D2D lines use orthogonal sets of subcarriers, it is evident that there will be no interference between the two types of links.

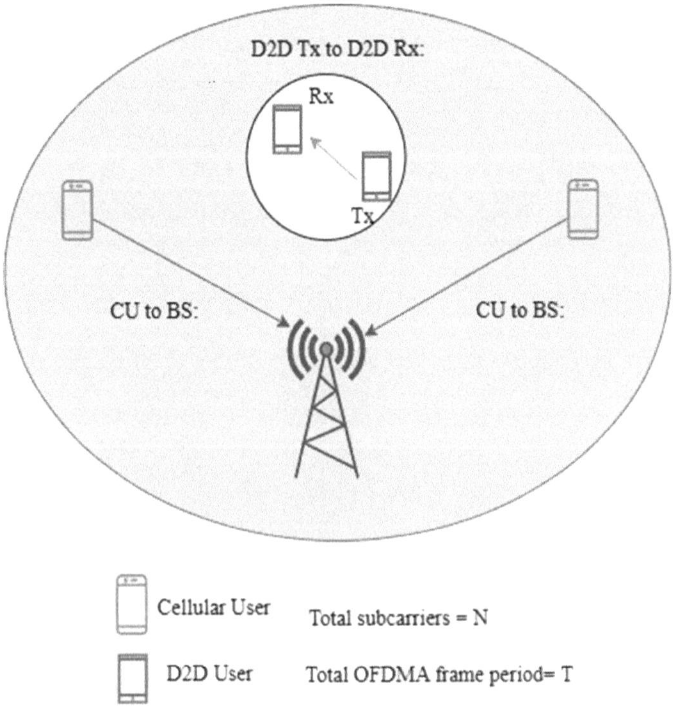

FIGURE 5.6 Overlay D2D communication.

In [47], a proposal is made for a spectrum-sharing protocol that would be used for D2D communication that would overlay cellular mode. According to [47], D2D users are able to support bidirectional transmission between the CUs and base station while simultaneously communicating with each other through a direct link. This is possible since the D2D users are connected directly to each other. In addition, an enhanced sum rate together with a power control strategy is offered for CUs and DUs. In [48], a technique based on random geometry is offered as a way for investigating how well D2D communication works over generalized fading channels. For the overlaid D2D network, closed-form formulas for SE and OP are developed. Nevertheless, the research presented in [47, 48] is restricted to the D2D communication that is superimposed on cellular networks. There has been no discussion of a comparison with the underlying or any other frameworks.

5.2.3 Cooperative D2D Communication

Recent research published in [49–53] presents the C-D2D model, which is a category of the D2D communication protocol. In order to obtain spatial performance benefits using the C-D2D model, a point-to-point D2D communication channel is divided into numerous time slots that are carried out by different entities in a coordinated manner. An increase in the sum rate of a C-D2D network can be accomplished by having a single node or many nodes function as an amplify-and-forward (AF) or decode-and-forward (DF) relay. In Figure 5.7 is a schematic representation of a C-D2D framework that is built on relays. The transmission over the cellular uplink is broken up into two parts in this instance. During the first time slot, the CU sends the messages to the BS, where it is picked up by the D2D Tx. For cellular transmission, D2D Tx can function as either a DF or AF relay. During the second time slot, the D2D Tx decodes the cellular signal that it has received and then transmits it to the BS using the fraction of the time and frequency resource block that is still available. The remaining portion of the time and frequency resource block is available for use by D2D Tx for D2D communication with D2D Rx. The authors of [49] present an adaptive mode selection strategy for D2D communication. The purpose of this scheme is to ensure performance improvement for users of both cellular and D2D networks. In terms of SE, the numerical findings shown in [49] explain that the C-D2D model performs better than the underlay and overlay D2D models presented in [54, 55]. In the event that the communication infrastructure fails, the paper [56] proposes a cooperative beamforming and relay selection technique that will make it easier for D2D communication to take place. On the other hand, the analysis presented in [49, 56] is restricted to numerical results.

In addition, each of the previously suggested methods can only support a single D2D pair. For the purpose of providing spatial diversity, traditional cooperative diversity network (CDN) typically consists of M UE-based relay nodes. However, in order to prevent interference with one another and to make the most of the resources that are available, each relay should send on orthogonal channels (i.e. on frequency/time slots). In [57], the outcomes were in an ineffective use of the available resources. It is possible to significantly improve the performance of the traditional CDN by including UE-based relay selection [58, 59]. Because of this, out of M UE relays, only the relay that is the most efficient will take part in the transmission of the source data to

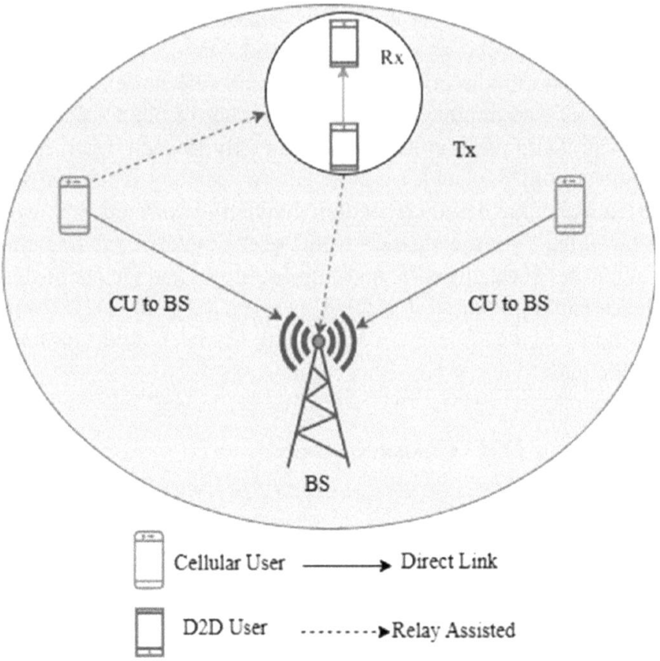

FIGURE 5.7 Cooperative D2D communication.

the destination. It has been suggested in [59] that a DF relay-based CDN with optimal relay selection would function best. They also obtain an analytical closed-form solution for the OP and EC. According to the findings, the optimal relay selection strategy is superior to the typical cooperative diversity networks in terms of performance at low SNR regime.

Within the context of C-D2D, it is possible to enhance the coverage area of an eNB by utilizing an in-coverage UE as a relay component. This relay node facilitates the transmission of information between the eNB and out-coverage UE [36]. This could be accomplished by selecting an in-coverage UE to act as a DF relay. [36] details the timing and synchronization signals that must be transmitted in order to accomplish the aforementioned goals in accordance with LTE-A releases 12 and 13. [60] proposes a power-effective relay selection issue for the relay-aided D2D framework as a means of reducing the number of resources needed for D2D communication and so optimizing network performance. However, it is expected that relay nodes are full-duplex, which reduces the significance of this finding in a practical sense. In [61], a framework for relay-assisted D2D communication is described. A D2D transceiver acts as a relay for the cellular link in this architecture. Additionally, it utilizes superposition coding for D2D transmission in the downlink system. The QoS of D2D communication is improved by a relay in [62]. Within a multi-relay OFDMA-based cellular network, to determine how resources should be allocated to each relay, an optimization challenge is provided. In contrast to an underlay D2D communication, the outcomes that were produced are examined. However, there is an additional cost associated with the set-up of additional relay in order to improve D2D communication because of the design. In addition, the scope of the investigation is restricted to relay-aided D2D communication; the effect of D2D channel performance on cellular network functionality is not taken into account.

5.3 SECURITY ASPECTS OF D2D AND C-D2D COMMUNICATIONS

In this part is the detailed summary of a few of the significant suggested improvements to the security of D2D communications made in [63]. In the meanwhile, the primary security criteria that should be met by D2D communications are described, and then a few possible research avenues based on D2D security and its requirements will be recommended, such as [63]:

- **Privacy.** User data privacy issues are mainly focused on the features of D2D services to restrict data leakage to any other parties other than a data owner. This is in contrast to the concerns regarding the confidentiality of data. This requirement has the overarching goal of mitigating the risks to users' privacy that are posed by security domains that involve mobile users.
- **Confidentiality and integrity.** In D2D communications, *confidentiality* is the safeguarding of data against leakage. It makes a guarantee that only approved machines can see each other's private information. Integrity guarantees that information is unaffected and constant at all times. It prevents accidental or illegal changes.
- **Availability and dependability.** D2D activities for verified users can only thrive if they are consistently accessible and reliable. The D2D systems must continue to function regardless of whether they are subjected to a denial-of-service (DoS) or distributed denial-of-service (DDoS) attack. D2D communications could be stymied by their inconsistent availability, which can frustrate users and slow their uptake.
- **Revocability.** If a user is found to be malicious, ineligible, harmful, or invalid, that user's access to the D2D service should be revoked as soon as possible. This stipulation is helpful in protecting against impersonation attacks. Group-based D2D communications necessitate careful consideration of revocability.
- **Authentication.** In order to guarantee that merely valid and authorized devices are involved, it is necessary for them to authenticate with each other. User authentication for identity in D2D communications is more distinct than conventional authentication techniques because D2D communications pertain to relay communications and fully distributed communications.
- **Secure routing.** Protect the routing backbone from attacks by implementing secure routing protocols. These kinds of attacks might break or reroute communication. For the sake of honest communication, this aspect is vital.
- **Physical layer security.** The study of physical layer security in D2D systems is currently undergoing development, with ongoing research efforts

focused on the exploration of novel strategies aimed at improving the security of wireless communications. The introduction of purposeful disturbance, sometimes referred to as artificial noise, has the potential to perplex anyone attempting to intercept communication and thereby augment the overall security of the communication channel. Beamforming techniques can be employed to effectively manipulate signal directionality and mitigate vulnerability to jamming attacks.

5.4 CONCLUSION

This chapter demonstrates that the C-D2D communication can enhance things like system capacity, coverage, and connectivity, as well as energy efficiency, support for mobility, privacy and security, and a lot of useful uses. It can also encourage more research and development in wireless communication systems. This chapter comprehensively provides a short overview of the latest developments in D2D and C-D2D communication systems. There is a thorough look at D2D communication, followed by an explanation of why and how C-D2D communication works. In this way, joint D2D communication is an interesting way to make wireless communication systems more reliable and open to more types of devices. C-D2D communication can improve coverage, increase capacity, and lower energy use by using the ability of close devices to work together. Because of the security needs, we looked at previous works to find open research questions and ideas for where future research should go. We found that there is a significant amount of work that needs attention and completion to fix the security problems in D2D networks. As we move forward with our study, we want to make AI-based C-D2D communication systems that will give the next generation of 6G networks intriguing opportunities as well as difficulties.

REFERENCES

[1] J. Laneman, D. Tse and G. Wornell, “Cooperative Diversity in Wireless Networks: Efficient Protocols and Outage Behavior,” in IEEE Transactions on Information Theory, vol. 50, no. 12, pp. 3062–3080, 2004.

[2] I. Budhiraja, N. Kumar, S. Tyagi, S. Tanwar, Z. Han, Md. J. Piran and D. Y. Suh, “A Systematic Review on NOMA Variants for 5G and Beyond,” in IEEE Access, vol. 9, pp. 85573–85644, 2021, doi: 10.1109/ACCESS.2021.3081601.

[3] G. Hou and L. Chen, “D2D Communication Mode Selection and Resource Allocation in 5G Wireless Networks,” in Computer Communications, vol. 155, pp. 244–251, 2020.

[4] L. Song et al., Wireless Device-to-Device Communications and Networks. Cambridge University Press, 2015, doi: 10.1017/CBO9781107478732.

[5] L. Wang and H. Tang, “Critical Technologies for D2D Communications,” in Device-to-Device Communications in Cellular Networks. Cham: Springer International Publishing, 2016, pp. 13–24, doi: 10.1007/978-3-319-30681-0_2.

[6] A. P. Chrysologou, N. D. Chatzidiamantis and G. K. Karagiannidis, “Cooperative Uplink NOMA in D2D Communications,” in IEEE Communications Letters, vol. 26, no. 11, pp. 2567–2571, 2022.

[7] R. Cheng, X. Zhou, H. Zhang, F. Fang and D. Yuan, “Non-Orthogonal Multiple Access in SWIPT Enabled Cooperative D2D Network,” in 2020 IEEE/CIC International Conference on Communications in China (ICCC), Chongqing, pp. 62–67, 2020.

[8] R. Dubey, P. K. Mishra and S. Pandey, "An Energy Efficient Scheme by Exploiting Multi-Hop D2D Communications for 5G Networks," in Physical Communication, vol. 51, p. 101576, 2022.
[9] Z. Aslam, A. Gupta, D. Deepak and N. Mawandia, "Power Saving in Multi-Hop Device to Device Communication," in 2023 4th International Conference for Emerging Technology (INCET), Belgaum, India, pp. 1–4, 2023, doi: 10.1109/INCET57972.2023.10170371.
[10] L. Militano, A. Orsino, G. Araniti, A. Molinaro and A. Iera, "A Constrained Coalition Formation Game for Multi-hop D2D Content Uploading," in IEEE Transactions on Wireless Communications, vol. 15, no. 3, pp. 2012–2024, 2016.
[11] S. Ahmed, "Study of Cluster-Based D2D Communication in Next Generation Mobile Network for the Post-Disaster Response," in 2023 25th International Conference on Advanced Communication Technology (ICACT), Pyeongchang, pp. 243–248, 2023.
[12] M. B. Yashoda and V. Shivashetty, "Bi-CRS: Bio-Inspired Cluster-Based Routing Scheme for D2D Communication in IoT," in R. P. Mahapatra, S. K. Peddoju, S. Roy, P. Parwekar and L. Goel (eds) Proceedings of International Conference on Recent Trends in Computing. Lecture Notes in Networks and Systems, vol. 341. Singapore: Springer
[13] S. K. Gupta, J. Y. Khan and D. T. Ngo, "Cluster-Based D2D Architecture for Safety Services in Vehicular Ad Hoc Networks," in 2018 IEEE Wireless Communications and Networking Conference Workshops (WCNCW), Barcelona, pp. 43–48, 2018.
[14] M. F. Kader, S. M. R. Islam and O. A. Dobre, "Device-to-Device Aided Cooperative NOMA Transmission Exploiting Overheard Signal," in IEEE Transactions on Wireless Communications, vol. 21, no. 2, pp. 1304–1318, 2022, doi: 10.1109/TWC.2021.3103607.
[15] M. F. Kader, S. M. R. Islam and O. A. Dobre, "Simultaneous Cellular and D2D Communications Exploiting Cooperative Uplink NOMA," in IEEE Communications Letters, vol. 25, no. 6, pp. 1848–1852, 2021, doi: 10.1109/LCOMM.2021.3062111.
[16] M. Le, Q.-V. Pham, H.-C. Kim and W.-J. Hwang, "Enhanced Resource Allocation in D2D Communications with NOMA and Unlicensed Spectrum," in IEEE Systems Journal, vol. 16, no. 2, pp. 2856–2866, 2022, doi: 10.1109/JSYST.2021.3136208.
[17] K. Lee, J. Lee, Y. Yi, I. Rhee and S. Chong, "Mobile Data Offloading: How Much Can Wi-Fi Deliver," in IEEE/ACM Transactions on Networking, vol. 21, no. 2, pp. 536–550, 2013.
[18] N. Ristanovic, J.-Y. Le Boudec, A. Chaintreau and V. Erramilli, "Energy Efficient Offloading of 3g Networks," in Proceeding of IEEE Conference on Mobile Ad-Hoc and Sensor Systems (MASS), pp. 202–211, 2011.
[19] Y. Choi, H.W. Ji, J. Yoon Park, H. Chul Kim and J. Silvester, "A 3W Network Strategy for Mobile Data Traffic Offloading," in IEEE Communication Magazine, vol. 49, no. 10, pp. 118–123, 2011.
[20] J. Korhonen, T. Savolainen, A. Ding and M. Kojo, "Toward Network-Controlled IP Traffic Offloading," in IEEE Communication Magazine, vol. 51, no. 3, pp. 96–102, 2013.
[21] P. Janis, C.-H. Yu, K. Doppler, C. Ribeiro, C. Wijting, K. Hugl, O. Tirkkonen and V. Koivunen, "Device-to-Device Communication Underlaying Cellular Communications Systems," in International Journal of Communications, Network and System Sciences, vol. 2, no. 3, pp. 169–178, 2009.
[22] P. Janis, V. Koivunen, C. Ribeiro, J. Korhonen, K. Doppler and K. Hugl, "Interference-Aware Resource Allocation for Device-to-Device Radio Underlaying Cellular Networks," in Proceeding IEEE Vehicular Technology Conference, pp. 1–5, 2009.
[23] B. Kaufman, J. Lilleberg and B. Aazhang, "Spectrum Sharing Scheme Between Cellular Users and Ad-Hoc Device-to-Device Users," in IEEE Transactions on Wireless Communications, vol. 12, no. 3, pp. 1038–1049, 2013.
[24] C.-H. Yu, K. Doppler, C. Ribeiro and O. Tirkkonen, "Resource Sharing Optimization for Device-to-Device Communication Underlaying Cellular Networks," in IEEE Transactions on Wireless Communications, vol. 10, no. 8, pp. 2752–2763, 2011.

[25] H. Min, W. Seo, J. Lee, S. Park and D. Hong, "Reliability Improvement Using Receive Mode Selection in the Device-to-Device Uplink Period Underlaying Cellular Networks," in IEEE Transactions on Wireless Communications, vol. 10, no. 2, pp. 413–418, 2011.
[26] D. H. Lee, K. W. Choi, W. S. Jeon and D. G. Jeong, "Two-Stage Semi-Distributed Resource Management for Device-to-Device Communication in Cellular Networks," in IEEE Transactions on Wireless Communications, vol. 13, no. 4, pp. 1908–1920, 2014.
[27] M. Tehrani, M. Uysal and H. Yanikomeroglu, "Device-to-Device Communication in 5G Cellular Networks: Challenges, Solutions, and Future Directions," in IEEE Communications Magazine, vol. 52, no. 5, pp. 86–92, 2014.
[28] Y. Li, Z. Wang, D. Jin and S. Chen, "Optimal Mobile Content Downloading in Device-to-Device Communication Underlaying Cellular Networks," in IEEE Transactions on Wireless Communications, vol. 13, no. 7, pp. 3596–3608, 2014.
[29] A. Sachan, S. Nigam and A. Bajpai, "An Energy Efficient Virtual-MIMO Communication for Cluster Based Cooperative Wireless Sensor Network," in 2018 9th International Conference on Computing, Communication and Networking Technologies (ICCCNT), Bengaluru, pp. 1–6, 2018, doi: 10.1109/ICCCNT.2018.8493985.
[30] A. Nosratinia, T. E. Hunter and A. Hedayat, "Cooperative Communication in Wireless Networks," in IEEE Communications Magazine, vol. 42, no. 10, pp. 74–80, 2004.
[31] P. K. Malik, D. S. Wadhwa and J. S. Khinda, "A Survey of Device to Device and Cooperative Communication for the Future Cellular Networks," in International Journal Wireless Information Networks, vol. 27, pp. 411–432, 2020.
[32] O. A. Amodu, M. Othman, N. K. Noordin and I. Ahmad, "Relay-Assisted D2D Underlay Cellular Network Analysis Using Stochastic Geometry: Overview and Future Directions," in IEEE Access, vol. 7, pp. 115023–115051, 2019, doi: 10.1109/ACCESS.2019.2933915.
[33] K. Xie, K. Xie, S. He, D. Zhang, J. Wen and J. Lloret, "Busy Tone-Based Channel Access Control for Cooperative Communication," in Transactions on Emerging Telecommunications Technologies, vol. 26, no. 10, pp. 1173–1188, 2015.
[34] K. R. Liu, A. K. Sadek, W. Su and A. Kwasinski, Cooperative Communications, and Networking. Cambridge: Cambridge University Press, 2009.
[35] S. Y. Lien, C. C. Chien, G. S. T. Liu, H. L. Tsai, R. Li and Y. J. Wang, "Enhanced LTE Device-to-Device Proximity Services," in IEEE Communications Magazine, vol. 54, no. 12, pp. 174–182, 2016.
[36] A. Al-Rimawi and D. Dardari, "Analytical Characterization of Device-to-Device and Cellular Networks Coexistence," in IEEE Transactions on Wireless Communications, no. 99, pp. 1–12, 2017.
[37] K. Doppler, M. Rinne, C. Wijting, C. Ribeiro and K. Hugl, "Device-to-Device Communication as an Underlay to LTE-Advanced Networks," in IEEE Communications Magazine, vol. 47, no. 12, pp. 42–49, 2009.
[38] F. H. Kumbhar, N. Saxena and A. Roy, "Reliable Relay: Autonomous Social D2D Paradigm for 5G LOS Communications," in IEEE Communications Letters, vol. PP, no. 99, p. 1, 2017.
[39] Y. L. Chung, "A Study of Advanced D2D Transmission for Use in 5G Cellular Systems," in 2017 International Conference on Information Networking (ICOIN), pp. 298–302, 2017.
[40] B. Chen, J. Zheng and Y. Zhang, "A Time Division Scheduling Resource Allocation Algorithm for D2D Communication in Cellular Networks," in IEEE International Conference on Communications (ICC) 2015, pp. 5422–5428, 2015.
[41] X. Lin, J. Andrews and A. Ghosh, "Spectrum Sharing for Device-to-Device Communication in Cellular Networks," in IEEE Transactions on Wireless Communications, vol. 13, no. 12, pp. 6727–6740, 2014.

[42] B. Sainath and N. Mehta, "Interference-Constrained Optimal Power-Adaptive Amplify-and-Forward Relaying and Selection for Underlay Cognitive Radios," in IEEE Transactions on Communications, vol. 62, no. 8, pp. 2709–2720, 2014.

[43] S. Stefanatos, A. Gotsis and A. Alexiou, "Operational Region of Overlay D2D Communications," in IEEE International Conference on Communications (ICC) 2015, pp. 2541–2547, 2015.

[44] P. Li, S. Guo, T. Miyazaki and W. Zhuang, "Fine-Grained Resource Allocation for Cooperative Device-to-Device Communication in Cellular Networks," in IEEE Wireless Communications, vol. 21, no. 5, pp. 35–40, 2014.

[45] S. G. Peng Li, Cooperative Device-to-Device Communication in Cognitive Radio Cellular Networks. New York: Springer International Publishing, 1st edition, 2014.

[46] H. Wang and X. Chu, "Distance-Constrained Resource-Sharing Criteria for Device-to-Device Communications Underlaying Cellular Networks," in Electronics Letters, vol. 48, no. 9, pp. 528–530, 2012.

[47] Y. Pei and Y. C. Liang, "Resource Allocation for Device-to-Device Communications Overlaying Two-Way Cellular Networks," in IEEE Transactions on Wireless Communications, vol. 12, no. 7, pp. 3611–3621, 2013.

[48] Y. J. Chun, S. Cotton, H. Dhillon, A. Ghrayeb and M. Hasna, "A Stochastic Geometric Analysis of Device-to-Device Communications Operating Over Generalized Fading Channels," in IEEE Transactions on Wireless Communications, vol. PP, no. 99, p. 1, 2017.

[49] Y. Cao, T. Jiang and C. Wang, "Cooperative Device-to-Device Communications in Cellular Networks," in IEEE Wireless Communications, vol. 22, no. 3, pp. 124–129, 2015.

[50] L. Jiang, C. Qin, X. Zhang and H. Tian, "Secure Beamforming Design for SWIPT in Cooperative D2D Communications," in China Communications, vol. 14, no. 1, pp. 20–33, 2017.

[51] Q. Sun, L. Tian, Y. Zhou, J. Shi and X. Wang, "Energy Efficient Incentive Resource Allocation in D2D Cooperative Communications," in 2015 IEEE International Conference on Communications (ICC), pp. 2632–2637, 2015.

[52] I. Krikidis, J. Laneman, J. Thompson and S. McLaughlin, "Protocol Design and Throughput Analysis for Multi-User Cognitive Cooperative Systems," in IEEE Transactions on Wireless Communications, vol. 8, no. 9, pp. 4740–4751, 2009.

[53] J. Yan, D. Wu, S. Sanyal and R. Wang, "Trust-Oriented Partner Selection in D2D Cooperative Communications," in IEEE Access, vol. 5, pp. 3444–3453, 2017.

[54] G. George, R. K. Mungara and A. Lozano, "An Analytical Framework for Device-to-Device Communication in Cellular Networks," in IEEE Transactions on Wireless Communications, vol. 14, no. 11, pp. 6297–6310, 2015.

[55] Y. Yang, Y. Zhang, L. Dai, J. Li, S. Mumtaz and J. Rodriguez, "Transmission Capacity Analysis of Relay-Assisted Device-to-Device Overlay/Underlay Communication," in IEEE Transactions on Industrial Informatics, vol. 13, no. 1, pp. 380–389, 2017.

[56] J. Z. Moghaddam, M. Usman, F. Granelli and H. Farrokhi, "Cognitive Radio and Device-to-Device Communication: A Cooperative Approach for Disaster Response," in 2016 IEEE Global Communications Conference (GLOBECOM), pp. 1–6, 2016.

[57] M. D. Selvaraj and R. K. Mallik, "Performance Analysis of a Multi-Relay Cooperative Diversity Network with Decode and Forward Relaying," in 2010 National Conference on Communications (NCC), pp. 1–5, 2010.

[58] A. Bletsas, A. Khisti, D. P. Reed and A. Lippman, "A Simple Cooperative Diversity Method Based on Network Path Selection," in IEEE Journal on Selected Areas in Communications, vol. 24, no. 3, pp. 659–672, 2006.

[59] S. S. Ikki and M. H. Ahmed, "Performance Analysis of Adaptive Decode-and-Forward Cooperative Diversity Networks With Best-Relay Selection," in IEEE Transactions on Communications, vol. 58, no. 1, pp. 68–72, 2010.

[60] B. Ma, H. Shah-Mansouri and V. W. S. Wong, "A Matching Approach for Power Efficient Relay Selection in Full Duplex D2D Networks," in 2016 IEEE International Conference on Communications (ICC), pp. 1–6, 2016.
[61] S. Shalmashi and S. B. Slimane, "Cooperative Device-to-Device Communications in the Downlink of Cellular Networks," in 2014 IEEE Wireless Communications and Networking Conference (WCNC), pp. 2265–2270, 2014.
[62] M. Hasan and E. Hossain, "Distributed Resource Allocation for Relay-Aided Device-to-Device Communication Under Channel Uncertainties: A Stable Matching Approach," in IEEE Transactions on Communications, vol. 63, no. 10, pp. 3882–3897, 2015.
[63] M. Wang and Z. Yan, "A Survey on Security in D2D Communications," Mobile Networks Applications, vol. 22, pp. 195–208, 2017. doi: 10.1007/s11036-016-0741-5.

6 UAV-Assisted Communication

Unmanned Aerial Vehicles in Vehicle-to-Everything Communication Networks

Kiran Kumar Godugu and Suseela Vappangi

6.1 INTRODUCTION AND LITERATURE REVIEW

Unmanned aerial vehicles (UAVs) garnered substantial consideration for their potential in revolutionizing various industries, including transportation, emergency response, and monitoring. Among their many applications, UAVs offer a promising solution to enabling efficient vehicle-to-everything (V2X) communication networks. However, several challenges need to be addressed to fully exploit the potential of UAV-based V2X communication networks [1]. V2X technology uses sensors, cameras, and wireless connectivity, including Wi-Fi, radio frequencies, Long-Term Evolution (LTE), and fifth-generation (5G), to allow automobiles to communicate with other vehicles, the people in them, and the environment around them. Trajectory planning and power allocation (PA) for an automobile network using a UAV acts as a relay for linking two independent distant automobiles [2].

6.1.1 Existing Works

Later, sixth-generation (6G) V2X services, like intended sharing, interactive gaming, and coordinated driving [3], should make self-driving cars more secure, more adaptable, as well as more effective [4]. The achievement of those programs depends on analyzing and combining huge amounts of information via devices placed in cars, on roadways, and in other kinds of infrastructure to make sure that users have a clear picture of their surroundings. The individual vehicle's potential to compute and perceive keeps getting better. But because robust CPUs and high-precision devices are expensive, the price-to-performance ratio of an intelligent car is low. The way people in the area think still limits the performance of 6G V2X applications, even though you do not look at the price performance proportion. Additionally, the range of the vehicles will be reduced by the power needed for sensing and data processing. Programs that need a lot of computing resources are a big problem for vehicular platforms that do not have a lot of resources. In [5], the authors are interested in bringing greater interest to the intriguing but difficult UAV-assisted 6G vehicular

DOI: 10.1201/9781032625034-6

edge computing (VEC) set-up. They call for additional investigations to have been done to solve the numerous challenges that need to be fixed before it can reach its maximum efficiency.

Recent breakthroughs in UAV technology have brought about a wide range of sectors where UAVs are deployed for intelligent transportation system (ITS) applications. Search-and-rescue activities [6] and out-of-sight indoor flights with tactile feedback [7] are two examples of the uses of UAVs. Other things that UAVs have been used for include transporting goods, packages, and people [8]. UAVs with cameras on board can be used in farmlands to keep an eye on the height and health of crops. Inter-vehicle contact can be added to UAVs to make them more useful. UAVs can have communication lines added to them so they can talk to each other or to other UAVs. A way for UAVs and CAVs to talk to each other is through dedicated short-range communication (DSRC). At the moment, Automatic Dependent Surveillance-Broadcast (ADS-B) is the standard way for the aerospace business to send location data. The writers of [9], on the other hand, say that ADS-B probably will not be able to handle low-altitude air traffic communication for much longer because there will be more of it. In [10], the writers looked into how DSRC can be used instead of ADS-B for UAV-to-UAV communication to avoid low-level collisions in the air. The writers of [11] looked into how a UAV-enabled ITS could be used and what problems it might cause in a smart city.

An in-depth look at how UAV channels are defined by using testing methods and statistically significant channel structures is given in [12]. In [13, 14], a full look at how UAVs will connect to wireless networks that go beyond 5G is given. We talked about the main points of the current study on UAV activities that use 5G methods from the physical layer, the network layer, and the sharing, computing, and caching layers all at the same time. There is a full lesson on how to use UAVs and what they can do for you in [15]. UAVs could also be used as moving base stations (BSs), servers, or switches in wireless networks to make them more reliable, cover more areas, and use less energy. The authors of [16] wrote a list of some of the most important technologies and innovative parts of 6G-V2X systems. To do this, they list the features and perks that these technologies could have, which are much more than what 5G can do. Machine learning (ML) can now be used better in 6G car networks, and we talked about some of those changes. We looked at a single-cell multi-UAV network in [17]. In this network, different UAVs send data to the base station using U2N and U2U signals. You can make these single-cell networks work with more than one cell by thinking about how the cells talk to each other and join with BS.

6.1.2 Motivations and Contributions

As UAVs are becoming more affordable, they will no longer limit mobile users to terrestrial mobile stations when it comes to future wireless networks. It is shown in Figure 6.1 that there are numerous possibilities for UAVs in wireless network, such as UAV swarm networks in natural disasters, UAV-aided V2X communications [1], UAV-enabled mobiles networks, ISP groups, V2X communication networks, and smart city corporate networks. By incorporating aerial and terrestrial network elements whose services are a bit more adaptable than the existing stationary ground telecommunications system, a three-dimensional (3D) structure will evolve into

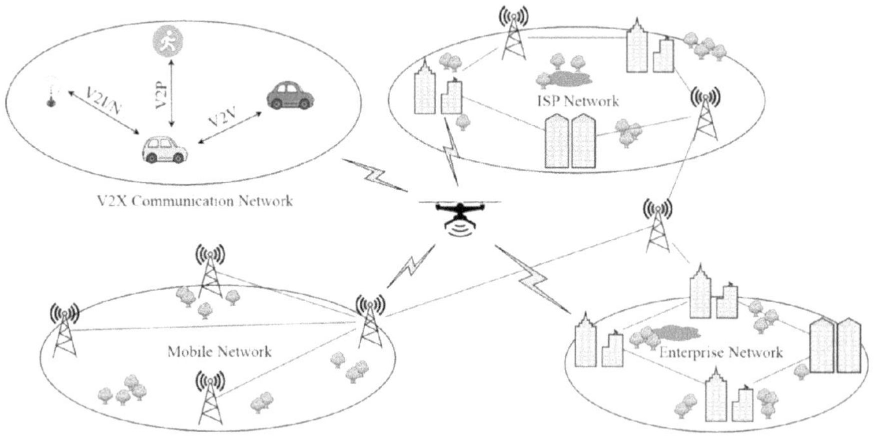

FIGURE 6.1 Applications of UAVs in wireless communication networks.

the wireless communications system. Spectrum access for UAV communications is important in the design and administration of the overall communications network, which is a key factor in the advancement of aerial platforms. Reliability of wireless transmissions is ensured by the minor signal attenuation, owing to less impediments in the way of aerial communications, including air-to air (A2A), air-to-ground (A2G), and ground-to-ground (G2G) communications. The signal in A2A communications travels through air at a speed like that in open space. The probability of a line-of-sight (LoS) or non-line-of-sight (NLoS) links between an UAV and a ground node varies depending on the terrain and the type of network being used.

In general, UAV transmissions occur in an unlicensed radio spectrum, such as the ISM bands (2.4 GHz and 5.8 GHz) [18]. Multi-hop 802.11 or Zigbee technologies are evaluated for small- and medium-range UAV applications based on its throughput and range requirements [19]. High-throughput systems necessitate the consideration of additional wireless technologies and spectrum. When using unlicensed spectrum, UAVs are more vulnerable to security risks and assaults that could compromise the transmission of sensitive data. Furthermore, the unlicensed spectrum is growing saturated, and UAVs may soon face spectrum shortages because of the dramatic expansion in the number of wireless devices (such as tablets, smartphones, and sensors) that also operate in unlicensed frequency.

In [20], the authors gave a comprehensive summary of the V2X architecture. Major security/privacy concerns, regulatory efforts, and defence methods currently proposed inside the V2X area were also discussed. The last step is to define potential open concerns and identify semantic gaps in current security remedies. In [21], the authors provided an in-depth look at the current developments in LTE and 5G that make possible V2X communications. They prove that LTE's established architectures cleared the way for V2X communications based on LTE thanks to its increased coverage, higher capacity, and widespread use. Under a realistic description of the channels used by aircraft, the authors of [22] propose a new system in which a UAV can aid in communication via V2X. For the V2X service, designers also tasked themselves with finding the sweet spot between maximizing SE and minimizing power consumption along the UAV's flight path. From a cross-layer layout perspective, the authors of [23] examine the benefits of media access control (MAC) protocols at the link layer and routing algorithms at the network layer. Centralized multi-hop wireless (CMW) is a suggested network design that makes use of UAV to provide the separated data–plane and network-control-plane broadcasts essential to multi-hop vehicle networks. Inspired by all the aforementioned works, we look forward to the integration of UAVs into V2X communication networks that hold tremendous promise for revolutionizing our transportation systems. The wider coverage, increased capacity, and freedom from geographical limits made possible by UAVs are just a few of the ways in which they might help usher in a more secure, more productive era.

The primary benefits this chapter adds are a broad look at the recent progress in science and technology that could affect how future wireless networks with UAVs helping V2X communication networks are designed. In contrast to [24], which looks at how the two main V2X technologies have changed over time, along with big data-driven Internet-of-Vehicle (IoV) and cloud-based IoV, we look ahead and include a lot of different points of view, such as new materials, algorithms, and system structures, in order to spark a lot of new research for 6G-V2X and beyond. We also want to stress how important ML will be for innovative usage occurrences for the upcoming vehicle communication networks.

6.1.3 Organization of the Chapter

The remaining parts of this chapter are broken down into the following sections. In Section 6.2, a summary of recent advancements in vehicle networks is provided, along with open difficulties and issues. In the final section, we will summarize our findings from the investigation.

6.2 PROPOSED UAV-AIDED V2X COMMUNICATION

6.2.1 Overview of V2X Communication

The potential of device-to-device (D2D) communication technology as a viable alternative for the next wireless networks has been widely acknowledged [25]. D2D approaches facilitate the potential uses by utilizing direct data exchange between devices in close proximity [26]. Numerous contemporary studies have explored the issue of resource sharing across a range of different approaches. The primary objective of these efforts is to improve the efficiency of frequency reuse when cellular users and D2D communications are permitted to utilize the same resources [27]. V2X has been seen as a distinct instance of D2D communications, characterized by varying network architectures. For instance, the implementation of vehicle-to-infrastructure (V2I), vehicle-to-network (V2N), and vehicle-to-vehicle (V2V) communication systems has the potential to enhance the safety and reliability of transportation.

Numerous standards have been created for LTE/LTE-Advanced (LTE-A) systems, such as the LTE-vehicle and LTE-Unlicensed, as referenced in [28, 29]. Despite the numerous benefits, the primary obstacle faced by V2X communication in IoT networks pertains to the control of interference among devices. Consequently, it is imperative to develop a suitable architecture that effectively mitigates mutual interference in order to guarantee the quality of service (QoS) for the entire network. Most existing studies on V2X communications have mostly focused on addressing interference management issues in the uplink data transmission. This emphasis is attributed to the challenges associated with establishing effective algorithms for resource reuse in the downlink data transmission. Hence, the allocation of spectrum among Internet of Things (IoT) devices remains a formidable task for upcoming wireless networks [30, 31].

In recent times, UAVs have garnered significant attention across various domains, including but not limited to package delivery, traffic management, agricultural practices, military operations, and photography [32–36]. Cellular wireless networks often face challenges, such as blind spots inside macrocells, high-speed data demand, low latency requirements, and the need for reliable data transfer. UAVs offer promising prospects in various applications, owing to their inherent advantages in terms of mobility, cost-effectiveness, and ability to be deployed in three-dimensional space. There are numerous challenges associated with the utilization of UAVs, which are counterbalanced by their advantages. These challenges include the need to maximize spectral efficiency (SE), achieve optimal two-dimensional (2D) or three-dimensional (3D) deployment, plan efficient flight paths, and minimize flight duration. This chapter examines the utilization of UAV relaying to enhance V2X communications, specifically focusing on optimizing the trajectory of the UAV to enhance downlink data transfer in the V2X network.

6.2.2 Evaluation and Classification of V2X Communication

Roads are a fundamental component of infrastructure and a crucial factor in facilitating the mobility and logistics of vehicles. The establishment of an interconnected network of cities, ports, airports, and countries serves as a fundamental foundation for fostering social and economic development. In recent years, there has been a significant rise in population growth and urbanization, leading to a rapid increase in car ownership across all nations [37]. According to the United Nations, the proportion of the global population residing in urban areas was reported to be 30% in 1950. Presently, this figure has escalated to 54%. According to the United Nations, it is projected that by the year 2050, over 66% of the global population would reside in urban areas [38–40]. The future of transportation will be influenced by a number of social developments, including but not limited to the following:

- A new wave of people moving to cities is putting stress on the transportation systems that are already in place.
- Even stricter rules and policies about emissions set by the government.
- There is a lot of demand on transportation systems and logistics/delivery facilities because new business ideas are coming up.
- Live and open data flow administration that works well.

So the straight way to deal with this scary problem is to improve the infrastructure and come up with different ways to make transportation smarter. These solutions must be sustainable on both an economic and environmental level. So here are some of the things we expect to happen in the transportation world over the next two to five years [41, 42]:

- Non-accidental traveling.
- For situations with more traffic, a flexible aiding system will be implemented.
- Increasing reliance on automobiles.
- Transportation that is both eco-friendlier and more efficient (electric cars).

In 1994, the United States Department of Transportation renamed intelligent transportation systems (ITS) from its predecessor as intelligent vehicle-highway systems (IVHS) for enhancing the services of the transport system. This was made possible by advances in IT, specifically sensors, geolocation, and communication technologies, which have enabled the invention of incredible new benefits that have offered a means of intelligently executing methods by taking into account context, in order to improve traffic flow, ITS facilitates communication between roads, pedestrians, vehicles, and urban infrastructure [43]. The primary goals of ITS are the review, growth and development, analysis, and integration of cutting-edge technology in order to maximize productivity, resource conservation, reduction in time, security, sustainability, and satisfaction among various traffic users [44–46]. Therefore, with the aid of ITS, the chaotic transportation environment may be converted into an intelligent and interconnected network. The International Society of Automotive Engineers (SAE) [47] has developed a six-stage plan for achieving intelligence and adaptive management. We may soon see how the industry responds to the trend toward autonomous vehicles. Business names like Waymo and Tesla, as well as others in the auto industry, are evolving to pave the way for the rise of autonomous vehicles. However, many automotive organizations and researchers still hope to one day realize the full potential of level 4 and level 5 vehicle automation. As a result, communication between all vehicles and road users is crucial if full transportation automation is to be achieved. In this setting, communication breathes new life into the ITS purpose by enabling cooperative vehicle in everything-based approaches to transportation. More security and interdependence on mobility through LOS and NLoS situations is possible through vehicle-to-vehicle (V2V), vehicle-to-pedestrian (V2P), vehicle-to-infrastructure (V2I), and vehicle-to-network (V2N) communication, as shown in Figure 6.2 [48].

6.2.3 Applications of V2X Communication

- Traffic monitoring and management
- Emergency response and disaster management
- Infrastructure inspection and maintenance
- Communication relay and network coverage enhancement
- Dynamic data collection for predictive analysis
- Precision agriculture and delivery services
- Simulation and testing for autonomous vehicles

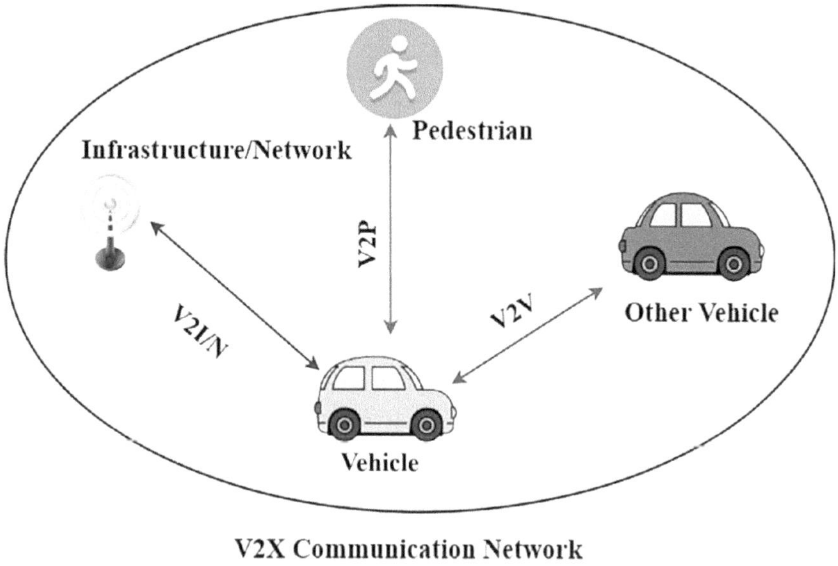

FIGURE 6.2 A typical V2X communication network.

Here we present a new paradigm for improving V2X communication with the use of UAVs. With multiple important contributions, our suggested paradigm is a major step forward in the field. Taking these issues into account, we propose a UAV-aided V2X communication system that can improve communication range, reduce congestion and latency, increase resilience to disruptions, maximize resource utilization, address privacy and security concerns, scale to meet increasing traffic demands, and adapt to changing traffic conditions, among other things. By resolving these problems, we want to usher in a new era of robust V2X communication that will aid in the development of IVS.

6.2.4 Integration of UAV-Aided V2X Communication Networks

We think about a V2X communication system, whereas a UAV assists a cellular downlink system with a BS which is placed at the cell centre. As shown in Figure 6.3, a BS with N antennas is used to support a vehicular user (VU) in a vehicle. UAV relays with a single antenna are utilized to aid the BS in data transmission, which boosts the V2X system's SE. To that end, the UAV can boost data rates for the VU so long as the overall throughput offered by that UAV does not exceed the throughput of the backhaul connection between the BS and the UAV.

The variant of the network that was looked at was suggested in [22]. It has a BS, a UAV-mounted relay, and two NOMA cars that are not connected to each other, as shown in Figure 6.4. These cars move along two different lanes of a street at speeds that stay at V1 and V2. There is not a straight link between the BS and the vehicles

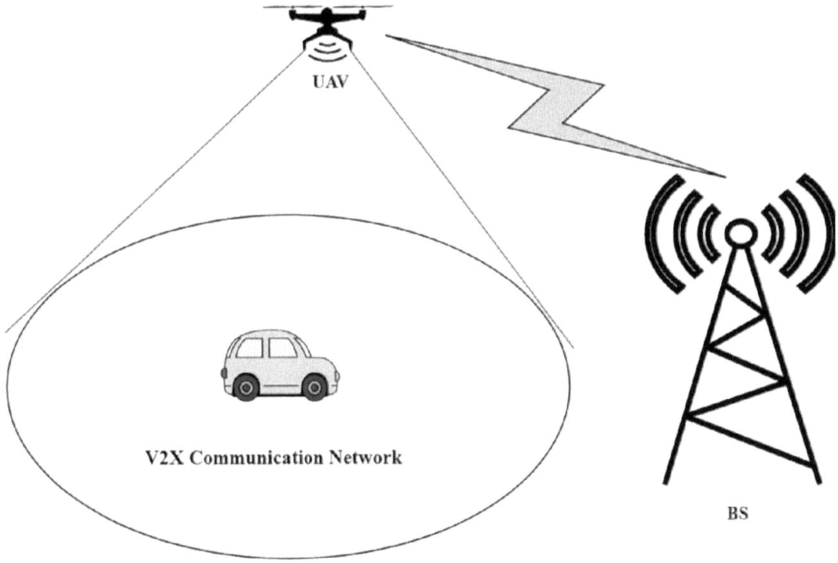

FIGURE 6.3 An illustration of UAV-assisted V2X communication network.

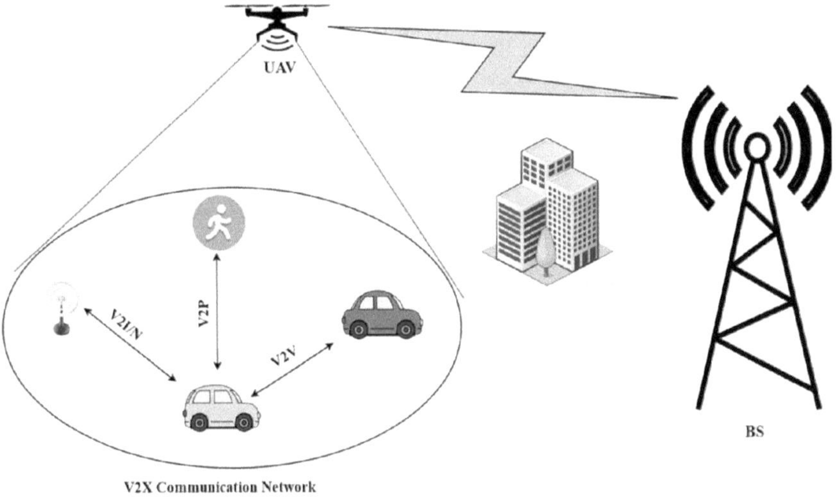

FIGURE 6.4 A system model for UAV-aided V2X communication network.

because of things like buildings that block the path and a lot of path loss. The UAV helps set up a communication link for these vehicles that are not connected. The relay server is thought to use the AF protocol and operate in half-duplex mode.

As UAVs have the ability to cover large areas, they could be used as radio entry points in the 6G–V2X communication systems. UAVs can help people in cars in a number of ways, including by transmitting, caching, and computing [49]. UAVs can work with static network infrastructure points (like BSs) to manage the wireless systems and make the user practice more effective. This is especially true in areas with a lot of vehicles. Since UAVs can move around in three dimensions almost without any limits, they can be used in a variety of V2X situations as a flying agent. For example, they may detect a road accident before the rescue team gets there, help police by keeping an eye on traffic violations, and send out warnings about road hazards that happened in an area that did not have an RSU already [11].

Even though UAV technology has come a long way, there are still some problems with V2X platforms that use UAVs. For instance, keeping reliable and fast wireless communication going between UAVs and vehicles on the ground is very hard because both the UAVs and the vehicles on the ground can move around, which makes channel features very unstable. LoS links between UAVs and ground vehicles are predicted, but measurements and models of these channels are still being made [50]. Other big problems are safety and rules, making sure that new networks work with old ones, and the fact that UAV batteries do not last very long. For 6G–V2X transmission, satellites are another possible way to talk over the air. Figure 6.4 shows an example of a V2X communication device that uses UAVs to help it work. Now, V2X guidelines only allow satellites to be used for locating devices. It is important to note that satellite transmission data rates have been going up a lot over the last few years.

Multibeam satellites [51], for instance, are widely used in satellite communication systems because they can speed up wireless data rates. So talking over a satellite could be a way for 6G–V2X to help the connection between a vehicle and a remote data server when there is no terrestrial coverage. A UAV can do computing and network management chores, just like a UAV-based V2X transmission. To make UAV-assisted V2X communications possible, a lot of research needs to be done to correctly model how the channels between satellites and high-mobility vehicles work. It will also be hard to combine the various ways of communicating used in V2X and UAV communications, such as the physical (PHY) or medium access control (MAC) layer transfer protocol.

Another important study question in the field of PHY is how to get the most efficient power transmission for high-speed, long-distance communications. This might not be possible with OFDM because it has a high peak-to-average power ratio (PAPR), which limits how far it can communicate. Researchers need to think again about, and try again, a few old modulation methods that work with steady signal patterns, like CPM [52]. One problem with traditional CPM is that it might not be able to handle high-speed transmission as well as OFDM does [53]. Multiple-input, multiple-output (MIMO) could be a solution to this issue, but CPM's constant phase constraint makes it impossible to easily combine with MIMO. For load-modulated MIMO [54, 55], a new phase modulation on the hypersphere (PMH) has been created not long ago. It is known that PMH can get the best power efficiency and is also getting close to the capacity of Gaussian inputs, which is a benefit. It is important to look into how PMH can be used for UAV-aided long-distance V2X.

6.2.5 Why UAVs in V2X Communication Networks

- UAVs can serve as aerial relay stations (ABS), transmitting data between vehicles and ground facilities across large distances.
- Because of their mobility, UAVs can bridge the distance between nodes in V2X networks.
- UAVs with sensors and cameras can be used for traffic monitoring.
- UAVs can be extremely useful in V2X networks because of the situational awareness they can provide to first responders during crises and catastrophes.
- UAVs serving as ABS must maintain constant and stable connections to moving cars and the current V2X network.

6.3 OPEN CHALLENGES AND FUTURE RESEARCH DIRECTIONS

The following represent a handful of the most significant topics for future study and open problems that need to be solved for UAV-based V2X communication networks to help them grow and be used more [1]:

- **Security and privacy threats.** Physical access to UAVs can lead to tampering, data theft, or disruption of communication systems, and secure communication protocols, implementation of GPS-based authentication, and employing anti-spoofing techniques to verify the identity of communicating UAVs are a must.
- **Resource allocation for UAV–V2X systems.** Efficient resource allocation in UAV–V2X systems requires a combination of intelligent algorithms, dynamic adaptation mechanisms, and compliance with regulatory standards. Continuous monitoring and optimization based on real-time data and system requirements are essential for maximizing the performance of these systems.
- **Beamforming or precoding techniques.** Beamforming or precoding techniques are used to focus the communication signals on specific directions, enhancing the signal quality at the intended receivers, with improved signal-to-noise ratio (SNR), increased coverage, and reduced interference.
- **Anti-jamming techniques.** Beamforming and precoding techniques are used to focus the communication signals on specific directions, enhancing the signal quality at the intended receivers, with improved signal-to-noise ratio (SNR), increased coverage, and reduced interference, whereas the design of jamming vectors is used to counteract potential jamming attacks from adversaries, frequency hopping, spread spectrum modulation, and adaptive power control.
- **UAV deployment.** UAVs can provide extended coverage and improved line-of-sight communication, and they can dynamically move to areas with high communication demand or act as relays in areas with poor connectivity.
- **V2X secrecy rates.** UAVs can be strategically positioned to minimize interference and enhance secrecy rate. UAVs can act as relays, enabling

multi-hop communication to improve coverage. But efficient relay strategies must be designed to maintain secrecy during multi-hop communication.

- **Network energy efficiency.** UAVs can be equipped with energy-efficient communication systems. Service suppliers must make effective use of the battery power available on UAVs because of the persistent energy constraints imposed by UAVs.
- **Trajectory design for UAVs.** When designing V2X safety devices, it is important to think about how legitimate cars and snoops move around, as well as how NLoS communications paths can take advantage of changes in elevation angles. Also, for V2X secure communications when UAV jammers set up security areas, the UAV's height and horizontal path must be planned together.

6.3.1 Future Research Directions

Additionally, we are interested in looking into smart techniques based on artificial intelligence (AI) in the future to make this design more useful and able to adapt to more complex and real-world V2X situations.

6.4 CONCLUSION

In this chapter, the integration of UAVs into V2X environment is primarily reliant on resolving communication reliability and power challenges. Academics and practitioners must collaborate to solve these challenges in UAV-based V2X communication networks in order to enable revolutionary applications and services in a number of industries. This chapter delivers a thorough summary of the recent features of UAV-based V2X communication networks for future wireless networks. It discusses their major issues, technological advancements, and potential applications. It also illustrates how UAV-based V2X communication networks could enhance their performance and usage in terms of improved SNR, greater coverage, and decreased interference in future. It is also shown where studies could go to improve their performance and use in terms of improved SNR, increased coverage, and reduced interference. As a result, the purpose of this chapter is to study the design and implementation issues of V2X communication networks based on UAVs in the context of ITS for a smart city.

REFERENCES

[1] B. Shang, L. Liu, J. Ma and P. Fan, "Unmanned Aerial Vehicle Meets Vehicle-to-Everything in Secure Communications," *in IEEE Communications Magazine*, vol. 57, no. 10, pp. 98–103, October 2019, doi: 10.1109/MCOM.001.1900170.

[2] S. Gyawali, S. Xu, Y. Qian and R. Q. Hu, "Challenges and Solutions for Cellular Based V2X Communications," *in IEEE Communications Surveys & Tutorials*, vol. 23, no. 1, pp. 222–255, 2021, doi: 10.1109/COMST.2020.3029723.

[3] J. Hu, C. Chen, T. Qiu and Q. Pei, "Regional-Centralized Content Dissemination for ev2x Services in 5G mm Wave-Enabled IOVs," *in IEEE Internet of Things Journal*, p. 1, 2020.

[4] S. Zhang, J. Chen, F. Lyu, N. Cheng, W. Shi and X. Shen, "Vehicular Communication Networks in the Automated Driving Era," *in IEEE Communications Magazine*, vol. 56, no. 9, pp. 26–32, September 2018.
[5] J. Hu, C. Chen, L. Cai, M. R. Khosravi, Q. Pei and S. Wan, "UAV-Assisted Vehicular Edge Computing for the 6G Internet of Vehicles: Architecture, Intelligence, and Challenges," *in IEEE Communications Standards Magazine*, vol. 5, no. 2, pp. 12–18, June 2021, doi: 10.1109/MCOMSTD.001.2000017.
[6] S. Waharte and N. Trigoni. "Supporting Search and Rescue Operations with UAVs," *in Proceedings of the 2010 International Conference on Emerging Security Technologies*, Canterbury, pp. 142–147, 6–7 September 2010.
[7] O. Kavas and H. Gurocak, "Haptic Interface with Linear Magnetorheological (MR) Brakes for Drone Control," *in Proceedings of the 2018 15th International Conference on Ubiquitous Robots (UR)*, Honolulu, HI, pp. 676–681, 26–30 June 2018.
[8] R. Kellermann, T. Biehle and L. Fischer, "Drones for Parcel and Passenger Transportation: A Literature Review," *in Transportation Research Interdisciplinary Perspectives*, vol. 4, p. 100088, 2020.
[9] A. Moore, S. Balachandran, S. D. Young, E. T. Dill, M. J. Logan, L. J. Glaab, C. Muñoz and M. Consiglio, "Testing Enabling Technologies for Safe UAS Urban Operations," *in 2018 Aviation Technology, Integration, and Operations Conference*, 2018. Reston, VA: American Institute of Aeronautics and Astronautics.
[10] C. Chakrabarty, A. Ippolito, J. Baculi, K. S. Krishnakumar and S. Hening, "Vehicle to Vehicle (V2V) Communication for Collision Avoidance for Multi-Copters Flying in UTM? TCL4," *in AIAA Scitech 2019 Forum*, 2019. Reston, VA: American Institute of Aeronautics and Astronautics.
[11] H. Menouar, I. Guvenc, K. Akkaya, A. S. Uluagac, A. Kadri and A. Tuncer, "UAV-Enabled Intelligent Transportation Systems for the Smart City: Applications and Challenges," *in IEEE Communications Magazine*, vol. 55, pp. 22–28, 2017.
[12] A. A. Khuwaja, Y. Chen, N. Zhao, M.-S. Alouini and P. Dobbins, "A Survey of Channel Modeling for UAV Communications," *in IEEE Communications Surveys & Tutorials*, vol. 20, no. 4, pp. 2804–2821, 2018, doi: 10.1109/COMST.2018.2856587.
[13] B. Li, Z. Fei and Y. Zhang, "UAV Communications for 5G and Beyond: Recent Advances and Future Trends," *in IEEE Internet of Things Journal*, vol. 6, no. 2, pp. 2241–2263, April 2019, doi: 10.1109/JIOT.2018.2887086.
[14] S. Poudel and S. Moh, "Task Assignment Algorithms for Unmanned Aerial Vehicle Networks: A Comprehensive Survey," *in Vehicular Communications*, vol. 35, p. 100469, June 2022, doi: 10.1016/j.vehcom.2022.100469.
[15] X. Gu and G. Zhang, "A Survey on UAV-Assisted Wireless Communications: Recent Advances and Future Trends," *in Computer Communications*, vol. 208, pp. 44–78, 2023, doi: 10.1016/j.comcom.2023.05.013.
[16] M. Noor-A-Rahim et al., "6G for Vehicle-to-Everything (V2X) Communications: Enabling Technologies, Challenges, and Opportunities," *in Proceedings of the IEEE*, vol. 110, no. 6, pp. 712–734, June 2022, doi: 10.1109/JPROC.2022.3173031.
[17] S. Zhang, H. Zhang, B. Di and L. Song, "Cellular UAV-to-X Communications: Design and Optimization for Multi-UAV Networks," *in IEEE Transactions on Wireless Communications*, vol. 18, no. 2, pp. 1346–1359, February 2019, doi: 10.1109/TWC.2019.2892131.
[18] Y. Saleem, M. H. Rehmani and S. Zeadally, "Integration of Cognitive Radio Technology with Unmanned Aerial Vehicles: Issues, Opportunities, and Future Research Challenges," *in Journal of Network and Computer Applications*, vol. 50, pp. 15–31, 2015.
[19] D. Lee, J. Lim and H. Baek, "An Airborne Communication Relay Scheme for IEEE 802.11 WLAN Based Network," *in 2018 International Conference on Information Networking (ICOIN)*, pp. 426–431, 2018.

[20] M. Hasan, S. Mohan, T. Shimizu and H. Lu, "Securing Vehicle-to-Everything (V2X) Communication Platforms," *in IEEE Transactions on Intelligent Vehicles*, vol. 5, no. 4, pp. 693–713, December 2020, doi: 10.1109/TIV.2020.2987430.

[21] O. Abbasi, H. Yanikomeroglu, A. Ebrahimi, N. Mokari and M. Alzenad, "Dynamic NOMA/OMA for V2X Networks with UAV Relaying," *in 2020 IEEE 92nd Vehicular Technology Conference (VTC2020-Fall)*, Victoria, BC, pp. 1–7, 2020, doi: 10.1109/VTC2020-Fall49728.2020.9348520.

[22] T. A. Kiet, N. T. L. Quyen, H. P. Hieu and V. T. Hung, "A Novel Approach for UAV-Aided Vehicle-to-Everything Networks," *in 2022 16th International Conference on Ubiquitous Information Management and Communication (IMCOM)*, Seoul, pp. 1–4, 2022, doi: 10.1109/IMCOM53663.2022.9721776.

[23] X. Ma, L. Wang, W. Han, X. Wang and T. Shang, "A UAV-Assisted V2X Network Architecture with Separated Data Transmission and Network Control," *in China Communications*, vol. 20, no. 6, pp. 260–276, June 2023, doi: 10.23919/JCC.2023.00.030.

[24] H. Zhou, W. Xu, J. Chen and W. Wang, "Evolutionary V2X Technologies Toward the Internet of Vehicles: Challenges and Opportunities," *in Proceedings of the IEEE*, vol. 108, no. 2, pp. 308–323, February 2020.

[25] M. N. Tehrani, M. Uysal and H. Yanikomeroglu, "Device-to-Device Communication in 5G Cellular Networks: Challenges, Solutions, and Future Directions," *in IEEE Communications Magazine*, vol. 52, no. 5, pp. 86–92, May 2014.

[26] "3rd Generation Partnership Project; Technical Specification Group SA; Study on Architecture Enhancements to Support Proximity Services (Prose) (Release 12)," 3GPP, Technical Report, 23.703 v2.0.0, March 2014.

[27] H. M. Kim, H. V. Nguyen, G.-M. Kang, Y. Shin and O.-S. Shin, "Device-to-Device Communications Underlaying an Uplink SCMA System," *in IEEE Access*, vol. 7, pp. 21756–21768, 2019.

[28] S. Chen, J. Hu, Y. Shi and L. Zhao, "LTE-V: A TD-LTE-Based V2X Solution for Future Vehicular Network," *in IEEE Internet of Things Journal*, vol. 3, no. 6, pp. 997–1005, 2016.

[29] K. Xiong, S. Leng, C. Huang, C. Yuen and Y. L. Guan, "Intelligent Task Offloading for Heterogeneous V2X Communications," *in IEEE Transactions on Intelligent Transportation Systems*, vol. 22, no. 4, pp. 2226–2238, 2021.

[30] B. Antonescu, M. T. Moayyed and S. Basagni, "mmWave Channel Propagation Modeling for V2X Communication Systems," *in Proceedings of the 2017 IEEE 28th Annual International Symposium on Personal, Indoor, and Mobile Radio Communications*, pp. 1–6, October 2017, doi: 10.1109/PIMRC.2017.8292718.

[31] R. I. Ansari, C. Chrysostomou, S. A. Hassan, M. Guizani, S. Mumtaz, J. Rodriguez and J. J. P. C. Rodrigues, "SG D2D Networks: Techniques, Challenges, and Future Prospects," *in IEEE Systems Journal*, vol. 12, no, 4, pp. 3970–3984, 2018.

[32] S. Zhang, H. Zhang and L. Song, "Beyond D2D: Full Dimension UAV-to-Everything Communications in 6G," *in IEEE Transactions on Vehicular Technology*, vol. 69, no. 6, pp. 6592–6602, 2020.

[33] M. Mozaffari, W. Saad, M. Bennis, Y. Nam and M. Debbah, "A Tutorial on UAVs for Wireless Networks: Applications, Challenges, and Open Problems," *in IEEE Communications Surveys & Tutorials*, vol. 21, no. 3, pp. 2334–2360, 2019.

[34] H. Wang, H. Zhao, J. Zhang, D. Ma, J. Li and J. Wei, "Survey on Unmanned Aerial Vehicle Networks: A Cyber Physical System Perspective," *in IEEE Communications Surveys & Tutorials*, vol. 22, no. 2, pp. 1027–1070, 2020.

[35] Y. Zeng, R. Zhang and T. J. Lim, "Throughput Maximization for UAV-Enabled Mobile Relaying Systems," *in IEEE Transactions on Communications*, vol. 64, no. 12, pp. 4983–4996, December 2016.

[36] X. Li, J. Tan, A. Liu, P. Vijayakumar, N. Kumar and M. Alazab, "A Novel UAV-Enabled Data Collection Scheme for Intelligent Transportation System Through UAV Speed Control," *in IEEE Transactions on Intelligent Transportation Systems*, vol. 22, no. 4, pp. 2100–2110, 2021.
[37] I. Soto, M. Calderon, O. Amador and M. Urueña, "A Survey on Road Safety and Traffic Efficiency Vehicular Applications Based on C-V2X Technologies," *in Vehicular Communications*, vol. 33, p. 100428, 2022.
[38] T. Buettner, "Urban Estimates and Projections at the United Nations: The Strengths, Weaknesses, and Underpinnings of the World Urbanization Prospects," *in Spatial Demography*, vol. 3, pp. 91–108, 2015.
[39] L. Mooren, "United Nations Road Safety Collaboration (UNRSC)," *in Journal of Road Safety*, vol. 25, no. 3, pp. 49–51, 2014.
[40] M. Aftabuzzaman, "Measuring Traffic Congestion - A Critical Review," *in 30th Australasian Transport Research Forum*, ETM GROUP, London, pp. 1–16, 2007.
[41] C. Systematics, "Traffic Congestion and Reliability: Trends and Advanced Strategies for Congestion Mitigation," Technical Report, United States Federal Highway Administration, 2005.
[42] T. Afrin and N. Yodo, "A Survey of Road Traffic Congestion Measures Towards a Sustainable and Resilient Transportation System," *in Sustainability*, vol. 12, no. 11, p. 4660, 2020.
[43] K. N. Qureshi and A. H. Abdullah, "A Survey on Intelligent Transportation Systems," *in Middle-East Journal of Scientific Research,* vol. 15, no. 5, pp. 629–642, 2013.
[44] B. Singh and A. Gupta, "Recent Trends in Intelligent Transportation Systems: A Review," *in Journal of Transport Literature*, vol. 9, pp. 30–34, 2015.
[45] A. Maimaris and G. Papageorgiou, "A Review of Intelligent Transportation Systems From a Communications Technology Perspective," *in 2016 IEEE 19th International Conference on Intelligent Transportation Systems (ITSC), IEEE*, pp. 54–59, 2016.
[46] Z. Xiong, H. Sheng, W. Rong and D. E. Cooper, "Intelligent Transportation Systems for Smart Cities: A Progress Review," *in Science China Information Sciences*, vol. 55, pp. 2908–2914, 2012.
[47] P. H. Feiler, B. A. Lewis and S. Vestal, "The SAE Architecture Analysis & Design Language (AADL) a Standard for Engineering Performance Critical Systems," *in 2006 IEEE Conference on Computer Aided Control System Design, 2006 IEEE International Conference on Control Applications, 2006 IEEE International Symposium on Intelligent Control, IEEE*, pp. 1206–1211, 2006.
[48] A. Vinel, N. Lyamin and P. Isachenkov, "Modeling of V2V Communications for C-ITS Safety Applications: A CPS Perspective," *in IEEE Communications Letters*, vol. 22, no. 8, pp. 1600–1603, 2018.
[49] X. Cao, P. Yang, M. Alzenad, X. Xi, D. Wu and H. Yanikomeroglu, "Airborne Communication Networks: A Survey," *in IEEE Journal on Selected Areas in Communications*, vol. 36, no. 10, pp. 1907–1926, September 2018.
[50] Y. Zeng, R. Zhang and T. J. Lim, "Wireless Communications With Unmanned Aerial Vehicles: Opportunities and Challenges," *in IEEE Communications Magazine*, vol. 54, no. 5, pp. 36–42, May 2016.
[51] O. Vidal, G. Verelst, J. Lacan, E. Alberty, J. Radzik and M. Bousquet, "Next Generation High Throughput Satellite System," *in Proceedings - 2012 IEEE 1st AESS European Conference on Satellite Telecommunications*, pp. 1–7, October 2012.
[52] J. B. Anderson, T. Aulin and C. E. Sundberg, Digital Phase Modulation. New York, NY: Plenum, 1986.
[53] M. P. Wylie-Green, E. Perrins and T. Svensson, "Introduction to CPM-SC-FDMA: A Novel Multiple-Access Power-Efficient Transmission Scheme," *in IEEE Transactions on Communications*, vol. 59, no. 7, pp. 1904–1915, July 2011.

[54] M. A. Sedaghat, R. R. Müller and C. Rachinger, "(Continuous) Phase Modulation on the Hypersphere," *in IEEE Transactions on Wireless Communications*, vol. 15, no. 8, pp. 5763–5775, August 2016.
[55] S. Bhat and A. Chockalingam, "Detection of Load-Modulated Multiuser MIMO Signals," *in IEEE Wireless Communication Letters*, vol. 7, no. 2, pp. 266–269, April 2018.

7 A Flight from 5G to 6G with UAV-Enabled Communications

A Case Study

Prangya Priyadarshini, Lopamundra Hota, and Arun Kumar

7.1 INTRODUCTION

UAVs, or drones, have sprung onto the consumer market due to their widespread use in various industries. UAV-aided communication boosts the efficiency of wireless communication networks. As aerial relays, UAVs can expand the range of cellular communications, cover areas with weak signals, and set up emergency networks in the aftermath of natural disasters [1]. Using drones makes various locations inaccessible due to a lack of infrastructure accessible. Drones are becoming increasingly prevalent in many industries, from agriculture and forestry to environmental protection and security, necessitating their usage in critical circumstances like rescue and search operations and transportation. In places where terrestrial networks are overloaded or unstable, UAVs can bridge the communication gap with fast speeds and minimal latency.

However, there are specific challenges associated with UAV-assisted communication. Due to the power constraint, the flight time of the UAVs is restricted to a few minutes. UAVs are susceptible to adverse weather conditions and may not be able to operate in all circumstances. Additionally, privacy concerns may be associated with UAVs, since an individual can pilot a UAV and pose as a mobile RSU while engaging in unlawful acts like spying on others. Even though UAV-assisted cellular networks are still in their infancy, researchers in communication, control, signal processing, computer science, and information theory have paid great attention to them because of their potential for supporting communication and sensing applications.

The most cost-effective way to deal with situations like sudden communication infrastructure failure due to natural disasters, communication equipment malfunction, or a sudden increase in the number of users demanding connectivity and a high data rate is to use UAV-assisted heterogeneous networks (Het-Nets). In the future, 5G and beyond 5G wireless connectivity will be the primary use of UAVs. Since 2000, scientists have been attempting to integrate UAVs with the cellular network [1]. The prototype was a GPRS-powered network connecting two UAVs to an

DOI: 10.1201/9781032625034-7

already-established ground operator. Third-generation partnership project (3GPP) participants provided several ideas for integrating UAVs into already-established LTE networks [2]. Different use cases in UAVs in wireless communication include:

- A portable base station may be set up in remote, high-demand, or infrastructure-deficient regions.
- In either store-and-forward (SF) or decode-and-forward (DF) modes, a single or fleet of UAVs can act as a flying relay node, connecting far-flung source and destination pairs.
- In the form of airborne user equipment (UE), drones can be used for last-mile deliveries and data collecting in remote locations.

However, there are specific challenges associated with UAV-assisted communication. In UAV-assisted communication, accurate modelling of the air-to-ground channel is essential. Channel modelling is more difficult in wireless communication because both the base station and the user are on the move, unlike in terrestrial communication, where only the user is on the move [3]. Unlike the 2D scenario of terrestrial BS placement, UAV BS installation is a 3D placement challenge. UAV BS installation must consider geographical specifics, user location, air-to-ground connection channel characteristics, hovering UAV battery time constraints, etc. In multi-UAV deployments, the 3D deployment strategy must account for UAV interference [4]. Trajectory planning is a challenging task for UAV-assisted communication, which considers various parameters like user demands, flight and energy constraints of UAVs, collision avoidance among UAVs, etc. Since communication networks facilitated by UAVs must coexist with the conventional terrestrial wireless communication network, scarce resources like spectrum must be shared. Resource management in UAV-assisted communication is challenging due to the unique characteristics of UAVs, such as their manoeuvrability, limited time and energy, etc. The availability of reliable backhaul connections is a crucial aspect of any communication network's blueprint. UAVs need wireless backhauling since they are airborne. Satellite links, Wi-Fi, millimetre waves, and free-space optical (FSO) network are all promising options for wireless backhauling [5]. UAVs and traditional wireless networks on the ground share bandwidth in UAV-assisted wireless communication. Practical resource allocation algorithms must be developed to optimize throughput and user coverage. Other important aspects of resource management include user association, power management, allocating bandwidth, and time slots.

7.2 VITAL NETWORK AND COMMUNICATION TECHNOLOGIES FOR UAV COMMUNICATION

Multiple network and communication technology forms are essential for drones' efficient and secure operation. Data transmission, navigation, control, and remote sensing are all made possible for UAVs by recent technological advancements. Figure 7.1 presents some of the most critical communication and network technologies for UAV communication.

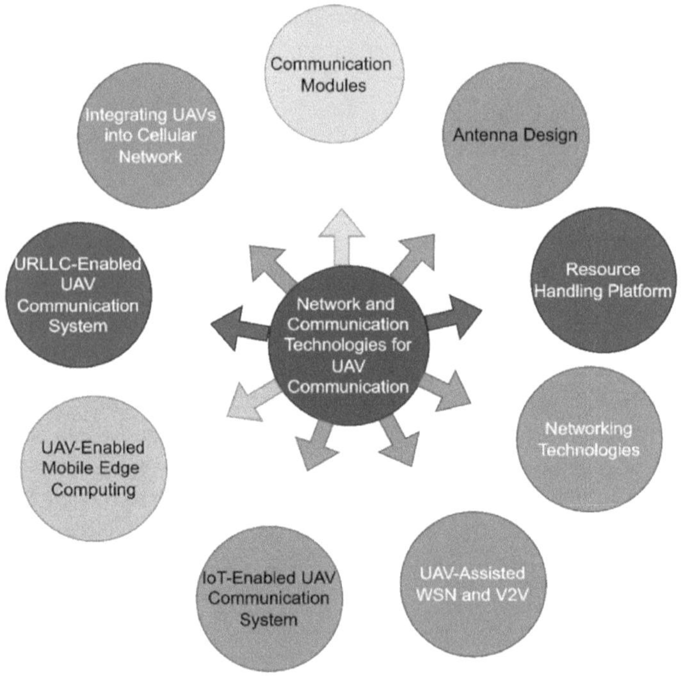

FIGURE 7.1 Network and communication technologies for UAV communication.

- **Communication modules.** Communication modules for UAVs are specialized hardware that allows for reliable and effective data exchange between UAVs and other devices or networks. Transceivers allow for bidirectional connection; antennas improve signal strength and coverage; modems encrypt and decrypt data for transmission; and telemetry systems report on the UAV's health in real time. Communication between UAVs relies heavily on accuracy and stability as key performance parameters. Orthogonal frequency-division multiplexing (OFDM) with multiple inputs and multiple outputs may accurately reconstruct transmitted data at the receiver end with minimal computing complexity and overhead [6]. Long-distance data link systems are also available, as are satellite communication modules for worldwide coverage and LTE/5G modules for integration with current mobile networks. Data integrity is protected by safeguards, including encryption methods, and compatibility with UAV avionics is achieved by adaptability to different communication protocols and standards.
- **Antenna design.** The goal of aerial design for UAV communication is to maximize signal transmission and reception while adhering to the physical limitations of UAVs. These antennas are crucial for the UAV's ability to communicate with ground stations, distant equipment, or other UAVs. Form factor, frequency band, radiation pattern, gain, and polarization are only a few factors considered while designing antennas for UAVs. Key

factors include using lightweight and long-lasting materials, compatibility with avionics equipment, and compliance with regulations. The authors in [7] have proposed a design considering bandwidth needs to achieve optimal antenna performance. A two-beam switching Yagi–Uda antenna was offered, which operates at 2.4 GHz to increase the radio control distance of a UAV controller [8]. An RF jammer uses a three-dimensional multiple-input, multiple-output (3D MIMO) radar to counteract surveillance [9]. The ability of UAVs to establish robust and reliable communication links is crucial to their ability to perform in various roles and adapt to the ever-changing conditions of the sky.

- **Resource handling platform.** UAV communication platforms are systems developed to efficiently handle and distribute the resources required for reliable data transfer between UAVs and other devices or networks. This system guarantees efficient communication with low interference and power consumption by optimizing bandwidth, quality of service, spectrum allocation, and latency control. Communication is optimized for various UAV uses, including traffic load balancing, network slicing, and adaptive protocols. The platform is designed to help with resource diversification and communication monitoring, and it is integrated with avionics systems to do so. A decentralized platform called AuRoRA has been employed as a ground station to deliver command signals to the servo motors of cars. This method avoided the problems that would have arisen from relying on only one computer to process the flight data and control signals [10]. Overall, this system improves UAV communication, allowing it to be used more effectively in various contexts, such as surveillance, mapping, and emergency response.
- **Networking technologies for UAV communication systems.** The success of UAV operations relies in part on several networking technologies. Short-range communications between ground stations and neighbouring drones are made possible via radio frequency (RF) alternatives, like Wi-Fi and Bluetooth. Third-generation (3G), fourth-generation (4G), and fifth-generation (5G) cellular networks provide coverage beyond the direct line of sight [11]. Satellite connections bring internet access to far places, although with increased lag. For UAVs to function effectively, ad hoc mesh networking must be used to set up these fluid connections. Software-defined radio (SDR) provides increased flexibility, while multi-channel communication and frequency hopping increase dependability and data throughput [12]. Optical and laser communication technologies provide safe, high-bandwidth transmission within line-of-sight ranges. Additionally, improved UAV communication is provided by directional antennas and dynamic spectrum access. Depending on the individual needs of the operation and the surrounding conditions, a mix of these technologies is typically necessary for effective UAV communication.
- **UAV-assisted WSN and vehicle-to-vehicle (V2V).** UAVs improve WSNs and V2V communication systems. As mobile base stations, UAVs can efficiently gather data from sensors in remote or inaccessible places for

WSNs [13]. UAVs fly over sensor nodes to collect and relay data, guaranteeing sensor coverage and energy savings. This allows accurate and rapid data collection for environmental monitoring, disaster management, and precision agriculture. UAVs operate as relays in vehicle-to-vehicle communication, expanding the range and assuring connectivity in low-line-of-sight situations. UAVs increase collision avoidance and cooperative awareness by conveying real-time signals between vehicles. This is useful in congested metropolitan areas and highways, where impediments might impair V2V communication. UAVs in WSNs and V2V systems have several benefits. For large-area monitoring applications, UAV-assisted WSNs improve data gathering and coverage. In addition, UAVs may be dynamically deployed to meet sensor node needs and data trends. In V2V communication, UAV relays improve resilience and enable essential information sharing, making transportation systems safer and more efficient [14].

- **IoT-enabled UAV communication system.** An IoT-enabled UAV communication system connects UAVs and linked objects using IoT technology. This integration includes UAV IoT devices, different communication protocols for real-time data transmission, on-board data processing for fast analysis, and multi-source data fusion. The system supports agricultural, surveillance, logistics, and disaster response by remotely monitoring and controlling UAVs, dynamic mission planning, and real-time data optimization. Protecting data and complying with rules require strong security measures. The authors of [15] proposed an adaptable bus-and-drones mobile infrastructure (AidLife) using current public transit to provide reliable communication during disasters. Device discovery, which comprises registering, deleting, querying, and forming routes for nodes, can lead to high energy usage and hinder network functioning. XML energy charts help identify energy-efficient devices for 5G IoT employing UAVs [16].
- **UAV-enabled mobile edge computing.** Mobile edge computing (MEC) moves computer resources to the network's periphery to improve data processing speed and lower application latency. When UAVs are included in this system, computer resources may be deployed on UAVs, enabling them to process data right where it was collected [17, 18]. This cooperation enables real-time data analysis; facilitates the use of technologies like video streaming, augmented reality, and data processing from the Internet of Things; and allows MEC services to be made available in new or changing geographic areas. Harnessing the promise of UAV-enabled mobile edge computing in areas like emergency response, remote sensing, and interactive media experiences presents several challenges, including optimizing resource allocation, controlling UAV mobility, and overcoming communication restrictions. Despite this, UAVs have limitations in terms of both flight time and battery life, and there are often many people who need to be reached inside the UAV's range of vision. Therefore, it is essential to develop effective methods of allocating resources in MEC networks that support multiple users and UAVs [19].

- **URLLC-enabled UAV communication system.** The term 'ultra-reliable and low-latency communications–enabled UAV communication system' describes a specialized arrangement in which unmanned aerial vehicles are outfitted with cutting-edge technology that guarantees extremely trustworthy and nearly immediate data transmission. Fifth-generation mobile networks, crucial for mission-critical applications like autonomous vehicles, will be made possible by URLLC's current wireless networking technology [20]. This system is designed where constant, real-time communication is crucial, focusing on dependability and minimizing communication delays. This configuration suits disaster response, remote surveillance, industrial automation, and healthcare support due to robust communication protocols, low-latency networks, and advanced error correction mechanisms. When dependable and immediate information transmission is of fundamental importance, the focus is on maintaining consistent connectivity, lowering communication latency, and assuring flawless operations.
- **Integrating UAVs into cellular network.** Connecting drones to existing mobile communication systems is an integral part of cellular network integration; this allows for more sophisticated protocols, such as 4G and 5G, allowing faster data transmission, broader coverage, and real-time communication. Field findings from a prototype LTE–UAV integrated network were reported in 2016 by the China Mobile Research Institute and Ericsson. They discuss the advantages of mobile technologies for the drone ecosystem, provide a high-level overview of the capabilities needed for drone applications, and evaluate the demands on mobile networks in terms of service in this prototype [21]. UAVs are revolutionizing industries, including surveillance, agriculture, and emergency response, thanks to advancements made possible by this integration, such as remote control, BVLOS operations, and access to real-time data. Safe and efficient UAV operations within cellular networks bring in a new era of superior aerial connectivity and applications, but only when hurdles like interference control, legal compliance, and network optimization are overcome.

7.3 CELLULAR-CONNECTED UAVS: 5G NR AND BEYOND

UAVs that are cellular-connected in the context of 5G NR (new radio) technology use fifth-generation (5G) cellular networks to set up dependable and high-speed communication linkages between ground-based networks and UAVs [22]. Flexible numerology (e.g. subcarrier spacing), better antenna technology and beamforming capabilities, a more sophisticated carrier aggregation method, and comprehensive bandwidth operations are all features of 5G NR. Meanwhile, research towards 5G improvements for UAVs was conducted in 3GPP Release 17 [23]. This research identifies new communication needs and key performance indicators (KPIs) for unmanned aerial vehicles (3GPP subscribers), such as communication service and command-and-control traffic KPIs. 5G NR's capabilities and characteristics make it possible to provide high-quality connections for UAVs, including 3D coverage, data

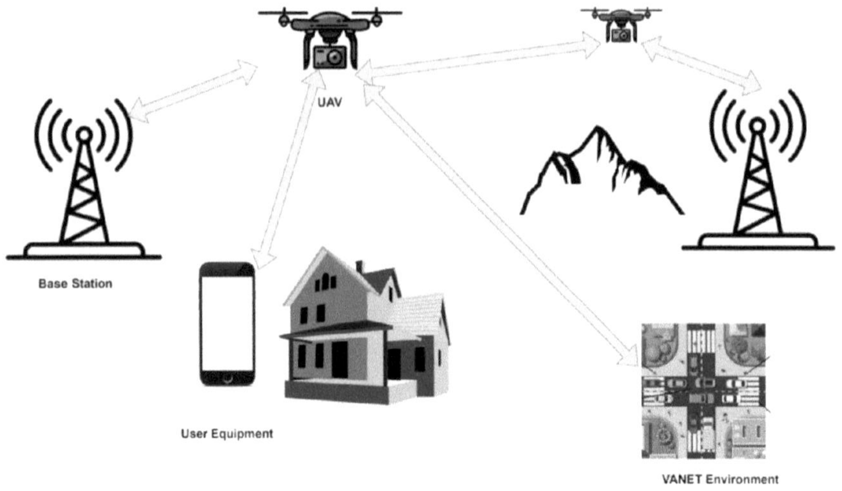

FIGURE 7.2 System model for cellular-connected UAV.

throughput, latency, and dependability. Figure 7.2 presents the system model for the cellular-connected UAV communication network.

The capacity of cellular-connected UAV systems is increased by 5G, which depends on cutting-edge antenna technologies, including improved beamforming and large multiple-input, multiple-output (MIMO) schemes while enabling single-user and multiuser MIMO [24]. Additionally, 5G uses narrow beam operations to take advantage of beam management and beam-sweeping characteristics, which can increase coverage and capacity while lowering interference for supporting UAVs.

Data transfer and real-time control of UAVs are made possible by 5G's low latency, ultra-high data speeds, and improved antenna technology. 5G allows UAVs to broadcast high-resolution images and video feeds to ground stations or remote operators with remarkable clarity and little latency, improving situational awareness in aerial photography and surveillance. Additionally, UAVs can quickly send enormous datasets back to analysts for real-time review, enabling speedy decision-making and issue detection, particularly for activities like infrastructure inspection. UAVs can process this data locally thanks to the integration of 5G with edge computing [25], which decreases latency and enables autonomous navigation and obstacle avoidance.

Unmanned aerial vehicles (UAVs) and 5G technologies work together to improve range and flexibility [26]. Regarding coverage, 5G's expansive reach enables UAVs to retain steady connectivity even in distant or underserved locations, allowing for beyond-line-of-sight (BLOS) missions and guaranteeing a durable link for rural applications. Combining 5G with satellite communication increases worldwide coverage, which is crucial for long-range missions and emergency response. Regarding mobility, 5G's support for high-speed data transmission and seamless handovers between cells allows UAVs to efficiently cover vast areas without losing connectivity, making it an excellent fit for urban air mobility and rapid response tasks that rely on constant, immediate interaction.

The foundation of 5G and UAV integration rests on the twin pillars of reliability and security. The 5G network's redundancy and sturdiness allow UAVs to maintain constant contact even in harsh conditions or during time-sensitive missions [27]. Congestion-related disruptions are avoided because network slicing in 5G allocates resources specifically for UAVs. At the same time, the UAV–5G ecosystem is protected from cyber threats and unauthorized access utilizing severe security measures, including end-to-end encryption, authentication procedures, and secure access control. This all-encompassing method guarantees secure and dependable UAV communication while protecting sensitive information. Together, these components strengthen the potential for 5G and UAVs to change industries and improve safety and efficiency in fields ranging from surveillance and emergency response to delivery services.

Unmanned aerial vehicles equipped with edge computing and 5G connectivity usher in a new era of innovation. UAVs can make split-second decisions because of the extremely low latency made possible by processing data locally at the network's edge. This is especially important for use cases like autonomous navigation and obstacle avoidance. In addition, 5G networks benefit greatly from the increased efficiency of the significantly reduced data transit needs made possible by this localized data processing. Scalability is improved to meet the rising demand for UAV applications, while privacy and security are enhanced since sensitive data stays inside the local network.

7.3.1 Cellular-Connected UAVs towards 6G

The capabilities of 5G are expected to be further enhanced by 6G [28], which is expected to provide considerable improvements in data throughput, dependability, latency, energy efficiency, connection density, quicker mobility, and network intelligence. It allows for various applications, such as continuous and constant connectivity, driverless vehicles, wireless XR apps, low-power gadgets, and holographic communications. Meanwhile, creating 6G calls for overcoming obstacles, including complex hardware design, integrated communication, control, and computing design, and the effective cohabitation of many different use cases. Space-air-ground integrated networks, intelligent surfaces, wireless power transfer, terahertz (THz) communications, cell-free massive MIMO, and AI-assisted designs are some of the enabling technologies and solutions that 6G can use to support new use cases with demanding performance requirements [29].

The projected ultra-high data speeds of 6G, which will much exceed those of 5G, will completely transform the capabilities of UAVs. 6G's terabit-per-second (Tbps) data transfer rates will allow UAVs to perform at previously unimaginable levels [30]. 8K video streaming in real time, ultra-high-resolution images, and instantaneous data interchange will become the norm, opening up new avenues for applications like surveillance, remote sensing, and emergency response. This enormous bandwidth significantly improves UAVs' situational awareness and mission efficiency since the data they collect may be quickly transmitted to ground stations or other UAVs.

When paired with UAVs, 6G's extraordinarily low latency – predicted to reach sub-millisecond levels – and its support for vast connection will be revolutionary.

With 6G's ultra-low latency, UAVs can make split-second choices, improving their ability to navigate, avoid obstacles, and fly autonomously in unpredictable settings. UAV fleets can collaborate on an unprecedented scale because of 6G's capacity to manage a substantially greater number of linked devices, a crucial characteristic known as massive machine-type communication (mMTC) [31]. Applications where several UAVs need to work together, such as extensive environmental monitoring, infrastructure inspection, and disaster response, would benefit significantly from this.

When used with UAVs, 6G's use of THz frequencies significantly improves over 5G's use of the millimetre wave band. THz frequencies provide even more expansive bandwidths, letting UAVs function with swifter and more stable connections. This is critical in congested metropolitan areas and busy skies, where UAV communication might be difficult. With 6G's THz technology, UAVs can access a broader spectrum, guaranteeing reliable, high-speed data transmission even in challenging circumstances [32]. To fully realize the promise of UAV applications like urban air mobility and delivery services, cutting-edge surveillance, and real-time data collection, this upgrade will be crucial, ushering in a whole new level of efficiency and dependability in UAV operations.

Regarding UAVs, quantum-secure communication is a crucial component of 6G technology [33]. The development of quantum computers poses a threat to the security of current techniques of encryption. In light of the possible dangers posed by quantum-based attacks on UAV communications, 6G seeks to provide encryption and security mechanisms immune to such attacks. This assures that, despite changing cryptographic problems, the data transferred between UAVs and ground control stations is secure and unmodifiable. Surveillance, monitoring of vital infrastructure, and military applications can all benefit from quantum-secure communication since it increases the robustness of UAV operations against new cyber threats and protects the confidentiality of sensitive data. Protecting UAV operations and expanding their usefulness across industries, this cutting-edge security feature of 6G technology ensures reliable communication routes.

The use of AI and machine learning in 6G-assisted UAV communication is pioneering [34]. UAVs analyze data on board with unparalleled efficiency and intelligence thanks to 6G and AI/machine learning algorithms. UAVs can recognize objects, identify abnormalities, and make real-time flying choices, including modifying flight routes or responding to changing environmental conditions. UAVs with AI-driven data analytics can instantly turn sensor data into usable information. Autonomous surveillance, disaster response, and precision agriculture benefit from UAVs with AI that quickly analyze and convey crucial information, improving decision-making and mission outcomes [35]. AI's capacity to adapt and learn makes UAV communication and coordination more sophisticated, enabling complex and adaptive UAV operations on 6G-powered networks.

Holographic communication in 6G-assisted UAVs might revolutionize data visualization and sharing [36]. UAVs can display real-time three-dimensional holographic pictures or information using 6G's predicted terahertz frequencies and ultra-high data speeds. This allows immersive, context-rich communication beyond 2D visual data transfer. A UAV with holographic communication can be used for search and rescue missions to give distant operators a 3D image of the catastrophe scene for better decision-making. Telemedicine uses UAV-streamed holographic pictures to help professionals diagnose and guide on-site medical personnel. This technology also allows remote professionals to do immersive, three-dimensional inspections in construction and infrastructure, boosting evaluation and decision support. Holographic communication in 6G-assisted UAVs makes data transfer vibrant and dynamic, improving UAV operations across sectors.

7.3.2 Advancing UAV Connectivity beyond 5G and 6G

To advance aerial technology, it is essential to expand UAV connectivity past 5G and 6G capabilities. By pushing the limits of networking and communication, UAVs can function more effectively, securely, and independently. The objective is to open up new avenues for UAV use in various fields, from improving surveillance and emergency response to revolutionizing agriculture and urban transportation. Advancing UAV connectivity beyond the capabilities of 5G and 6G relies on several enabling technologies:

- **Massive MIMO.** Ground BSs may conduct adaptive 3D beamforming with narrow beams by utilizing massive antenna arrays to effectively serve UAVs in the sky and terrestrial users. Massive MIMO may improve cellular-connected UAV networks' spectral efficiency and coverage performance. Massive MIMO with null-steering capabilities in cellular-connected UAV networks may dramatically reduce inter-cell interference [37]. The benefit of 3D massive MIMO over a fixed antenna tilting technique is the ability to minimize inter-cell interference while dynamically flexibly regulating beams in 3D. However, its performance may be constrained if the number of UAVs grows and broadband services are required for high-altitude UAVs. To fulfil the performance needs of UAVs and guarantee effective coexistence with terrestrial UEs, specialized up-tilted BS antennas may be required. Another exciting technique is cell-free massive MIMO, which uses many dispersed antennas to provide UAVs with dependable, cost-effective, and widespread support [38].
- **High-frequency operation.** THz and millimetre wave communications are potential options for providing high-data-rate services [39]. Operating at such high frequencies provides wide bandwidth availability. It permits the use of small antennas in massive MIMO systems to address UAV communications' high-data-rate needs. Nevertheless, transmission range is limited while working at such high frequencies. UAV networks may optimize LoS connections and reduce the impact of shadowing with the use of high-gain narrow beams, multi-hop communications, and combined UAV mobility and scheduling.
- **Terahertz (THz) communication.** Communication technologies like 6G and 5G might use terahertz frequencies in the future. The bandwidth of THz waves is far greater than that of millimetre wave frequencies utilized in 5G. Terahertz (THz) communication is a state-of-the-art wireless technology that might dramatically alter the future of high-speed data transfer and network connectivity since it operates in the terahertz frequency range

between 0.1 and 10 THz [35]. Data rates of terabits per second are made possible, making it ideal for data-intensive applications like real-time 3D mapping and high-definition video streaming, for which it is uniquely suited. With its directional and highly secure nature, THz communication shows promise for applications ranging from secure military communications to novel use cases in UAVs, where rapid, high-quality data exchange is paramount. However, THz communication faces challenges, such as limited propagation range, due to atmospheric absorption and line-of-sight requirements.

- **Quantum communication.** The state-of-the-art field of quantum communication uses quantum mechanics' underlying principles to ensure that messages cannot be intercepted or altered in transit. Using entanglement and quantum bits (qubits), it generates cryptographic keys that are impossible to crack and ensures the highest level of security throughout data transmission [40]. Quantum communication uses quantum superposition and entanglement to make it impossible to overhear or eavesdrop on a conversation without disrupting the quantum state, which would immediately warn the persons involved. The ramifications of this technique for secure data transfer are far-reaching, ranging from quantum-resistant encryption approaches for the post-quantum future to secure government communications. In the case of military or critical infrastructure use, this technique might be implemented to encrypt data sent between UAVs and base stations.
- **Swarm communication and coordination.** A paradigm change in robotics and autonomous systems, swarm communication, and coordination involves several autonomous agents working together to complete a job. Drones, robots, and even tiny sensors may create these swarms, and they can all talk to one another to share data and coordinate their efforts. Swarm systems, which use decentralized algorithms and peer-to-peer communication, may self-organize to accomplish complicated goals like search and rescue, environmental monitoring, and disaster response [34]. Swarm systems have enormous promise for improving efficiency, scalability, and resilience in various applications, from precision agriculture to exploring challenging and dangerous settings, thanks to their capacity to function in a coordinated manner without centralized control [41].
- **Satellite integration.** The combination of satellites and UAVs, driven by 5G and 6G technology, creates a game-changing synergy. UAVs can attain unprecedented levels of connection and performance by combining the worldwide coverage of satellites with the high-speed, low-latency capabilities of 5G and 6G networks. By combining terrestrial 5G/6G networks with satellite connectivity, UAVs can maintain constant, high-quality communication anywhere. Its high data transmission rates, low latency, and extensive communication footprint allow UAVs to perform complicated jobs, like remote sensing, disaster management, and precision agriculture [40]. In addition, merging these technologies expands the possibilities for UAVs in many other sectors and mission-critical applications, such as autonomous, long-range missions, real-time data processing, and collaborative operations over vast landscapes.

- **Intelligent surfaces.** Reconfigurable intelligent surfaces (RISs) can enhance propagation conditions and communication connection quality by intelligently reflecting electromagnetic signals in a desired direction [42]. RISs provide efficient control over the phase and amplitude of incoming signals, significantly improving signal quality with minimal additional power consumption; because of their adaptability and LoS channels, RISs and UAVs work together to improve 6G networks' power efficiency. Meanwhile, RISs may be used by cellular-connected UAV systems to improve coverage by optimally reflecting transmitted signals towards airborne UAVs, hence reducing downlink interference.
- **Power transfer, energy harvesting, and wake-up radio.** Energy harvesting and wireless power transfer are promising options for enabling battery-limited gadgets as technology progresses towards 6G. Likewise, energy harvesting (such as solar or wind) and wireless power transmission through ground BSs can be helpful options for UAVs with limited battery life. To further reduce the UAV's overall energy footprint, near-zero power wake-up radio technology can be used in UAV communications. However, there are design issues when considering UAV operation, given the uncertainty in energy harvesting, the need for long-distance and dependable power transmission for fast-moving UAVs, and the requirement for sensitive wake-up radios for long-distance UAV communications.
- **3D cellular network.** To provide 3D connectivity in the future 6G network, aerial BSs installed on drones or tethered balloons offer a promising option. In particular, this makes it possible to design a 3D cellular network infrastructure with various airborne platforms, including flying base stations, backhauling nodes, and unmanned aerial vehicle user equipment (UAV-UEs) [43]. A 3D cellular network can provide dependable, low-latency, and high-data-rate services to UAV–UEs in the air.
- **Integrated sensing and communication (ISAC).** ISAC is an essential enabling technology for 6G since it enables communication and sensing functions on the same shared frequency and hardware platform [44]. Since UAVs operate at low altitudes, are very mobile, and are sometimes too tiny to be spotted by radars and surveillance systems, ISAC can help make monitoring and managing them more accessible. UAV identification, high-precision localization, and monitoring are all possible thanks to the dependable wireless communication provided by ISAC systems based on cellular technology. For UAVs with cellular connectivity, ISAC technology may be used for obstacle identification to aid navigation and route planning.

7.4 NEXT-GEN WIRELESS CONNECTIVITY WITH UAV-ENHANCED RIS TECHNOLOGY

Data transmission may be significantly improved using RIS, a device built from inexpensive metasurfaces that can reflect or refract the signals appropriately [45]. Relays provide a similar job, but with varying capabilities. The radio signals are amplified with the additional power in relays to travel greater distances. RIS, on the other hand, is a passive device that reflects rather than amplifies the signal. The RIS is the most promising technology for boosting data transmission efficiency. It allows for two-way communication without fostering the production or amplification of the noise signal.

UAVs offer several benefits, including coverage, data collecting, communication, deployment, and precise monitoring. On the other hand, RIS technology enhances existing networks and devices by improving signal capacity and channel gain while being low-cost and deployable on any surface. When used together, RIS and UAVs open up a new frontier for improving 5G networks and beyond, allowing them to provide customers with unprecedented benefits. RISs may accomplish passive beamforming by intelligently adjusting the reflecting components, while UAVs rely on their excellent mobility to create LoS-dominated transmission links with the ground users. For instance, with the aid of an RIS or many RISs, a UAV may be set up as a mobile base station to facilitate communication with devices on the ground. In addition to enhancing the quality of wireless channels, incorporating RIS during transmission may help deal with problems like LOS and energy consumption at UAVs [46].

Multiple antennas may considerably improve a device's performance, but adding more antennas to UAVs introduces weight, size, and power consumption limitations. Adjusting the phase shift of the metasurfaces in the RIS allows for increased gain without installing numerous antennas in UAVs. Since the user located closer to RIS will be able to receive the signal from RIS rather than the UAV, including RIS in the UAV system aids in reducing the trajectory of the UAV. The RIS will be able to reflect the signals with the aid of the UAV to improve signal transmission to far users who can receive it with a reasonable data rate. Figure 7.3 presents the RIS-assisted UAV cellular communication scenario.

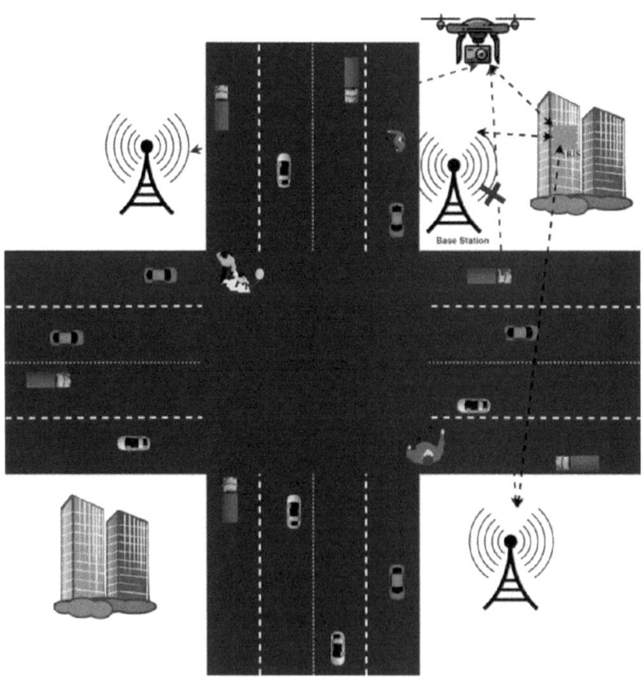

FIGURE 7.3 RIS-assisted UAV cellular communication.

Combining UAVs and RIS technology creates more efficient, adaptive, and durable wireless communication networks. This paradigm shift might impact telecommunications, agriculture, transportation, and more. A beacon of innovation, UAV-assisted RIS ushers in the future of wireless communications as demand for high-quality, ubiquitous connection grows.

7.5 APPLICATIONS OF UAV-AIDED COMMUNICATION

Several applications make use of the UAVs' communication network capabilities. Here are some examples:

1. **Surveillance and security.** UAVs with communication systems play a crucial role in surveillance operations, helping to monitor public events, essential infrastructure, and border crossings. They improve situational awareness by sending control centres real-time data [47].
2. **Search and rescue.** UAV communication makes locating missing people or catastrophe survivors much easier. They create communication channels between the victims and the rescuers, facilitating effective coordination and raising the possibility of successful rescues [48].
3. **Disaster response.** UAVs with communications capabilities help set up networks for emergency response teams in disaster zones. They make it easier to coordinate relief activities by improving communication, data exchange, and allocating supplies [49].
4. **Precision agriculture.** Insights on crop health, soil conditions, and irrigation demands are available to farmers through UAVs outfitted with communication facilities. By better allocating public resources, agricultural productivity is increased [43].
5. **Environmental monitoring.** UAVs help monitor air and water quality and wildlife habitats. UAVs with the ability to communicate can provide data for immediate processing and decision-making [50].
6. **Infrastructure inspection.** UAV communication makes inspecting power lines, pipelines, and bridges easier. They transmit high-resolution pictures and sensor data for remote inspection and upkeep preparation [51].
7. **Telecommunication support.** UAVs can serve as makeshift cell towers in places without human inhabitants or with damaged infrastructure. Connectivity in disaster-stricken areas is restored as they set up lines of communication [52].
8. **Scientific research.** Scientists use UAVs to gather information in difficult or dangerous terrain. Ecology, geology, and climate science researchers might benefit from UAVs outfitted with communication technologies since the data they collect can be transmitted in real time [53].
9. **Media and journalism.** Media coverage is improved when UAVs are used, since they provide aerial views of events and locales. UAVs with built-in communications send live footage to media outlets, increasing the accuracy and breadth of their coverage [54].
10. **Logistics and delivery.** UAVs help logistics teams coordinate supplies and find the most efficient routes through better communication. Communication technologies facilitate interactions between UAVs and control centres.

11. **Mapping and surveying.** UAVs with data collection and transmission capabilities are used in mapping and surveying. They stream geographic information in real time, allowing for more precise cartography and modelling [55].
12. **Law enforcement and public safety.** UAV communication is used by law enforcement organizations for airborne patrol and investigation. UAVs with the ability to communicate aid in emergency response.

7.6 IMPLEMENTATION CHALLENGES FOR UAV

UAVs are rapidly becoming an integral feature of 5G and beyond wireless networks. The field is expanding rapidly, and researchers are making significant strides. While there are many benefits of using UAVs for wireless communication, several obstacles must be overcome before UAVs can be used effectively. Some of the challenges are addressed in the following.

7.6.1 Positioning a UAV Base Station in Three Dimensions

In contrast to terrestrial communication, where BS placement is solely in a 2D plane, aided communication BS placement needs to be handled in a 3D plane, where the altitude is a crucial design factor [56]. The placement of mobile bases is influenced by several variables, such as user location, air-to-ground channel characteristics, local geography, UAV energy and battery limitations, and flying duration restrictions.

7.6.2 Resource Management

As UAVs share resources and the current ground terrestrial communication network, resource management in UAV-aided wireless communication networks is a crucial and complex challenge that has to be solved [57]. The management of resources is made more difficult by the unique characteristics of UAVs, such as energy and time constraints, path planning, interference from air-to-ground and air-to-air connections, manoeuvrability, and mobility aspects. Therefore, effective resource-sharing plans must be implemented to create an optimized network.

7.6.3 Channel Modelling

Channel properties between the transmitter and receiver significantly influence the communication quality of any given communication system. In a UAV-assisted communication system, the air–ground link's channel characteristics are quite different from those in a terrestrially deployed communication system, where only ground-to-ground communication connections are used, the base stations are immovable, and the users are all mobile. However, in UAV-assisted communication, UAV-deployed BSs are also movable, just like the user; therefore, a slight change in direction can significantly impact the air-to-ground channel characteristics. The location, altitude, type, and elevation angle of the deployed UAV, as well as other channel variables, all affect the characteristics of the UAV-assisted communication channel [58].

7.6.4 Interference Management

The challenge of managing interference is crucial in the deployment of many UAVs. UAVs must be used to maximize their footprint while minimizing interference with one another. To maximize throughput and coverage on the network, interference mitigation must be considered.

7.6.5 Trajectory Planning

Another significant obstacle to communication aided by UAVs is trajectory planning. Factors, including flight duration energy constraints, collision avoidance, user requirements, and others, must be considered when determining the best UAV path. Finding infinite no-optimization variables (the position of the UAV) is a challenging job in trajectory planning.

7.6.6 Energy Limitations

For the adoption of UAV-assisted communication in numerous circumstances, UAVs' limited on-board energy is a significant barrier. Typically, UAVs run on batteries, which are a finite resource. For various functions, including data processing, mobility management, and transmission control, on-board energy is required. The physical factors of an object, such as the specifics of a UAV mission, the state of the environment, the path planning strategies of the UAVs, etc., also affect energy consumption. The lifetime of a UAV is directly defined by its energy constraint. The UAV lifetime must thus be extended by creating effective energy utilization methods.

7.6.7 Backhaul Connectivity

Another crucial and challenging aspect of UAV-assisted communication is wireless backhaul connectivity. Due to the flying nature of UAVs, wireless backhauling is necessary rather than cable. Satellite networks, Wi-Fi, mmWave, and FSO communication are several choices for UAV wireless backhauling. Even though satellite links have a higher backhaul capacity than Wi-Fi links, Wi-Fi remains a top contender for wireless backhauling in UAV-assisted communication due to its lower cost and shorter latency. Due to the LoS link maintenance function of UAV-aided communication, mmWave and FSO are sources of extended backhauling.

7.6.8 Security

A broadcasting scenario using UAVs is constantly in control without a human pilot. As a result, hackers might likely target open communication channels. The communication channel must be protected from eavesdropping, man-in-the-middle assaults, and other malicious attempts. Security is one of the most important aspects of integrated UAV-assisted communication networks and is also challenging to maintain amongst nodes with different characteristics due to the multi-layer architecture of these networks. The multi-layer network structure is easily vulnerable to malicious assaults. SDN controllers oversee the management of resources in multi-layer integrated networks. Keeping SDN controllers secure is also crucial for UAV-assisted communication. Several studies in recent years have focused on fixing the security issues with communicating with UAVs. A safe UAV-assisted communication experience is presented using AI-based solutions [59]. To improve the physical layer security of UAV-assisted communication, the author in [51] employs multi-antenna, friendly-jamming, and relay selection–based strategies. In [60], the authors present a method of secure transmission using two UAVs, wherein one UAV is used to jam several ground eavesdroppers so that the desired nodes can continue their safe connection.

7.7 CONCLUSION

UAVs are transitioning from 5G to 6G, which represents a step towards a time when UAVs are more than just communication tools but also intelligent, autonomous entities that can easily integrate into various sectors, from healthcare and agriculture to disaster relief and infrastructure inspection. The cutting-edge AI and other technological advancements allow UAVs to perform tasks previously only possible in science fiction, giving them more versatility and importance than ever. This chapter presents the UAV communication mechanism in detail, along with 5G and 6G modes of communication, providing high data rate, low latency, energy minimization, and secure communication. With next-generation connectivity, UAV-enhanced RIS technology, energy efficiency, e3Dnhanced LOS, improved mobility, remote connectivity, and many other use cases can be attained. Further, the applications and implementation challenges are discussed in detail.

REFERENCES

[1] Shahzadi, Rizwana, Mudassar Ali, Humayun Zubair Khan, and Muhammad Naeem. "UAV assisted 5G and beyond wireless networks: A survey." Journal of Network and Computer Applications 189 (2021): 103114.

[2] Sharma, Abhishek, Pankhuri Vanjani, Nikhil Paliwal, Chathuranga M. Wijerathna Basnayaka, Dushantha Nalin K. Jayakody, Hwang-Cheng Wang, and P. Muthuchidambaranathan. "Communication and networking technologies for UAVs: A survey." Journal of Network and Computer Applications 168 (2020): 102739.

[3] Zheng, Yi, Yuwen Wang, and Fanji Meng. "Modeling and simulation of path loss and fading for air-ground link of HAPs within a network simulator." In 2013 International Conference on Cyber-Enabled Distributed Computing and Knowledge Discovery, pp. 421–426. IEEE, 2013.

[4] Arshad, Qazi Kamal Ud Din, Ahsan Ullah Kashif, and Ijaz Mansoor Quershi. "A review on the evolution of cellular technologies." In 2019 16th International Bhurban Conference on Applied Sciences and Technology (IBCAST), pp. 989–993. IEEE, 2019.

[5] Nguyen, Tri Minh, Wessam Ajib, and Chadi Assi. "A novel cooperative NOMA for designing UAV-assisted wireless backhaul networks." IEEE Journal on Selected Areas in Communications 36, no. 11 (2018): 2497–2507.

[6] Vahidi, Vahid, and Ebrahim Saberinia. "A low complexity and bandwidth efficient procedure for OFDM data reconstruction in DSC 5G networks." In 2018 15th IEEE Annual Consumer Communications & Networking Conference (CCNC), pp. 1–4. IEEE, 2018.

[7] Zabihi, Roshanak, and Rodney G. Vaughan. "Monopole and conformal PIFA for small cylindrical ground plane mounting." In 2017 IEEE International Symposium on Antennas and Propagation & USNC/URSI National Radio Science Meeting, pp. 1487–1488. IEEE, 2017.

[8] Ngamjanyaporn, P., C. Kittiyanpunya, and M. Krairiksh. "A switch-beam circular array antenna using pattern reconfigurable Yagi-Uda antenna for space communications." In 2017 International symposium on antennas and propagation (ISAP), pp. 1–2. IEEE, 2017.

[9] Multerer, Thomas, Alexander Ganis, Ulrich Prechtel, Enric Miralles, Askold Meusling, Jan Mietzner, Martin Vossiek, Mirko Loghi, and Volker Ziegler. "Low-cost jamming system against small drones using a 3D MIMO radar based tracking." In 2017 European radar conference (EURAD), pp. 299–302. IEEE, 2017.

[10] Pizetta, Igor Henrique Beloti, Alexandre Santos Brandao, and Mário Sarcinelli-Filho. "A hardware-in-the-loop platform for rotary-wing unmanned aerial vehicles." Journal of Intelligent & Robotic Systems 84 (2016): 725–743.

[11] Nayak, B. P., L. Hota, A. Kumar, A. K. Turuk, and P. H. Chong. "Autonomous vehicles: Resource allocation, security, and data privacy." IEEE Transactions on Green Communications and Networking 6, no. 1 (2021): 117–131.

[12] Akeela, Rami, and Behnam Dezfouli. "Software-defined radios: Architecture, state-of-the-art, and challenges." Computer Communications 128 (2018): 106–125.

[13] Kerrache, Chaker Abdelaziz, Abderrahmane Lakas, Nasreddine Lagraa, and Ezedin Barka. "UAV-assisted technique for detecting malicious and selfish nodes in VANETs." Vehicular Communications 11 (2018): 1–11.

[14] Barka, Ezedin, Chaker Abdelaziz Kerrache, Nasreddine Lagraa, and Abderrahmane Lakas. "Behavior-aware UAV-assisted crowd sensing technique for urban vehicular environments." In 2018 15th IEEE Annual Consumer Communications & Networking Conference (CCNC), pp. 1–7. IEEE, 2018.

[15] Narang, Mamta, William Liu, Jairo Gutierrez, and Luca Chiaraviglio. "A cyber-physical buses-and-drones mobile edge infrastructure for large scale disaster emergency communications." In 2017 IEEE 37th International Conference on Distributed Computing Systems Workshops (ICDCSW), pp. 53–60. IEEE, 2017.

[16] Sharma, Vishal, Fei Song, Ilsun You, and Mohammed Atiquzzaman. "Energy efficient device discovery for reliable communication in 5G-based IoT and BSNs using unmanned aerial vehicles." Journal of Network and Computer Applications 97 (2017): 79–95.

[17] Wu, Gaoxiang, Yiming Miao, Yu Zhang, and Ahmed Barnawi. "Energy efficient for UAV-enabled mobile edge computing networks: Intelligent task prediction and offloading." Computer Communications 150 (2020): 556–562.

[18] Zhou, Yi, Cunhua Pan, Phee Lep Yeoh, Kezhi Wang, Maged Elkashlan, Branka Vucetic, and Yonghui Li. "Secure communications for UAV-enabled mobile edge computing systems." IEEE Transactions on Communications 68, no. 1 (2019): 376–388.

[19] Zhou, Fuhui, Rose Qingyang Hu, Zan Li, and Yuhao Wang. "Mobile edge computing in unmanned aerial vehicle networks." IEEE Wireless Communications 27, no. 1 (2020): 140–146.

[20] Han, Aitong, Tiejun Lv, and Xuewei Zhang. "UAV beamwidth design for ultra-reliable and low-latency communications with NOMA." In 2019 IEEE International Conference on Communications Workshops (ICC Workshops), pp. 1–6. IEEE, 2019.

[21] Muruganathan, Siva D., Xingqin Lin, Helka-Liina Määttänen, Jonas Sedin, Zhenhua Zou, Wuri A. Hapsari, and Shinpei Yasukawa. "An overview of 5g release-15 study on enhanced LTE support for connected drones." IEEE Communications Standards Magazine 5, no. 4 (2021): 140–146.

[22] Gapeyenko, Margarita, Dmitri Moltchanov, Sergey Andreev, and Robert W. Heath. "Line-of-sight probability for mmWave-based UAV communications in 3D urban grid

deployments." IEEE Transactions on Wireless Communications 20, no. 10 (2021): 6566–6579.

[23] Rahman, Imadur, Sara Modarres Razavi, Olof Liberg, Christian Hoymann, Henning Wiemann, Claes Tidestav, Paul Schliwa-Bertling, Patrik Persson, and Dirk Gerstenberger. "5G evolution toward 5G advanced: An overview of 3GPP releases 17 and 18." Ericsson Technology Review 2021, no. 14 (2021): 2–12.

[24] Pratschner, Stefan, Stefan Schwarz, and Markus Rupp. "Single-user and multi-user MIMO channel estimation for LTE-Advanced uplink." In 2017 IEEE International Conference on Communications (ICC), pp. 1–6. IEEE, 2017.

[25] Hassan, Najmul, Kok-Lim Alvin Yau, and Celimuge Wu. "Edge computing in 5G: A review." IEEE Access 7 (2019): 127276–127289.

[26] Chiputa, M., M. Zhang, G. M. N. Ali, P. H. J. Chong, H. Sabit, A. Kumar, and H. Li. "Enhancing handover for 5G mmwave mobile networks using jump Markov linear system and deep reinforcement learning". Sensors 22, no. 3 (2022): 746.

[27] Si-Mohammed, Samir, Maha Bouaziz, Hamed Hellaoui, Oussama Bekkouche, Adlen Ksentini, Tarik Taleb, Lechoslaw Tomaszewski, Thomas Lutz, Gokul Srinivasan, Tanel Jarvet, and Pawel Montowtt. "Supporting unmanned aerial vehicle services in 5G networks: New high-level architecture integrating 5G with U-space." IEEE Vehicular Technology Magazine 16, no. 1 (2020): 57–65.

[28] Chen, Zhi, Boyu Ning, Chong Han, Zhongbao Tian, and Shaoqian Li. "Intelligent reflecting surface assisted terahertz communications toward 6G." IEEE Wireless Communications 28, no. 6 (2021): 110–117.

[29] Angjo, Joana, Ibraheem Shayea, Mustafa Ergen, Hafizal Mohamad, Abdulraqeb Alhammadi, and Yousef Ibrahim Daradkeh. "Handover management of drones in future mobile networks: 6G technologies." IEEE Access 9 (2021): 12803–12823.

[30] Meng, Xi, Nan Zhang, Mengnan Jian, Michel Kadoch, and Dacheng Yang. "Channel modelling and estimation for reconfigurable intelligent surface based 6G SAGIN IoT." IEEE Internet of Things Journal 10 (2022): 9273–9282.

[31] Taneja, Ashu, Adi Alhudhaif, Shtwai Alsubai, and Abdullah Alqahtani. "A Novel Multiple Access Scheme for 6G Assisted Massive Machine Type Communication." IEEE Access 10 (2022): 117638–117645.

[32] Vaigandla, Karthik Kumar, S. Bolla, and R. Karne. "A survey on future generation wireless communications-6G: Requirements, technologies, challenges and applications." International Journal 10, no. 5 (2021).

[33] Ahmed, Abdulsattar M., Sayf A. Majeed, and Younis S. Dawood. "A survey of 6G mobile systems, enabling technologies, and challenges," International Journal of Electrical and Electronic Engineering & Telecommunications 12, no. 1 (2023): 1–21.

[34] Zuhair, Mohd, Farnazbanu Patel, Dhruvin Navapara, Pronaya Bhattacharya, and Deepti Saraswat. "BloCoV6: A blockchain-based 6G-assisted UAV contact tracing scheme for COVID-19 pandemic." In 2021 2nd International Conference on Intelligent Engineering and Management (ICIEM), pp. 271–276. IEEE, 2021.

[35] Naeem, Faisal, Georges Kaddoum, Saud Khan, Komal S. Khan, and Nadir Adam. "IRS-empowered 6G networks: Deployment strategies, performance optimization, and future research directions." IEEE Access 10 (2022): 118676–118696.

[36] Bhattacharya, Pronaya, Deepti Saraswat, Amit Dave, Mohak Acharya, Sudeep Tanwar, Gulshan Sharma, and Innocent E. Davidson. "Coalition of 6G and blockchain in AR/VR space: Challenges and future directions." IEEE Access 9 (2021): 168455–168484.

[37] Garcia-Rodriguez, Adrian, Giovanni Geraci, David López-Pérez, Lorenzo Galati Giordano, Ming Ding, and Emil Bjornson. "The essential guide to realizing 5G-connected UAVs with massive MIMO." IEEE Communications Magazine 57, no. 12 (2019): 84–90.

[38] Sudheesh, P. G., Mohammad Mozaffari, Maurizio Magarini, Walid Saad, and P. Muthuchidambaranathan. "Sum-rate analysis for high altitude platform (HAP) drones with tethered balloon relay." IEEE Communications Letters 22, no. 6 (2017): 1240–1243.
[39] Rao, P. V., M. Anand, J. Alfred Daniel, C. B. Sivaparthipan, S. Stewart Kirubakaran, Lydia J. Gnanasigamani, and P. Punitha. "Millimeter assisted wave technologies in 6G assisted wireless communication systems: A new paradigm for 6G collaborative learning." Wireless Networks (2023): 1–20.
[40] Kumar, Prabhat, Randhir Kumar, AKM Najmul Islam, Sahil Garg, Georges Kaddoum, and Zhu Han. "Distributed AI and blockchain for 6G-assisted terrestrial and non-terrestrial networks: Challenges and future directions." IEEE Network 37, no. 2 (2023): 70–77.
[41] Christensen, Anders Lyhne, Sancho Oliveira, Octavian Postolache, Maria Joao de Oliveira, Susana Sargento, Pedro Santana, Luis Nunes, Fernando Velez, Pedro Sebastião, Vasco Costa, Miguel Duarte, Jorge Gomes, Tiago Rodrigues, Fernando Silva. "Design of communication and control for swarms of aquatic surface drones." In International Conference on Agents and Artificial Intelligence, vol. 2, pp. 548–555. Scitepress, 2015.
[42] Taneja, Ashu, Shalli Rani, and Norbert Herencsar. "Energy aware solution for IRS-aided UAV communication in 6G wireless networks." Sustainable Energy Technologies and Assessments 58 (2023): 103318.
[43] Phani Praveen, S., Mohammed Hasan Ali, Muhammad Aslam Jarwar, Chander Prakash, Chavva Ravi Kishore Reddy, L. Malliga, and C. Chandru Vignesh. "6G assisted federated learning for continuous monitoring in wireless sensor network using game theory." Wireless Networks (2023): 1–27.
[44] Liu, Fan, Yuanhao Cui, Christos Masouros, Jie Xu, Tony Xiao Han, Yonina C. Eldar, and Stefano Buzzi. "Integrated sensing and communications: Toward dual-functional wireless networks for 6G and beyond." IEEE Journal on Selected Areas in Communications 40, no. 6 (2022): 1728–1767.
[45] Ge, Yimeng, Jiancun Fan, and Jinbo Zhang. "Active reconfigurable intelligent surface enhanced secure and energy-efficient communication of jittering UAV." IEEE Internet of Things Journal 10 (2023): 22386–22400.
[46] Hou, Zhifeng, Jin Chen, Yuzhen Huang, Yijie Luo, Ximing Wang, Jiangchun Gu, Yifan Xu, and Kailing Yao. "Joint trajectory and passive beamforming optimization in IRS-UAV enhanced anti-jamming communication networks." China Communications 19, no. 5 (2021): 191–205.
[47] Ullah, Zaib, Fadi Al-Turjman, and Leonardo Mostarda. "Cognition in UAV-aided 5G and beyond communications: A survey." IEEE Transactions on Cognitive Communications and Networking 6, no. 3 (2020): 872–891.
[48] Fontanesi, Gianluca, Hamed Ahmadi, and Anding Zhu. "Over the sea UAV based communication." In 2019 European Conference on Networks and Communications (EuCNC), pp. 374–378. IEEE, 2019.
[49] Do-Duy, Tan, Long D. Nguyen, Trung Q. Duong, Saeed R. Khosravirad, and Holger Claussen. "Joint optimisation of real-time deployment and resource allocation for UAV-aided disaster emergency communications." IEEE Journal on Selected Areas in Communications 39, no. 11 (2021): 3411–3424.
[50] Pan, Miaoxin, Chongcheng Chen, Xiaojun Yin, and Zhengrui Huang. "UAV-aided emergency environmental monitoring in infrastructure-less areas: LoRa mesh networking approach." IEEE Internet of Things Journal 9, no. 4 (2021): 2918–2932.
[51] Li, Bin, Zesong Fei, Yan Zhang, and Mohsen Guizani. "Secure UAV communication networks over 5G." IEEE Wireless Communications 26, no. 5 (2019): 114–120.
[52] Mayor, Vicente, Rafael Estepa, Antonio Estepa, and German Madinabeitia. "Deploying a reliable UAV-aided communication service in disaster areas." Wireless Communications and Mobile Computing 2019 (2019).

[53] Wang, Zhe, Lingjie Duan, and Rui Zhang. "Adaptive deployment for UAV-aided communication networks." IEEE Transactions on Wireless Communications 18, no. 9 (2019): 4531–4543.
[54] Sharma, Bhisham, Mohammad S. Obaidat, Vinay Sharma, and Kuei-Fang Hsiao. "Routing and collision avoidance techniques for unmanned aerial vehicles: Analysis, optimal solutions, and future directions." International Journal of Communication Systems 33, no. 18 (2020): e4628.
[55] Gesbert, David, Omid Esrafilian, Junting Chen, Rajeev Gangula, and Urbashi Mitra. "UAV-aided RF mapping for sensing and connectivity in wireless networks." IEEE Wireless Communications 30 (2022): 116–122.
[56] Constantinides, Chris, and Paul Parkinson. "Security challenges in UAV development." In 2008 IEEE/AIAA 27th Digital Avionics Systems Conference, p. 1-C. IEEE, 2008.
[57] Verdone, Roberto, and Silvia Mignardi. "Joint aerial-terrestrial resource management in UAV-aided mobile radio networks." IEEE Network 32, no. 5 (2018): 70–75.
[58] Zeng, Yong, Rui Zhang, and Teng Joon Lim. "Wireless communications with unmanned aerial vehicles: Opportunities and challenges." IEEE Communications Magazine 54, no. 5 (2016): 36–42.
[59] Challita, Ursula, Aidin Ferdowsi, Mingzhe Chen, and Walid Saad. "Machine learning for wireless connectivity and security of cellular-connected UAVs." IEEE Wireless Communications 26, no. 1 (2019): 28–35.
[60] Cai, Yunlong, Fangyu Cui, Qingjiang Shi, Minjian Zhao, and Geoffrey Ye Li. "Dual-UAV-broad enabled secure communications: Joint trajectory design and user scheduling." IEEE Journal on Selected Areas in Communications 36, no. 9 (2018): 1972–1985.

8 Development and Analysis of an RF-Powered Rechargeable Wireless Sensor Network for LR-WPAN

Chandrima Thakur and Sudipta Chattopadhyay

8.1 INTRODUCTION

Industrial Internet of Things (IIoT) consist of ultra-low battery-powered automated smart sensors to sense, collect, process, and communicate temporal events in industrial environment [1, 2]. Classic IIoT sensor nodes (motes) are mostly equipped with non-rechargeable (NR) batteries and perform energy-intensive network operations [3]. For stable network operation, the sensor batteries require frequent recharging/replacement after a specific time, which is expensive in terms of both labour and cost. Also, the battery lifetime of the sensor motes must be monitored continuously for better maintenance. Energy is an essential component in smart production environments as a key facilitator in the advancement of Industry 4.0, and energy management is a crucial factor to be considered when the lifetime of a sensor node needs to be improved [4]. This motivates robust research to develop radio frequency (RF) energy harvesting (EH) technique to acquire energy from ambient sources or other dedicated sources and utilize the harvested energy (electrically converted) to power the sensor motes and improve their operational capability and lifetime [5]. RF wireless power transfer (RF-WPT) technology has emerged as a viable, attractive green technology to power the sensors in a smart factory, where ambient RF cannot deliver sufficient energy to power the sensors [6]. The application of wireless recharge mechanisms to WSNs is referred to as rechargeable wireless sensor network (R-WSN) [7], where numerous sensors harvest energy emitted by a single or multiple wireless power transmitters to recharge their batteries or supercapacitors.

DOI: 10.1201/9781032625034-8

8.1.1 Prior Works and Motivation

Low-rate wireless personal area network (LR-WPAN) [8] has emerged as an appealing communication system in IIoT. ZigBee/IEEE 802.15.4 [9] is a low-cost WSN for industrial application that operates in the unlicensed RF worldwide (915 MHz Americas, 2.4 GHz global, or 868 MHz Europe). EH using dedicated RF source has gained popularity due to its ability to provide stable energy supply to the sensor motes [10]. Here, multiple sensors are configured with an EH module and a rechargeable lithium-ion (Li-ion) battery. Four battery chemistries were investigated in [11] to determine the best battery type for IoT applications. Energy provisioning methods include renewable EH (REH) [12], where the sensors recharge their batteries by capturing renewable energy (solar, wind, etc.) from the environment, and WPT [13], where batteries are recharged using electrical power from wireless chargers. The authors in [14] have proposed a cooperative integration of ambient RF EH with dedicated WPT that can prove to be an effective approach for different power-constrained sensor networks. However, the amount of harvested energy might be relatively low, which severely restricts the success of ambient RF for R-WSNs in practice [15]. Additionally, the deployment environment affects the quantity of energy harvested via REH. Moreover, appropriate circuit design is required to improve the RF-to-DC efficiency, keeping the harvester small in size. In contrast, RF-WPT technology [16, 17] has gained significant interest as a prospective technology that provides stable energy supply to the sensors through transmitting RF energy and has a powerful advantage in LR-WPAN as it exploits low-level, ambient RF power to sustain autonomous operation of sensor motes. Authors in [18] have outlined a range of EH technologies suitable for industrial automation. In [19], the total harvested energy is typically modelled by the linear model. In practice, however, EH circuits produce a non-linear end-to-end WPT. In [20], a non-linear EH model is built on a logistic function that reflects the dynamics of RF energy conversion efficiency for various input power levels. In contrast to [20], the authors in [21] proposed a quadratic RF EH efficiency model that obtained better results. Meanwhile, the authors in [22] designed a piecewise linear approximation model and obtained better performance, but they have not analyzed the battery recharging time. One of the commercially available evaluation kits allowing wireless energy transmission is the Bluetooth-enabled PowerSpot (PS) RF Wireless Power Development Module (P1110-EVAL-PS) [23] to charge various sensors and multiple consumer devices using the harvested RF power. Experimental results in [24] found P2110 harvester to be more efficient for long-range applications, whereas P1110 showed higher efficiency for short-range applications. In contrast to existing simulators like Castalia, EKHO, and COOJA that do not have energy modelling metrics table, NetSim v13.1 [25] provides EH model where the residual energy, as well as energy consumption in each battery node, is estimated by battery model metrics table. Rechargeable or EH mode represents the harvesting sub-module that periodically increases the battery's remaining energy. On the other hand, each node in the network is powered completely by its own battery in NR or no-harvesting mode. In [26], an agricultural simulation environment of WSN has been modelled with solar EH model in NetSim, where the lifetime of the sensor network with and without EH feature is estimated using the recharge state and recharge current duration. Authors in [27, 28] have found out the network lifetime curves of the sensor nodes deployed in an IoT environment using NetSim and have evaluated the throughput and battery life.

To the best of our knowledge, none of the aforementioned articles have analyzed the battery recharging time and network lifetime (NL) for any R-WSN in IIoT environment using RF-WPT. Figure 8.1 presents the flow chart of the proposed work.

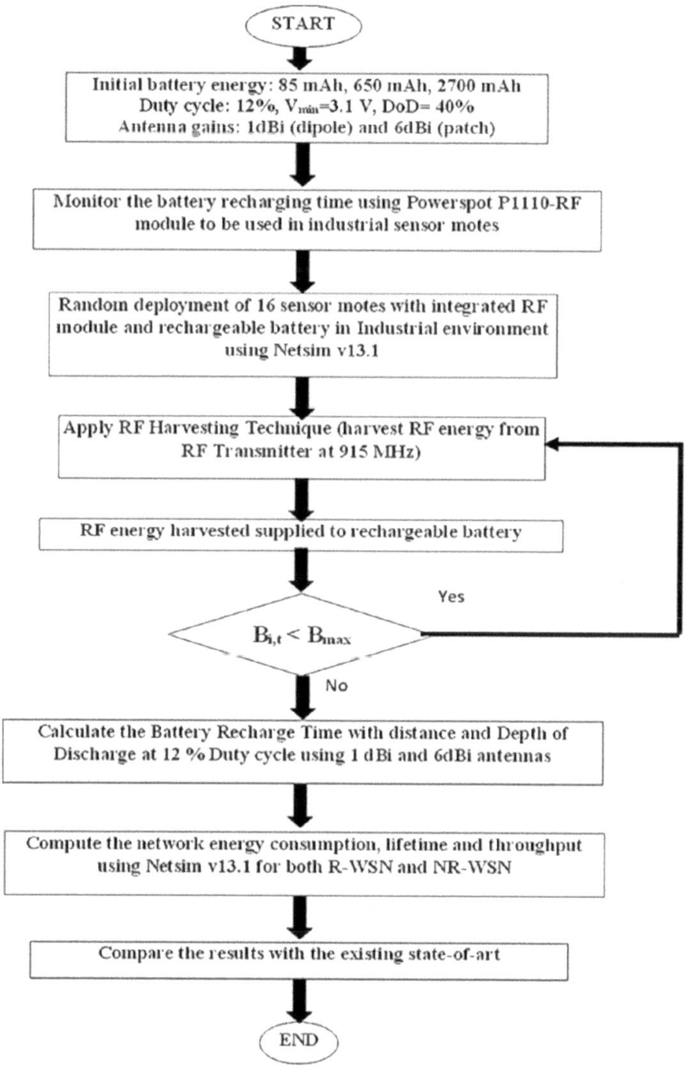

FIGURE 8.1 Flow chart of the proposed work.

8.1.2 Major Contributions

The main contributions are listed as follows:

- In the first instance, in a laboratory environment, we have set up an advanced energy measurement system with PowerSpot (PS) RF Wireless Power Development module to monitor the battery recharging time of the available on-board rechargeable batteries. This RF harvester-cum-storage module can be integrated in any sensor mote to enable true self-powering functionality for long-term operation in any indoor IIoT environment.
- Secondly, we characterize the effect of antenna gain on the battery recharge time with respect to variation in distance, number of RF transmitters, and depth of discharge (DoD).
- Rather than exploring each sensor mote individually, we have modelled the operation of an entire network comprising of 16 sensor motes (with inbuilt rechargeable batteries) using NetSim v13.1. Here, a realistic virtual simulation IIoT environment has been introduced and tested with the available on-board rechargeable batteries to find out the network energy consumption. Moreover, a comparative study of the proposed network has been carried out in terms of consumed, harvested, remaining energy for both R-WSN and NR-WSN, where the specifications of all three on-board batteries are entered in NetSim Battery Model metrics.
- Furthermore, the simulated values of energy consumption are processed in MATLAB to compare the network lifetime (NL) and throughput with non-rechargeable WSN (NR-WSN).
- Finally, our results are also compared with some of the existing published articles in terms of harvested energy.

8.2 NETWORK MODEL

Figure 8.2 presents an R-WSN based on LR-WPAN architecture deployed for environmental monitoring in an industrial warehouse (30 × 40 m^2). Here, 16 sensor motes (S_1, . . ., S_{16}), say, Zolertia mote (also known as Z1 mote [29]), working at 915 MHz radios are installed at different locations in the warehouse, and they report data to an access point (AP) within a period of 4 seconds. Using internet connection, the data received at AP gets stored in a database that runs on a remote server which can be accessed by the user anytime using special Java-based application. For simplicity, in this chapter, we assume that there is only one transmitter located at a depot within the monitoring area of the WSN. Each mote comprises of a wireless EH module (P1110), an energy storage module (preferably a rechargeable battery), and a sensor module. To keep the sensor motes working perpetually, they are assumed to have EH capabilities. Each Z1 mote has an on-board rechargeable battery (i.e. Li-ion or AAA) that stores the energy charged by the EH module. The harvesting module powers the sensor module using an analogue-to-digital converter (ADC). The motes adopt TI MSP430 as an MCU and TI CC2420 as an RF transceiver that is responsible for data transmission and reception. The sensed data is sent periodically to a cloud platform

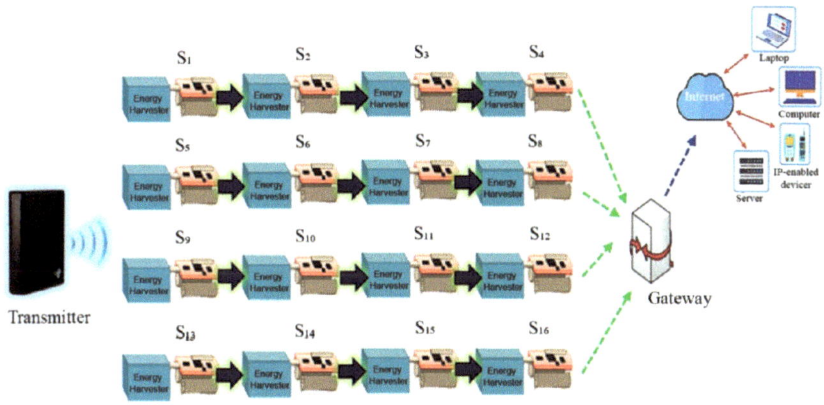

FIGURE 8.2 Proposed framework for R-WSN in LR-WPAN.

through the internet and then to a Wi-Fi router or GSM mobile. The user can view the data using a monitoring and control program that runs on their mobile device or personal computer. The assumptions for the constructed network model are listed in the following:

- The amount of the available RF energy is the same throughout the WSN.
- The motes are static and homogeneous, with the same initial battery energy, the same hardware; transmit/receive similar number of packets; and are distributed in the industry to monitor variables, such as humidity, temperature, sound, luminosity, etc.
- Each of the energy-constrained motes is powered by rechargeable batteries and fitted with an RF harvester that harvests energy from a dedicated power beacon.
- Multiple sensor motes are managed by one gateway.
- The rechargeable batteries of the motes are subjected to recharge via radiative WPT.

8.3 MEASUREMENT METHODOLOGY

In this section, we have given a detailed description of measurement tools and procedure to calculate the battery charging time of on-board rechargeable batteries. RF module P1110-EVAL-PS is used for measurement of battery recharging time. Software tools used for modelling and simulation are NetSim v13.1 and MATLAB R2022b. Measurements are taken at different distance, and DoD is varied using vertically polarized PCB dipole (1 dBi) and patch (6.1 dBi) antennas.

8.3.1 Measurement Tools

Table 8.1 shows the specifications of three widely used rechargeable test batteries to drive the sensor motes, namely, one Li-ion 2032 coin cell battery (Bat 1), a single

TABLE 8.1
Various On-Board Rechargeable Battery Specifications

Battery type	NiMH 900 mAh 10.5 mm × 44.5 mm	Li-MN 650mAh 14 mm × 50 mm	Li-ion coin cell 85 mAh, 20 mm × 3.2 mm
Make	Duracell	EFest	Route JD
Model no.	DX2400	IMR14500	PD2032C1
Characteristics of a single storage element			
Nominal voltage (V)	1.2	3.6	3.7
Current rating (mAh)/capacity	900 mAh	650 mAh	85 mAh
Self-discharge (per month)	30	<10	<10
Charge–discharge efficiency (%)	66	95	99.9
Memory effect	No	No	No
Charging method	Trickle/Pulse	Pulse	Pulse
Charging time (in hours)	4–9	4–5	2–3
Recharge cycles	1,000	1,200	1,200

Source: [23].

FIGURE 8.3 PS development board with battery holder for harvesting energy to charge the on-board batteries [23].

AA lithium manganese Li-Mn (Bat 2), and 3 AAA nickel metal hydride (Ni-MH) in series (Bat 3). We have tested our developed model (shown in Figure 8.1), selecting all the three batteries one at a time, and have analyzed their performance using NetSim modeller tool. The runtime of a battery is dictated by the battery capacity, expressed in milliampere per hour (mAh) or amp-hours (Ah). Z1 motes can be powered by any one of the given three batteries [29]. Figure 8.3 depicts the evaluation board (EVB) to charge the on-board batteries [23].

8.3.2 Block Diagram of WPT Model

Figure 8.4 depicts the block diagram of the WPT model. Left part of the diagram depicts power beacon (TX91501–3W-ID), and the right part shows a sensor mote powered by P1110-EVAL-PS power harvester module [23]). The WPT system is made up of two channels: RF energy transfer and data communication. There is no interference between these channels, as they use different frequency bands. The RF energy harvested by the P1110 RF module gets stored in a rechargeable battery or capacitor. An internal RF sensor in this module measures the received power using ADC. The beacon and EVB exchange information by means of IEEE 802.15.4 standard [9]. Any battery-powered sensor mote can wirelessly receive RF energy from the power beacon using WPT when it is attached to a power source. It is made up of a signal generator, which produces a weak continuous-wave RF source signal, and an amplifier, which amplifies the weak signal. The amplifier draws DC power from the power supply and converts it to RF. The amplified high-power signal is sent out by the transmit antenna. The harvester in the sensor mote receives an RF signal from the beacon via the receive antenna and converts it to DC using a rectifier. The DC power is then stored in the storage module, that is, a rechargeable battery.

8.3.3 Measurement Procedure

Figure 8.5 shows the experimental set-up for the determination of battery charging time using PS development module consisting of a transmitter TX91501–3W and the EVB. The former is a wireless dedicated 915 MHz transmitter with 3 watts equivalent isotopically radiated power (EIRP) fitted with an inbuilt patch antenna (8 dBi gain). The latter is a power harvester EVB with on-board rechargeable batteries which can harvest energy from RF signal radiated by the transmitter to provide intermittent/pulsed power output. In the current setting, we have not considered the operation of Z1 mote, as we have mainly focussed on battery charging time using P1110-EVAL-PS module. P1110 IC converts 915MHz RF received using a dipole or patch antenna to DC and stores it in any one of the selected on-board rechargeable batteries. The harvester collects the RF input power in the range of –5 to +20 dBm. The circuitry required to activate the PS transmitter via Bluetooth

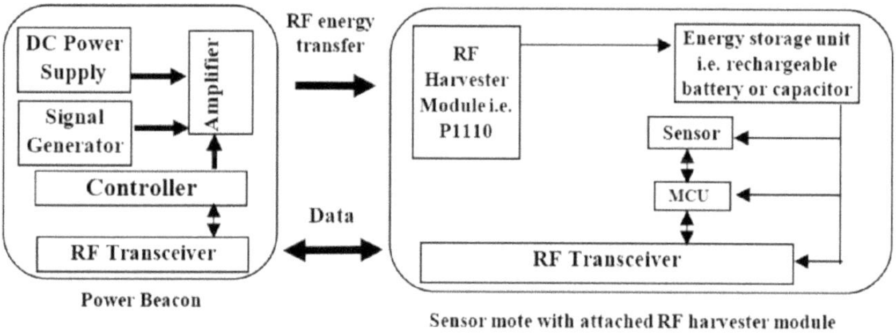

FIGURE 8.4 Block diagram.

FIGURE 8.5 Experiment setting for monitoring battery recharge time at 915 MHz.

Low Energy (BLE) communication is included in EVB. The board's circuitry keeps track of the received signal strength indication (RSSI) of the RF field received from the PS transmitter, the amount of current supplied to recharge the selected battery, and the voltage of the selected battery being recharged. Via Bluetooth, the data is transmitted to a smartphone and is displayed in the PS Dev app. A sliding switch on the EVB allows the user to select any one of the three batteries. The Li-ion battery is connected to BAT 1 pin, LiMn to Bat 2, and AAA to Bat 3. When any battery voltage goes below a predetermined threshold (3.1V), the selected battery will recharge again, and the cycle will repeat. When the charge threshold ≥3.1V is achieved, P1110 enables the voltage output. The output voltage from P1110 is applied to the BAT pin. The RSSI is monitored using RSSI pin on the board or from PS Dev app in the smartphone. After reaching the max voltage of 3.6 V or 3.7 V, P1110 shuts off its output and prevents the battery from getting overcharged. During the experiment, the transmitter is fixed at a position and the harvester board is varied with ten steps (10 cm, 20 cm, 30 cm, and so on). To find out the battery recharge time, measurement is taken at different distances using dipole antenna attached to the harvester circuit, as shown in Figure 8.5. Similar experiment is carried out with patch antenna. As per the recharge time of batteries, we can use them in Z1 motes. Initially, we monitor the battery recharge time for Bat 1 with a capacity of 85 mAh and nominal voltage of 3.6 V. We repeat the process with the other on-board batteries and compare all the three with respect to distance and DoD.

8.4 SIMULATION NETWORK SCENARIO OF THE PROPOSED MODEL

8.4.1 Simulation Model and Network Set-Up

To test the feasibility and performance of our proposed R-WSN described in Figure 8.2, we create a virtual IIoT scenario using NetSim v13.1 in Figure 8.6, comprising of similar rechargeable sensor motes with P1110 EH modules for monitoring environmental parameters. Z1 mote requires 36 mJ at 3.6 V of the harvested voltage

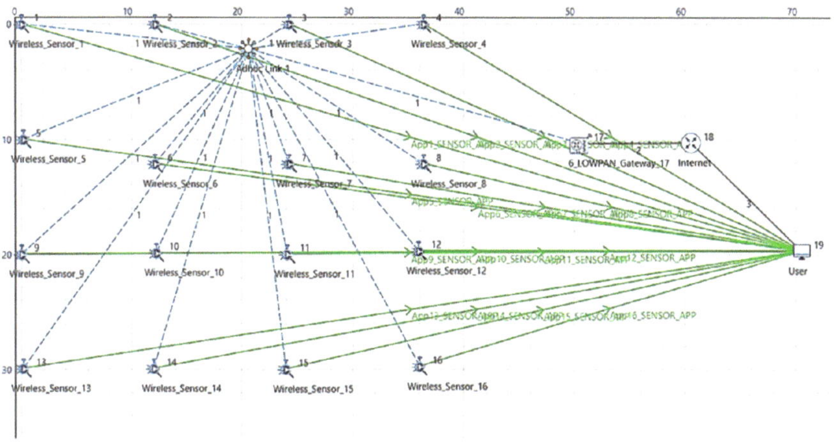

FIGURE 8.6 RF simulation module implementation using NetSim modeller tool R-WSN comprising of 16 battery-operated RF-powered sensor motes.

to switch on the mote for data connection and activation of the microcontroller MSP430F2617. Z1 can be powered in several ways: battery-powered (using AA or AAA or Li-ion coin cell in the battery-holder) or USB-powered. It is flexible in the sense that it can switch to any battery as per the required application. The measured and processed data from the sensor motes is transmitted to the gateway. Finally, the transmitted data is forwarded to the user through the internet. Simulation is carried out in NetSim modeller to find the network energy consumption. The 16 RF-powered Z1 motes can operate as both with (rechargeable) or without (non-rechargeable) RF harvesting feature. For real-time IIoT applications, the harvested power needs to be deterministic and consistent. Also, proper placement of RF source and sensor motes is required to obtain fixed amount of harvested power. In our network, if any sensor's battery voltage goes below the crucial minimum voltage level (Vmin), which is 3.1 V, then it is considered to be 'non-operational' and requires re-powering by EH.

8.4.2 Design and Implementation

To model the circuit described in Figure 8.6, we have used the sensor battery/energy model that is a part of WSN/IoT networks in the simulator. The ZigBee interface feature of the sensor motes contains the energy model, which is user-configurable. We analyze the network performance in terms of energy consumption by entering the specifications of all the three batteries one at a time. Energy consumption is calculated individually for each battery. According to the results of the simulations and computations, NetSim offers a detailed battery model metrics table that provides the energy consumption of each mote with respect to transmission, reception, idle mode, sleep mode. NetSim also supports an abstract EH model where a specified amount of energy (calculated from recharging current and voltage specified) is added to the remaining energy of the node periodically to replenish the battery energy.

8.5 THEORETICAL ANALYSIS

We consider a R-WSN consisting of a set of rechargeable homogeneous sensor motes $S = \{s_1, s_2, ...s_n\}$ deployed in a rectangular area $L \times L$ for environmental monitoring; n is the total number of sensor motes. The coordinate of the ith sensor mote s_i is taken as (x_i, y_i). The energy of the RF transmitter is assumed to be infinite and is located at a depot within the monitoring area of the network. Sensor s_i includes a sensing module, a processing unit, an RF transceiver, an EH module, rechargeable battery for storage, and a power management unit. Sensory data is generated by s_i with a data rate R_i and fed to the processing unit. After processing, the data is forwarded to the gateway using the transceiver unit via single-hop or multi-hop communication [26]. $B_{i,t}$ represents the remaining energy of s_i at some time point t, and $B_{\max}$ is the maximum energy level of each sensor mote. When $B_{i,t} <= B_{\max}$, then s_i starts harvesting energy from the RF transmitter to continue operation of the network. For NR-WSN, when $B_{i,t} <= B_{\max}$, then s_i becomes non-operational.

8.5.1 Battery Charging Time

According to the omnidirectional WPT model, the Friis transmission equation is expressed as [10]

$$P_R = \frac{P_T \Delta \eta_t |h|^2 |s|^2}{d^{\beta}} \quad (8.1)$$

where P_R represents the power received by a sensor mote, P_T is the power transmitted by the RF transmitter, s is the transmitted signal, and h is the channel coefficient between the beacon and the harvester. $|s|^2$ and $|h|^2$ denote the powers of s and h, respectively. β is the path loss exponent; d indicates distance between the transmitter and harvester board. η_t and Δ are the transmitter efficiency and the large-scale propagation effect parameter, respectively.

$$\Delta = \frac{G_T G_R}{L_P} \left(\frac{\lambda}{4\pi} \right)^2 \quad (8.2)$$

Here, Δ is dependent on transmitter and receiver antenna gains (G_T and G_R), wavelength of transmitted signal (λ), and polarization loss (L_p).

Let θ represent the charging orientation angle between sensor and transmitter. According to directional WPT model [10], the Friis transmission equation is expressed as

$$P_R = \frac{P_T A_R A_T \Delta \eta_t |h|^2 |s|^2 (\cos\theta + c)}{d^{\beta}} \quad (8.3)$$

Where A_T and A_R represent the maximum power transferring area and effective power receiving area of transmitter and sensor, respectively; c is a constant.

The received RF input power is converted to DC output power by the RF EH circuit. The efficiency of the P1110 EH module used here is approximately 80% at 915 MHz, given as

$$\eta_{EH} = \frac{P_{DC}}{P_R} \tag{8.4}$$

where P_{DC} indicates DC output power.

The harvested energy by the harvesting circuitry is expressed as

$$E_H = P_{DC} T_{Bat} \tag{8.5}$$

Where T_{Bat} is the time to recharge the battery, as expressed in the following:

$$T_{Bat} = \frac{E_H}{\sum_{n=1}^{N} \left(\eta P_{T,n} G_{T,n} G_R \left(\frac{\lambda}{4\pi d_n} \right)^2 \right)} \tag{8.6}$$

Now, E_H amount of harvested energy is transferred to a rechargeable battery with a certain capacity and voltage and stored to power the sensor motes. Thus, the expression for battery recharging time is given as

$$T_{Bat} = \frac{C_B D_B V_B^2}{2\sum_{n=1}^{N} \left(\eta P_{T,n} G_{T,n} G_R \left(\frac{\lambda}{4\pi d_n} \right)^2 \right)} \tag{8.7}$$

where V_B is the charging voltage, C_B is the battery capacity, and D_B is the DoD of the battery. From equation (8.7), the recharging time is inversely proportional to the gain of antenna, number of transmitters, and directly proportional to the battery capacity, voltage, distance, and DoD.

8.5.2 Energy Consumption (E_C)

The E_C of a sensor mote occurs during four modes: idle listening (E_{Idle}) and sleep mode (E_{Sleep}), data transmission (E_{T_X}), and data reception (E_{R_X}). The sum of all the consumed energies gives the total energy consumed by a mote. E_{T_X} depends on the amount of data transmitted by one mote to other motes. In the reception mode, the amount of data received by the sensor motes from the gateway is a measure of the total energy consumed. In idle mode, the motes consume energy by turning their transceivers in 'ON' position without any data transmission or reception. During sleep mode, the sensors switch off their receiver, and the energy consumption is negligible, hence ignored in calculation. The energy consumption is computed by the following equation [25]:

$$E = V \times I \times \tau \tag{8.8}$$

where I is the current drawn at different modes, V is the battery voltage of the mote (voltage is 3.6 V for Z1 mote), and τ is the time period when the mote is 'ON'.

The energy consumed during transmission is expressed as

$$E_{Tx} = V \times I_{Tx} \times \tau_{Tx} = V \times I_{Tx} \times \frac{l_{beacon}}{R} \tag{8.9}$$

where l_{beacon} represents length of beacon packet and R is the data rate.

Energy consumption in reception mode is given as

$$E_{Rx} = V \times I_{Rx} \times \tau_{Rx} = V \times I_{Rx} \times \frac{l_{beacon}}{R} \tag{8.10}$$

where I_{Rx} denotes current drawn during data reception and τ_{Rx} is the receiving time. For idle mode, the energy consumed during is given as

$$E_{Idle} = V \times I_{Idle} \times \tau_{Idle} \tag{8.11}$$

where I_{Idle} is the idle-mode current and τ_{Idle} represents time in idle mode.

The energy consumed by a single sensor mote is calculated as

$$E_C = E_{Tx} + E_{Rx} + E_{Idle} \tag{8.12}$$

8.5.3 Network Lifetime (NL)

In a WSN, the NL is defined as the time interval from the start time to the moment the network stops functioning [26]:

$$T_{Net} = \frac{E_{initial} - E_W}{E_C + \chi . E_S} \tag{8.13}$$

Where

T_{Net} is the average network lifetime (in days), $E_{initial}$ is the initial network energy (joules), E_W is the expected wasted (unused) energy in the network when it dies (joules), E_C is the network's energy consumption per day (joules), E_S is the expected energy consumption (joules), χ is packet arrival rate (bps).

8.5.4 Initial Network Energy $E_{Initial}$ Calculation

We assume each mote operates with a single AA LiMn (Bat 2) rechargeable battery. The initial energy of a single AA 3.7 V battery having capacity of 650 mAh in joules is obtained as 0.65 Ah × 3.7 × 3,600 = 8,658 J. Thus, initial energy of the whole network with 16 Bat 2-driven motes is obtained as 8,560 J × 16 = 138,528 J. Similarly, we can find the initial energy of the network with the other two test batteries (Bat 1 and Bat 3), as presented in Table 8.2.

TABLE 8.2
On-Board Battery Energy Calculation

Sl. no.		1 mote			Whole network (comprising of 16 motes)		
	Parameters	Bat 3	Bat 2	Bat 1	Bat 3	Bat 2	Bat 1
1	Initial energy (in J)	34,992	8,658	1,101.6	559,872	138,528	17,625.6
2	Unused or wasted energy (in J)	24,494.4	6,060.59	771.12	391,910.4	96,969.6	12,337.9

8.5.5 Unused Network Energy E_W Calculation

We presume that the battery state-of-charge (SoC) becomes 'zero' for NR-WSN when the maximum rated energy of the battery is reduced to 70%. Here, one sensor mote has maximum energy of 8,658 J. Thus, when the battery energy reaches less than 70% of 8,658 J, that is, 6,060.59 J, each sensor mote is considered 'non-operational'. E_W for the entire network is 6,060.59 J × 16 = 96,969.59 J.

If any mote energy goes below the predefined threshold level, the battery recharging for the motes occurs autonomously for R-WSN. After battery charging using RF harvesting, when the battery SoC of the energy constrained motes is more than 70%, they become operational and automatically participate in the network. Thus, the NL becomes infinite (ideally), and the network never stops completely. In traditional NR-WSNs, very low duty cycle is kept by the designers to save the energy consumption of the motes. But in RF EH, as the motes are not energy-limited, the duty cycle can be increased. Here we have taken 5% DoD and 12% duty cycle for R-WSN, that is, the network sends one message to the gateway node after every 8.3 sec for better monitoring in industrial application.

Let the expected energy consumption E_S by a sensor mote be 16 mW or 16 mJ/s. For 16 motes, it becomes 16 mW × 16 = 256 mW, or 256 mJ/s. The number of secs-rounds/day is 24 hrs × 60 mins × 60 secs = 86,400s. For 1% duty cycle, a sensor mote operates for 864 sec out of 86,400 sec in 24 hrs. Thus, the expected energy consumption by a single mote E_{S_1} is 16 mJ/s × 864 s = 13.82 J/day. For 16 motes, E_S becomes 13.82 J × 16 = 221.184 J/day. Thus, the NL is calculated as given in equation (8.13). Using the same procedure, the NL for various duty cycles is calculated and shown in Table 8.3.

8.5.6 Throughput

The rate of successful transmission of data (in kbps) within the network for a specified duration of simulation time defines the network throughput, expressed as

$$\text{Throughput} = SRP \times \text{packet size} / T_s \tag{8.14}$$

where SRP denotes successful reception of data packets and T_s is the simulation time.

TABLE 8.3
Duty Cycle versus Network Lifetime

Duty cycle (in %)	Energy consumption/ day for 1 mote (in J/day)	Energy consumption/ day for the network (comprising of 16 motes) (in J/ day)	Network lifetime (no. of days)		
			Bat 1	Bat 2	Bat 3
1	13.82	221.184	24.59	187.934	253.18
5	69.12	1,105.92	4.92	37.58	151.88
10	138.24	2,211.84	2.46	18.79	76.07
12	162.12	2,593.92	2.03	16.23	64.75
20	276.48	4,423.68	1.19	9.39	27.97
50	691.2	11,059.2	0.5	3.75	5.06

8.6 PERFORMANCE EVALUATION

In this section, we evaluate the recharging time by varying the DoD for all the batteries. The distance between the RF transmitter and EVB was not shortened by less than 10 cm during measurement. Table 8.4 shows the measurement results of recharging time versus DoD for Bat 1, 2, and 3 at 915 MHz and 10 cm distance.

The recharging time values with varying distance (in cm) for Bat 1, 2, and 3 using patch and dipole antennas at 915 MHz and 5% DoD are listed in Table 8.5. It is evident that recharging time increases with increasing distance. Moreover, recharging time is faster with patch antenna compared to dipole. In addition, Bat 1 gives the shortest recharging time value at 10 cm distance with 6.1 dBi patch antenna at 5% DoD as compared to Bat 2 and 3. For Bat 1, the shortest charging time is found to be 0.24 hrs (14.4 min) at a distance of 10 cm, and the longest charging time was determined as 0.672 hrs (40.32 min) at a distance of 50 cm for 6.1 dBi patch antenna. Moreover, using RF harvesting, the network throughput is improved by 1.08, 1.07, and 1.02 times for Bat 1, 2, and 3, respectively, and the maximum charging time was determined to be 1.92 hrs at a distance of 50 cm with a 1 dBi dipole antenna. Similarly, the recharging times for Bat 2 and 3 are also recorded and shown in Table 8.5.

Table 8.6 shows variation in recharging time with increase in number of RF transmitters (N) for Bat 1, 2, and 3 using PCB patch and dipole. Due to resource constraint, only two transmitters are used in this work. It is observed that the recharging time is faster with increase in number of transmitters. Using 6.1 dBi PCB patch, the recharging time for Bat 1 is found to be 0.672 hrs (40.32 min) for $N = 1$ that reduces to 0.432 hrs (25.92 minutes) for $N = 2$. Meanwhile, the recharging time using 1 dBi PCB dipole for Bat 1 is found to be 1.92 hrs for $N = 1$ that decreases to 0.96 hrs (57.6 min) for $N = 2$. Similarly, the recharge times for Bat 2 and 3 are also recorded and shown in Table 8.6.

TABLE 8.4
Test Results of Recharging Time (in hrs) versus DoD at a Distance of 10 cm

Battery type	Parameters	Patch	Dipole
	Received RF power (mW)	**8.12**	**2.56**
	Harvested DC power (mW)	**5.33**	**1.43**
	Conversion efficiency (in %)	**65.7**	**55.7**
	Discharge depth (in %)	**Recharging time (in hours)**	
Bat 1	5	0.24	0.37
	10	0.43	0.84
	15	0.56	1.05
	20	0.67	1.18
	25	0.76	1.28
	30	0.85	1.41
	35	0.91	1.55
	40	0.97	1.69
Bat 2	5	0.6	1.70
	10	1.52	2.33
	15	1.62	2.86
	20	1.87	3.30
	25	2.09	3.68
	30	2.32	4.05
	35	2.47	4.36
	40	2.65	4.7
Bat 3	5	0.97	1.95
	10	1.68	2.76
	15	1.89	3.38
	20	2.19	3.90
	25	2.46	4.5
	30	2.70	4.82
	35	2.90	5.2
	40	3.08	5.51

TABLE 8.5
Recharging Time versus Distance Using Patch and Dipole Antennas

	Recharging time (in hrs)					
Distance	**Bat 1**		**Bat 2**		**Bat 3**	
(in cm)	Patch	Dipole	Patch	Dipole	Patch	Dipole
10	0.24	0.32	0.6	0.72	0.97	2.88
20	0.312	0.44	0.96	2.64	3.84	11.04
30	0.432	0.72	1.93	5.04	8.4	20.88
40	0.6	1.2	3.12	8.64	13.5	35.76
50	0.672	1.92	4.56	13.92	18.72	50.2

TABLE 8.6
Recharging Time for Bat 1, 2, and 3 for Single and Double RF Transmitters at 50 cm Distance and 5% DoD

No. of transmitters (N)	Recharging time (in hrs)					
	Patch			Dipole		
	Bat 1	Bat 2	Bat 3	Bat 1	Bat 2	Bat 3
1	0.672	4.56	18.72	1.92	13.92	50.6
2	0.432	2.64	9.04	0.96	6.72	15.08

To evaluate the performance of the overall WSN using Bat 1, 2, and 3, a simulation environment is designed as shown in Figure 8.6 using NetSim, comprising of a set of 16 WSN nodes, one gateway, one router, and one user application. The 16 sensor nodes are uniformly deployed in the network area, and the initial energy for each node is varied according to the battery specifications given in Table 8.1. Physical quantity like temperature, light, humidity, etc. is sensed by a sensing agent in the industrial environment that sends data to all the motes. The simulation parameters are listed in Table 8.7.

TABLE 8.7
Simulation Parameters

Performance Parameters	Values
Simulation time	10,000
Frequency	915 MHz
Sensor mote	Z1 mote
No. of sensors	16
Initial battery capacity	1,101 J; 8,658 J; 34,992 J
Recharging current (mA)	50
Transmitting current (mA)	8.8
Idle mode current (mA)	3.3
Receiving current (mA)	9.6
Voltage (V)	3.7
Mobility model	No mobility
No. of sink (gateways)	1
Type of harvesting source	RF
Area of deployed nodes	30×40 m^2
Transmission rate	250 kbps
Data size	800 bits
Duty cycle (%)	12

Two sets of simulation results are obtained in the pre-analysis, for only battery-powered (i.e. NR-WSN) and another with included EH feature (i.e. R-WSN). Tables 8.8 and 8.9 show the simulated results generated in NetSim using Bat 1.

Table 8.10 presents a detailed analysis on the total energy consumption of the network driven by Bat 1 for NR-WSN and R-WSN scenarios. Here, the transmission mode energy consumed is 24,312.8331 mJ for NR-WSN, and 67,397.6475 mJ

TABLE 8.8
Results for NR-WSN Using Bat 1 without EH Feature

Device name	Consumed energy (mJ)	Remaining energy (mJ)	Harvested energy (mJ)	Transmitting energy (mJ)	Receiving energy (mJ)	Idle energy (mJ)
WIRELESS_SENSOR_1	1,101,600	0	0	24,312.8331	23,470.78487	1,053,816.382033
WIRELESS_SENSOR_2	1,101,600	0	0	24,312.8331	23,470.78487	1,053,816.382033
WIRELESS_SENSOR_3	1,101,600	0	0	24,312.8331	23,470.78487	1,053,816.382033
WIRELESS_SENSOR_4	1,101,600	0	0	24,312.8331	23,470.78487	1,053,816.382033
WIRELESS_SENSOR_5	1,101,600	0	0	24,312.8331	23,470.78487	1,053,816.382033
WIRELESS_SENSOR_6	1,101,600	0	0	24,312.8331	23,470.78487	1,053,816.382033
WIRELESS_SENSOR_7	1,101,600	0	0	24,312.8331	23,470.78487	1,053,816.382033
WIRELESS_SENSOR_8	1,101,600	0	0	24,312.8331	23,470.78487	1,053,816.382033
WIRELESS_SENSOR_9	1,101,600	0	0	24,312.8331	23,470.78487	1,053,816.382033
WIRELESS_SENSOR_10	1,101,600	0	0	24,312.8331	23,470.78487	1,053,816.382033
WIRELESS_SENSOR_11	1,101,600	0	0	24,312.8331	23,470.78487	1,053,816.382033
WIRELESS_SENSOR_12	1,101,600	0	0	24,312.8331	23,470.78487	1,053,816.382033
WIRELESS_SENSOR_13	1,101,600	0	0	24,312.8331	23,470.78487	1,053,816.382033
WIRELESS_SENSOR_14	1,101,600	0	0	24,312.8331	23,470.78487	1,053,816.382033
WIRELESS_SENSOR_15	1,101,600	0	0	24,312.8331	23,470.78487	1,053,816.382033
WIRELESS_SENSOR_16	1,101,600	0	0	24,312.8331	23,470.78487	1,053,816.382033

TABLE 8.9
Results for R-WSN Using Bat 1 with EH Feature

Device name	Consumed energy (mJ)	Remaining energy (mJ)	Harvested energy (mJ)	Transmitting energy (mJ)	Receiving energy (mJ)	Idle energy (mJ)
WIRELESS_SENSOR_1	3,014,622.019	62,886,273.879	64,799,295.89	67,397.6475	53,149.85061	2,894,074.521
WIRELESS_SENSOR_2	3,014,622.019	62,886,273.879	64,799,295.89	67,397.6475	53,149.85061	2,894,074.521
WIRELESS_SENSOR_3	3,014,622.019	62,886,273.879	64,799,295.89	67,397.6475	53,149.85061	2,894,074.521
WIRELESS_SENSOR_4	3,014,622.019	62,886,273.879	64,799,295.89	67,397.6475	53,149.85061	2,894,074.521
WIRELESS_SENSOR_5	3,014,622.019	62,886,273.879	64,799,295.89	67,397.6475	53,149.85061	2,894,074.521
WIRELESS_SENSOR_6	3,014,622.019	62,886,273.879	64,799,295.89	67,397.6475	53,149.85061	2,894,074.521
WIRELESS_SENSOR_7	3,014,622.019	62,886,273.879	64,799,295.89	67,397.6475	53,149.85061	2,894,074.521
WIRELESS_SENSOR_8	3,014,622.019	62,886,273.879	64,799,295.89	67,397.6475	53,149.85061	2,894,074.521
WIRELESS_SENSOR_9	3,014,622.019	62,886,273.879	64,799,295.89	67,397.6475	53,149.85061	2,894,074.521
WIRELESS_SENSOR_10	3,014,622.019	62,886,273.879	64,799,295.89	67,397.6475	53,149.85061	2,894,074.521
WIRELESS_SENSOR_11	3,014,622.019	62,886,273.879	64,799,295.89	67,397.6475	53,149.85061	2,894,074.521
WIRELESS_SENSOR_12	3,014,622.019	62,886,273.879	64,799,295.89	67,397.6475	53,149.85061	2,894,074.521
WIRELESS_SENSOR_13	3,014,622.019	62,886,273.879	64,799,295.89	67,397.6475	53,149.85061	2,894,074.521
WIRELESS_SENSOR_14	3,014,622.019	62,886,273.879	64,799,295.89	67,397.6475	53,149.85061	2,894,074.521
WIRELESS_SENSOR_15	3,014,622.019	62,886,273.879	64,799,295.89	67,397.6475	53,149.85061	2,894,074.521
WIRELESS_SENSOR_16	3,014,622.019	62,886,273.879	64,799,295.89	67,397.6475	53,149.85061	2,894,074.521

for R-WSN. Similarly, receiving mode energy consumed is 23,470.78487 mJ without harvesting and 53,149.85061 mJ with EH. In idle mode, energy consumption for NR-WSN is 1,053,816.382 mJ, and for R-WSN is 2,894,074.521 mJ. Energy consumption for sleep mode is assumed 'zero' for both the cases. Finally, the total energy consumed by NR-WSN and R-WSN in all modes of operation is found to be 17,625,600 mJ and 48,233,952.3 mJ, respectively, using Bat 1. It can be observed

TABLE 8.10
Network Energy Consumption for NR-WSN and R-WSN Using Bat 1

Packets generated	Parameters	NR-WSN (A) (in mJ)	R-WSN (B) (in mJ)	Improvement factor (B/A)
10,000	Remaining energy	0	62,886,273.88	NA
	Harvested energy	0	64,799,295.89	NA
	Transmitting mode energy consumed	24,312.8331	67,397.6475	2.77
	Receiving mode energy consumed	23,470.78487	53,149.85061	2.26
	Idle mode energy consumed	1,053,816.382	2,894,074.521	2.74
	Energy consumption (1 mote)	1,101,600	3,014,622.019	2.73
	Network energy consumption	17,625,600	48,233,952.3	2.73

that although the total network energy consumption increases for R-WSN, at the same time, the remaining energy does increase from 0 to 64,799.295 J.

Similarly, Tables 8.11 and 8.12 present a detailed analysis on the total energy consumption of the network driven by Bat 2 and 3, respectively, for both the scenarios. It is observed that the energy consumption of the network powered by rechargeable Li-ion battery, LiMn AA, and 3 AAA is increased by 2.73, 3.46, and 1.00 times with RF harvesting. With addition of harvested energy in the battery, the remaining energy for Bat 1, 2, and 3 is found to be 62,886.273 J, 45,284.922 J, and 89,900.888 J, respectively.

The throughput comparison values for R-WSN and NR-WSN is depicted from Table 8.13. With RF harvesting, the network throughput is boosted by 1.08, 1.07, and 1.02 times for Bat 1, 2, and 3, respectively.

TABLE 8.11
Network Energy Consumption for NR-WSN and R-WSN Using Bat 2

Packets generated	Parameters	NR-WSN (A)	R-WSN (B)	Improvement factor (B/A)
10,000	Remaining energy	0	45,284,922.22	NA
	Harvested energy	0	66,599,271.68	NA
	Transmitting mode energy consumed	39,499.46914	138,514.0865	3.50
	Receiving mode energy consumed	32,343.60269	89,393.6576	2.76
	Idle mode energy consumed	8,586,156.928	29,744,441.71	3.46
	Energy consumption (1 mote)	8,658,000	29,972,349.46	3.46
	Network energy consumption	138,528,000	479,557,591.36	3.46

TABLE 8.12
Network Energy Consumption for NR-WSN and R-WSN Using Bat 3

Packets generated	Parameters	NR-WSN (A)	R-WSN (B)	Improvement factor (B/A)
10,000	Remaining energy	25,101,629.101703	89,900,888.189864	3.58
	Harvested energy	0	64,799,291.362509	NA
	Transmitting mode energy consumed	134,680.677481	134,646.316201	0.99
	Receiving mode energy consumed	108,858.17088	108,938.63808	1.00
	Idle mode energy consumed	9,646,832.049846	9,646,818.21807	0.99
	Energy consumption (1 mote)	9,890,370	9,890,403.172301	1.00
	Network energy consumption	158,245,920	158,246,450.75	1.00

TABLE 8.13
Throughput Comparison for R-WSN and NR-WSN

	Throughput (in kbps)	
Battery type	R-WSN	NR-WSN
Bat 1	175.88	162.64
Bat 2	174.87	162.48
Bat 3	165.07	162.19

8.7 RESULTS AND DISCUSSIONS

Figure 8.7 highlights a plot of recharging time (t_C) versus DoD for Bat 1, 2, and 3 at a fixed distance of 40 cm using PCB patch and dipole antennas, respectively. From both the figures, we observe a close resemblance between measured (experimental) and simulated values. It is shown that the recharging time increases with increase in DoD. Table 8.14 presents the % deviation calculation in recharging time of a single sensor mote driven by Bat 1, 2, and 3 using patch antenna at varying distances (in cm) and DoD (in %). The shortest recharging time is found to be 0.24 hrs (14.4 mins) for Bat 1 at a distance of 10 cm. The % deviation is lowest for Bat 3 (0.12) at a distance of 50 cm and 5% DoD, and highest for Bat 1 (6.06) at a distance of 100 cm and 20% DoD, as highlighted in blue colour.

The effect of antenna gains on the recharging time for Bat 1, 2, and 3 at different distances is shown in Figure 8.8. Based on the charging time values given in Table 8.5, the % improvement in recharging time with increasing distance has been calculated for Bat 1, 2, and 3 with respect to antenna gains. For a fixed distance of 10 cm, it was found that patch antenna shortened the recharging time by 25.34%, 16.67%, and 66.32% for Bat 1, 2, and 3, respectively, compared to dipole antenna. On further increasing the distance to maximum 50 cm, the recharging time for patch antenna is improved by 65%, 67.24%, and 62.96% for Bat 1, 2, and 3, respectively, as compared to dipole.

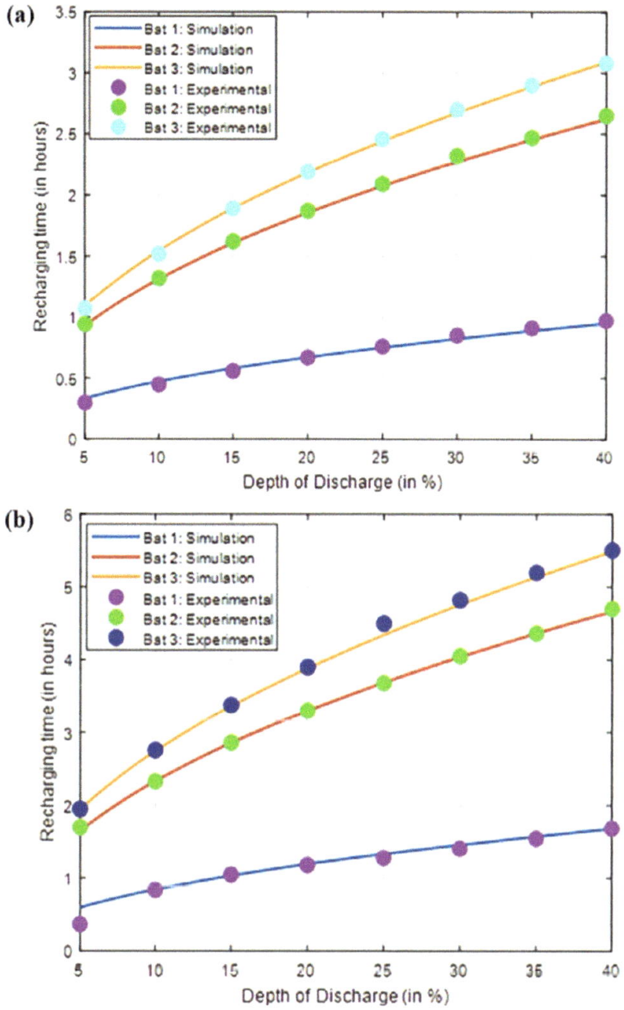

FIGURE 8.7 Recharging time versus DoD: (a) patch antenna, (b) dipole antenna.

Based on the charging time values given in Table 8.6, the percentage improvement in recharging time with increasing number of RF transmitters has been calculated for Bat 1, 2, and 3 with respect to antenna gains, as depicted in Figure 8.9. It is found that dual RF transmitters decrease the recharging time by 35.7%, 42.1%, and 51.7% for Bat 1, 2, and 3, respectively, compared to single RF transmitter using 6.1 dBi patch. Figure 8.10 compares the impact of transmit power on the harvested energy for proposed and existing baseline models (linear [19], constant-linear (CL) [22], and constant-linear-constant (CLC) [22]). Accordingly, the percentage improvement in harvested energy for our proposed work over the existing works has been presented in Table 8.15. For a fixed transmit power of 28 dBm, the % improvement of our work over linear [19], CL [22], and CLC [22] is found to be 99.3%, 89.3%, and 84%, respectively.

TABLE 8.14
Percentage Deviation Calculation in Recharging Time of a Single Sensor Mote Driven by Bat 1, 2, and 3 Using Patch Antenna at Varying Distances (in cm) and DoD (in %)

Distance (in cm)	Battery type	Recharging time (in hrs)								
		DoD 5%			DoD 10%			DoD 20%		
		Measured	Simulated	% deviation	Measured	Simulated	% deviation	Measured	Simulated	% deviation
10	Bat 1	0.24	0.239	0.416	0.432	0.42	2.7	0.69	0.67	3.1
	Bat 2	0.6	0.591	1.5	1.52	1.554	1.87	1.85	1.87	2.67
	Bat 3	0.97	0.964	0.62	1.68	1.56	1.56	2.12	2.024	0.95
20	Bat 1	0.312	0.309	0.96	0.862	0.84	2.5	1.73	1.68	2.9
	Bat 2	0.96	0.954	0.625	4.26	4.2	1.41	8.376	8.16	2.57
	Bat 3	3.84	3.85	0.25	17.38	17.28	0.57	33.96	33.6	1.06
50	Bat 1	0.672	0.661	1.63	1.25	1.22	2.12	2.592	2.52	2.78
	Bat 2	4.56	4.584	0.52	9.24	9.12	1.29	18.22	18	1.18
	Bat 3	18.74	18.72	0.12	37.63	37.44	0.6	75.8	74.88	1.20
100	Bat 1	7.52	7.2	4.25	11.04	10.8	4.34	15.84	14.88	6.06
	Bat 2	18.72	21.84	3.19	31.68	30.24	4.54	44.88	42.72	5.34
	Bat 3	25.68	25.2	1.84	36.48	35.52	2.63	52.56	50.4	4.10

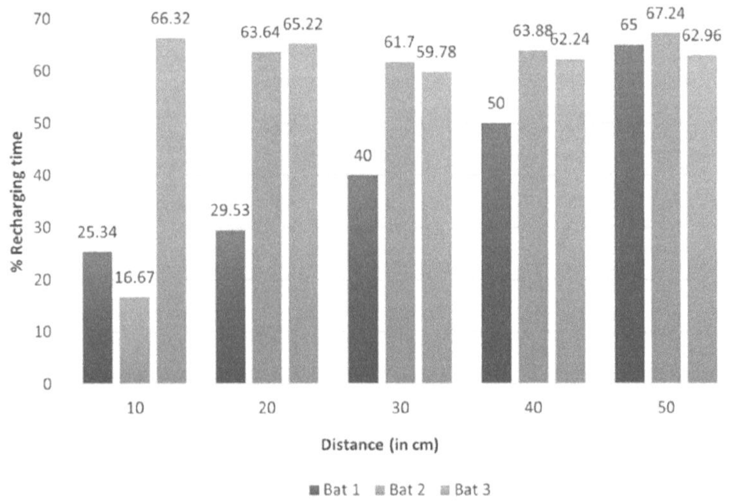

FIGURE 8.8 Effect of antenna gain on recharging time for Bat 1, 2, and 3 at different distance, expressed as percentage.

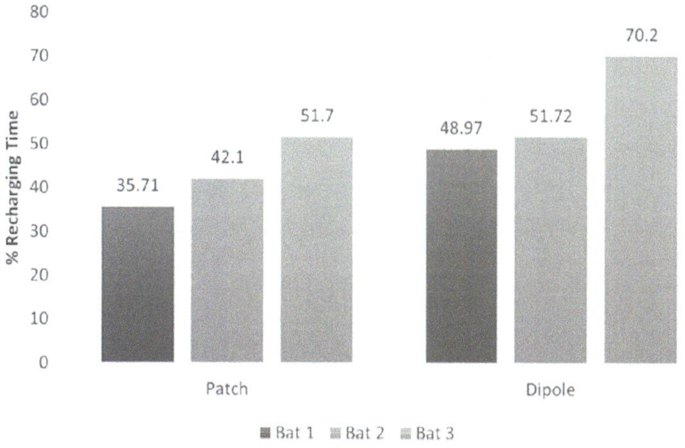

FIGURE 8.9 Percentage improvement in recharging time for Bat 1, 2, and 3 for single and dual RF transmitters with 6.1 dBi patch.

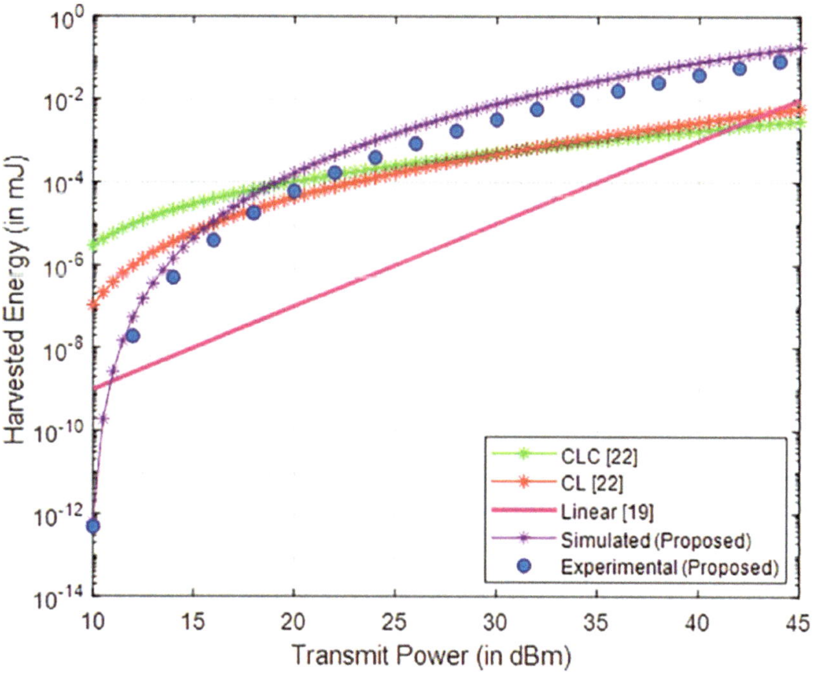

FIGURE 8.10 Comparison of harvested energy for proposed and existing works.

TABLE 8.15
Percentage Improvement in Harvested Energy

Parameter	Harvested energy (in mJ)				% harvested energy		
Transmit power (in dBm)	CLC [22]	CL [22]	Linear [19]	Proposed	Proposed over CLC [22]	Proposed over CL [22]	Proposed over linear [19]
26	2.9×10^{-5}	6.1×10^{-6}	10–7	4.3×10^{-6}	85.2 (deterioration)	29.5 (deterioration)	97.6
27	9.9×10^{-5}	4.2×10^{-5}	10–6	1.6×10^{-4}	38.13	73.75	99.35
28	2.4×10^{-4}	1.6×10^{-4}	10–5	1.5×10^{-3}	84	89.3	99.33
29	5.1×10^{-4}	4.8×10^{-4}	10–4	7.7×10^{-3}	93.4	93.7	98.75
31	0.011	0.003	0.001	0.078	85.9	96.15	98.7

Figure 8.11 highlights a comparison of throughput and NL for NR-WSN and R-WSN using Bat 1. For Bat 1, beyond 34 hrs, the throughput goes below 162.64 Kbps and keeps on decreasing, as is evident from the figure. For NR-WSN, the NL for Bat 1 is 59 hrs. This signifies that network is operational for 59 divided by 24 hrs, that is, 2.45 days approximately. The NL for R-WSN is 10,000 sec and extends towards infinite time. Since the NetSim simulation software cannot display simulations beyond 10,000 and outcomes for an unlimited duration, the maximum limit is displayed only up to 10,000. We can say that the network is operational for 10,000 divided by 24 hrs, that is, 416 days (1.13 years approx.). Similarly, Figures 8.12 and 8.13 highlight a comparison of throughput and NL for NR-WSN and R-WSN using Bat 2 and 3, respectively. The NL for NR-WSN using Bat 2 and 3 is around 390 hrs and 1,559 hrs, respectively, that is, the network is operational for 16.25 days and 64.95 days approximately. The average network throughput for R-WSN using Bat 1, 2, and 3 is 175.88, 174.87, and 165.07 kbps as opposed to 162.64, 162.48, and 162.19 Kbps for NR-WSN.

The key findings from the analysis are summarized here:

- Bat 1 gives the shortest recharging time value at 10 cm distance with 6 dBi patch antenna at 5% DoD as compared to Bat 2 and 3. The recharging time for Bat 1 (Li-ion coin cell) is much faster (14.4 min) using 6.1 dBi patch at 10 cm as compared to Li-Mn (Bat 2) and 3 AAA (Bat 3).
- 6.1 dBi patch shortens the recharging time by 65%, 67.24%, and 63% for Bat 1, 2, and 3, respectively, as compared to 1 dBi dipole.
- Dual RF transmitters reduce the recharging time by 35.7%, 42.1%, and 51.7% for Bat 1, 2, and 3, respectively, compared to single RF transmitter with 6.1 dBi patch.
- The energy consumption for RF-powered WSN driven by Li-ion battery, LiMn AA, and 3 AAA is increased by 2.73, 3.46, and 1.00 times. At the same time, the remaining energy for Bat 1, 2, and 3 is found to be 62,886.273

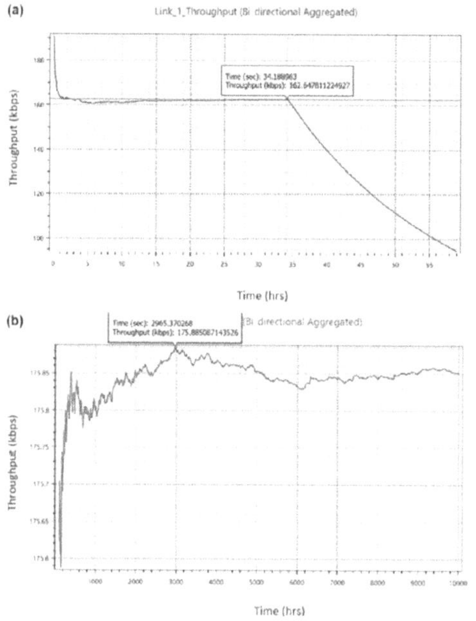

FIGURE 8.11 Comparison plot of throughput and network lifetime with Bat 1: (a) NR-WSN, (b) R-WSN.

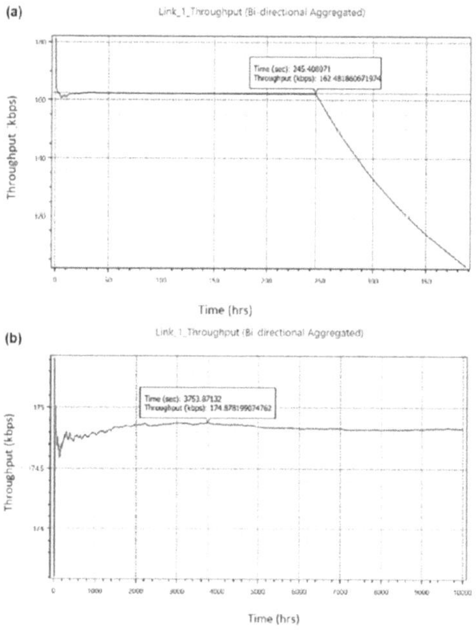

FIGURE 8.12 Comparison plot of throughput and network lifetime with Bat 2: (a) NR-WSN, (b) R-WSN.

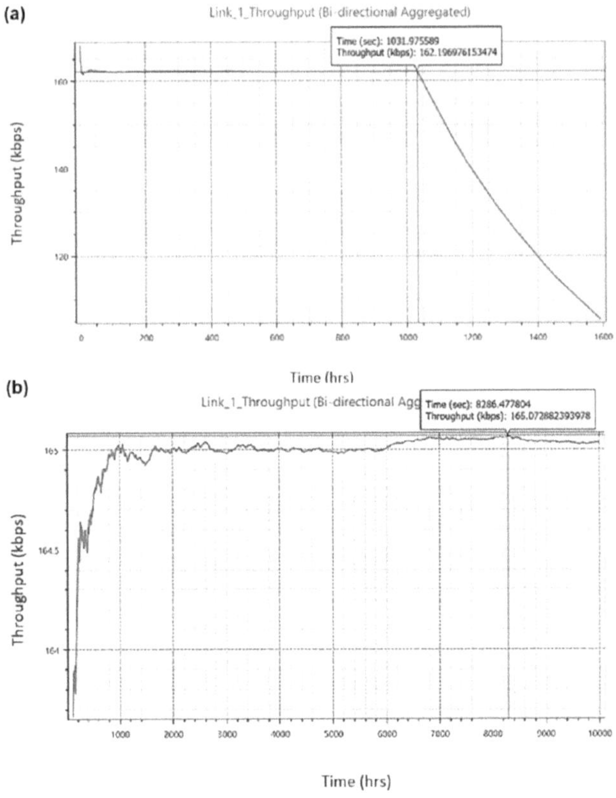

FIGURE 8.13 Comparison plot of throughput and network lifetime with Bat 3: (a) NR-WSN, (b) R-WSN.

J, 45,284.922 J, and 89,900.888 J, respectively. Thus, a conclusion can be made that with EH, the lifetime of a wireless sensor node can be extended with proper selection of correct parameters.

- The NL at 12% duty cycle and 5% DoD is extended up to 1.13 years and higher (ideally, up to infinite time), as compared to only 2.45, 16.45, and 66.67 days by RF harvesting. Besides, the network throughput is increased up to 1.08, 1.07, and 1.02 times for Bat 1, 2, and 3 with RF harvesting.

8.8 CONCLUSION AND FUTURE SCOPE

Our chapter investigates the impact of RF EH on a battery-driven R-WSN for IIoT application to enhance the throughput and network lifetime. P1110 Powerharvester® Receiver IC converts RF energy into DC power and charges the on-board batteries. The simulation findings show that RF energy harvesting can extend the network's life span to a theoretically unlimited amount of time. In addition, from the analysis, RF harvesting can be considered a cost-efficient strategy, as the battery of the sensor motes need not be replaced frequently to keep the network operational. The performance of the proposed model is compared with existing state-of-art network models. The results obtained prove the superiority of the proposed model in terms of harvested energy.

Future work is to make the network more efficient by recharging the batteries at a faster rate. Also, we intend to investigate the impact of RF EH on a battery-less R-WSN for IIoT application and compare the charging time and network lifetime. Our system model can be a reference for new RF-powered IIoT designs and can be relevant for planning and deploying IEEE 802.15.4–based sensor networks with appropriate battery system and specific performance demands.

8.8.1 Acknowledgements

This research work was supported by the Department of Science and Technology and Biotechnology, Government of West Bengal (Sanction Order no. 459 (Sanc.)/ STBT-11012(16)/29/2021-ST SEC).

REFERENCES

[1] Yang, D., Mahmood, A., Hassan, S. A., & Gidlund, M. (2022). Guest editorial: Industrial IoT and sensor networks in 5G-and-beyond wireless communication. IEEE Transactions on Industrial Informatics, 18(6), 4118–4121.

[2] Rahmani, H., Shetty, D., Wagih, M., Ghasempour, Y., Palazzi, V., Carvalho, N. B., . . . & Grosinger, J. (2023). Next-generation IoT devices: Sustainable eco-friendly manufacturing, energy harvesting, and wireless connectivity. IEEE Journal of Microwaves, 3(1), 237–255.

[3] Sherazi, H. H. R., Grieco, L. A., Imran, M. A., & Boggia, G. (2020). Energy-efficient LoRaWAN for industry 4.0 applications. IEEE Transactions on Industrial Informatics, 17(2), 891–902.

[4] Benbuk, A. A., Kouzayha, N., Costantine, J., & Dawy, Z. (2023). Charging and wake-up of IoT devices using harvested RF energy with near-zero power consumption. IEEE Internet of Things Magazine, 6(1), 162–167.

[5] Sanislav, T., Mois, G. D., Zeadally, S., & Folea, S. C. (2021). Energy harvesting techniques for internet of things (IoT). IEEE Access, 9, 39530–39549.

[6] Oruganti, S. K., Khosla, A., & Thundat, T. G. (2020). Wireless power-data transmission for industrial internet of things: Simulations and experiments. IEEE Access, 8, 187965–187974.

[7] Lee, D., Lee, C., Jang, G., Na, W., & Cho, S. (2022). Energy-efficient directional charging strategy for wireless rechargeable sensor networks. IEEE Internet of Things Journal, 9(19), 19034–19048.

[8] Tanash, R., AlQudah, M., & Al-Agtash, S. (2023). Enhancing energy efficiency of IEEE 802.15.4-based industrial wireless sensor networks. Journal of Industrial Information Integration, 33, 100460.

[9] Infrastructure, L. E. C., & Layer, M. L. P. (2015). IEEE standard for low-rate wireless networks. IEEE Stand, 2015, 1–708.

[10] Sharma, P., & Singh, A. K. (2023). A survey on RF energy harvesting techniques for lifetime enhancement of wireless sensor networks. Sustainable Computing: Informatics and Systems, 37, 100836.

[11] Ali, O., Ishak, M. K., Ooi, C. A., & Bhatti, M. K. L. (2022). Battery characterization for wireless sensor network applications to investigate the effect of load on surface temperatures. Royal Society Open Science, 9(2), 210870.
[12] Sah, D. K., & Amgoth, T. (2020). Renewable energy harvesting schemes in wireless sensor networks: A survey. Information Fusion, 63, 223–247.
[13] Kaswan, A., Jana, P. K., & Das, S. K. (2022). A survey on mobile charging techniques in wireless rechargeable sensor networks. IEEE Communications Surveys & Tutorials, 24(3), 1750–1779.
[14] Zhang, H., Guo, Y. X., Zhong, Z., & Wu, W. (2019). Cooperative integration of RF energy harvesting and dedicated WPT for wireless sensor networks. IEEE Microwave and Wireless Components Letters, 29(4), 291–293.
[15] Huang, J., Zhou, Y., Ning, Z., & Gharavi, H. (2019). Wireless power transfer and energy harvesting: Current status and future prospects. IEEE Wireless Communications, 26(4), 163–169.
[16] Van Leemput, D., Sabovic, A., Hammoud, K., Famaey, J., Pollin, S., & De Poorter, E. (2023). Energy harvesting for wireless IoT use cases: A generic feasibility model and trade-off stud. IEEE Internet of Things Journal, 10(17), 15025–15043.
[17] Niotaki, K., Carvalho, N. B., Georgiadis, A., Gu, X., Hemour, S., Wu, K., . . . & Shinohara, N. (2023). RF energy harvesting and wireless power transfer for energy autonomous wireless devices and RFIDs. IEEE Journal of Microwaves, 3(2), 763–782.
[18] Tang, X., Wang, X., Cattley, R., Gu, F., & Ball, A. D. (2018). Energy harvesting technologies for achieving self-powered wireless sensor networks in machine condition monitoring: A review. Sensors, 18(12), 4113.
[19] Yang, Z., Xu, W., Pan, Y., Pan, C., & Chen, M. (2017). Energy efficient resource allocation in machine-to-machine communications with multiple access and energy harvesting for IoT. IEEE Internet of Things Journal, 5(1), 229–245.
[20] Boshkovska, E., Ng, D. W. K., Zlatanov, N., & Schober, R. (2015). Practical non-linear energy harvesting model and resource allocation for SWIPT systems. IEEE Communications Letters, 19(12), 2082–2085.
[21] Xu, X., Özçelikkale, A., McKelvey, T., & Viberg, M. (2017, May). Simultaneous information and power transfer under a non-linear RF energy harvesting model. In 2017 IEEE International Conference on Communications Workshops (ICC Workshops) (pp. 179–184). IEEE.
[22] Alevizos, P. N., & Bletsas, A. (2018). Sensitive and nonlinear far-field RF energy harvesting in wireless communications. IEEE Transactions on Wireless Communications, 17(6), 3670–3685.
[23] Powercast Corporation. (2015). [Online]. Available: www.powercastco.com/products/development-kits/#P1110-EVAL-PS.
[24] Ijemaru, G. K., Ang, K. L. M., & Seng, J. K. (2022). Wireless power transfer and energy harvesting in distributed sensor networks: Survey, opportunities, and challenges. International Journal of Distributed Sensor Networks, 18(3), 15501477211067740.
[25] Kamalinejad, P., Mahapatra, C., Sheng, Z., Mirabbasi, S., Leung, V. C., & Guan, Y. L. (2015). Wireless energy harvesting for the Internet of Things. *IEEE Communications Magazine*, 53(6), 102–108. Available: www.tetcos.com/index.html.
[26] Sharma, H., Haque, A., & Jaffery, Z. A. (2019). Maximization of wireless sensor network lifetime using solar energy harvesting for smart agriculture monitoring. Ad Hoc Networks, 94, 101966.
[27] Mallick, S., Habib, A. Z. S. B., Ahmed, A. S., & Alam, S. S. (2017, December). Performance appraisal of wireless energy harvesting in IoT. In 2017 3rd International Conference on Electrical Information and Communication Technology (EICT) (pp. 1–6). IEEE.

[28] Habaebi, M. H., Basaloom, A. A. S., Islam, M. R., Yacine, M., Mesri, M., Elsheikh, E. A., & Suliman, F. M. (2021). Energy harvesting effect on prolonging low-power lossy networks lifespan. International Journal of Distributed Sensor Networks, 17(8), 15501477211028549.
[29] "Zolertia re-mote," [Online]. Accessed 7 January 2019. Available: https://github.com/Zolertia/Resources/wiki/RE-Mote.

9 Cooperative NOMA with Hybrid PTS-Based Energy Harvesting Relay

Priyanka Ghosh, Sanjay Dhar Roy, and Sumit Kundu

9.1 INTRODUCTION

In comparison with previous generations, the fifth generation of wireless technology has several notable benefits, such as increased data speeds, high reliability, reduced latency, and better coverage [1]. The 5G network uses some of the latest communication technologies, like non-orthogonal multiple access (NOMA), which makes 5G function more easily than do earlier networks. Multiple users can use the same time-frequency resources with the help of NOMA, which improves spectral efficiency and expands wireless networks' capacity [2].

Cooperative NOMA, which integrates cooperative communication strategies with NOMA principles, is an extension of the NOMA concept [3]. Users with better channel conditions act as relays to support users with worse channel conditions in cooperative NOMA [4–6]. Wireless devices can now transfer data and power simultaneously by utilizing a technique called simultaneous wireless information and power transmission (SWIPT) [7]. The information-decoding and energy-harvesting functions are separated by SWIPT by the application of two widely used techniques: power splitting (PS) [8] and time switching (TS) [9].

The power of the received signal is divided into two sections at the receiver's antenna, one for information decoding and the other for energy harvesting, using the power splitting protocol. The receiver may therefore gather energy and receive data at the same time. Another method for enabling independent time allocation for energy harvesting and information decoding in SWIPT systems is the time switching protocol. When it comes to data reception and power harvesting, time switching uses time-division multiplexing to manage these tasks at various time intervals, in contrast to the power splitting protocol, which divides power at the antenna concurrently [9]. To achieve a balance between energy harvesting and information decoding efficiency, the power splitting and time switching protocols have been combined into the hybrid power-time switching (PTS) protocol [10]. A relaying-assisted ultra-reliable low-latency communication network operating with finite block-length codes has been proposed in [11] to improve reliability performance by introducing a trade-off between the PS and TS protocols. A hybrid power-time splitting (PTS) protocol has been utilized for evaluating the robustness of a non-linear energy harvesting (EH)–based full-duplex (FD) Internet of Things (IoT) network with infinite and

 DOI: 10.1201/9781032625034-9

finite block-length codes in [12]. The effect of the harvested power at relay on the outage and throughput of an energy harvesting–based cooperative network in terms of time and power factors using the proposed PTS-based relaying protocol has been investigated in [13]. The performance of energy harvesting of a hybrid PTS-based non-linear NOMA network is compared with the same for TSR- and PSR-based system in [14]. The performance of an FD multiuser network based on non-linear hybrid PTS-based EH system is analyzed in [15]. Several researchers have analyzed outage performance of hybrid PTS-based NOMA [16, 17]. In this work, we study a cooperative NOMA system where energy is harvested at the near user by using the (1) TS protocol and the (2) hybrid PTS protocol to decode signals for both near and far users. The near user relays the signal for the far user using harvested energy [18]. This model with TSR is already investigated by us in [9] using MATLAB-based simulation, and the PSR scheme of this model is also studied in [19]. The main contributions of the chapter are as follows:

1. Presenting analytical framework of outage probability of a cooperative NOMA network aided by energy harvesting relay under (a) TSR scheme and simulation results for (b) hybrid PTS scheme of energy harvesting.
2. Development of closed-form analytical expressions on outage probability under the TSR scheme.
3. Indicating the influence of time switching parameter, joint impact of time, power splitting parameters (in PTS) on system outage.
4. MATLAB-based simulation results validating our analytical formulation.

9.2 TWO-USER NOMA MODEL, NOMA SYSTEM MODEL

The EH-assisted cooperative NOMA network consists of one source terminal (S), one far user (U_1), and one near user (U_1), as shown in Figure 9.1. Each node has been given a single antenna that operates in half-duplex mode. The nearby user U_2

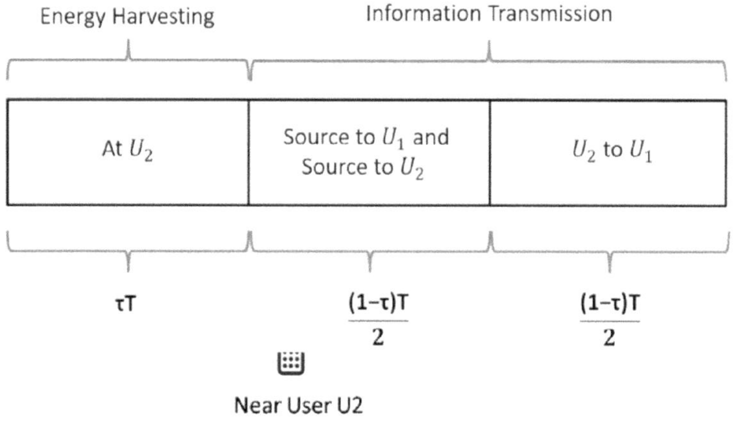

FIGURE 9.1 System model for two-user cooperative NOMA.

performs the relay operation to assist the far user U_1 in signal transmission and energy harvesting. The channels between S and U_1, S and U_2, and U_2 and U_1 have fading channel coefficients of h_1, h_2, and h_3, respectively. All the channels are considered to be independent, to be uniformly distributed, and to experience Rayleigh fading. The channel gains $|h_1|^2$, $|h_2|^2$, and $|h_3|^2$ will therefore have exponential PDF. The means of the channel gains are $\frac{1}{\lambda_1}$, $\frac{1}{\lambda_2}$, and $\frac{1}{\lambda_3}$. At both users, the additive white Gaussian noise (AWGN) has a variance of σ^2 and a mean of zero. Figure 9.2 and Figure 9.3 show the time frame for information transmission and energy harvesting for TSR model and PTS model, respectively.

9.3 COOPERATIVE NOMA AIDED BY A RELAY ENERGIZED BY SWIPT FOLLOWING

9.3.1 TSR Scheme of Energy Harvesting

In TSR scheme, energy is harvested for τT time, where τ is the time switching factor, and the rest of the time is divided into two equal halves and utilized for information transmission from source to users in two hops through an intermediate relay.

9.3.2 PTS Scheme of Energy Harvesting

In case of hybrid PTS, energy harvesting is allotted for both τT time and half of the rest of the time, that is, $\frac{(1-\tau)T}{2}$ time. In $\frac{(1-\tau)T}{2}$ time, energy harvesting is done through power splitting (PS), that is, the power is divided into two parts to harvest energy with a PS factor ρ and to transmit signal with rest of the power (1 – ρ).

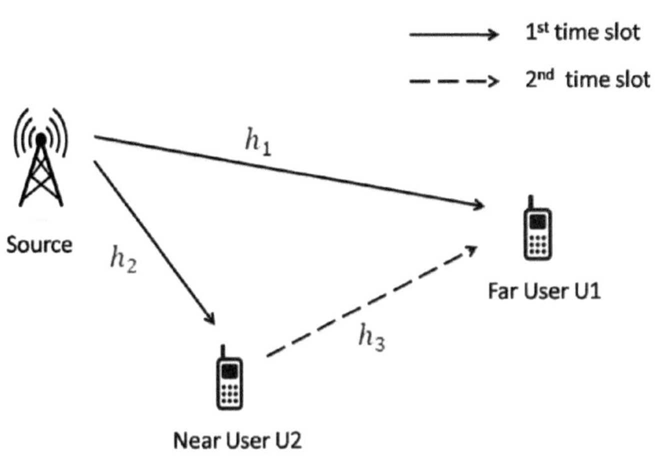

FIGURE 9.2 Time frame of TSR model.

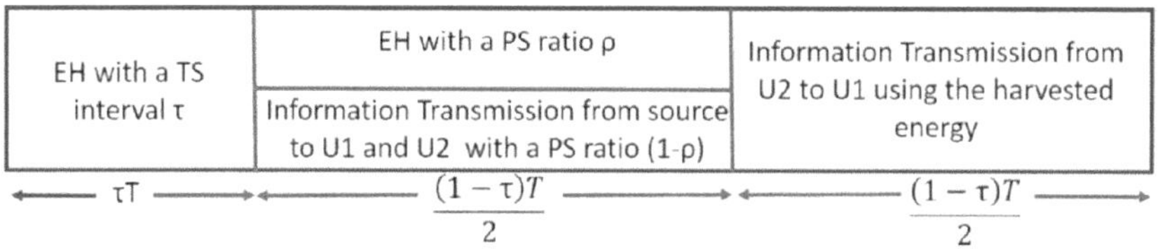

FIGURE 9.3 Time frame of the PTS model.V

During a τ fraction of time, that is, τT, only the harvester at the relay is kept to harvest energy from the transmitted signal of the two sources. The signal that the source terminal simultaneously delivered to the near user (U_2) and far user (U_1) in accordance with the NOMA principle is then expressed as

$$x(t) = s_1(t)\sqrt{P_s a_1} + s_2(t)\sqrt{P_s a_2} \tag{9.1}$$

where a_1 and a_2 are power allocation coefficients for U_1 and U_2, respectively; $a_1 > 0$, $a_2 > 0$, and $a_1 + a_2 = 1$; s_1(t) and s_2(t) are transmitted signals with unit power for U_1 and U_2, respectively; and P_s is the transmit power. In the downlink NOMA, the far user is assigned the higher power. U_1 and U_2 receive the transmitted signal in order to decode the data. For the TSR and PTS schemes, the signal received at user U_2 is stated as:

$$y_{SU_2} = h_2\left(s_1(t)\sqrt{P_s a_1} + s_2(t)\sqrt{P_s a_2}\right) + n_1(t) \tag{9.2}$$

and

$$y_{SU_2} = \sqrt{(1-\rho)}h_2\left(s_1(t)\sqrt{P_s a_1} + s_2(t)\sqrt{P_s a_2}\right) + n_1(t) \tag{9.3}$$

The signals from the near user and the far user are received by U_2. Therefore, by executing perfect SIC at U_2, both the signals $s_1(t)$ and $s_2(t)$ are decoded in accordance with the fundamental idea of NOMA. U_2 considers its own signal to be noise when decoding the far user signal s_1(t). Next, to obtain the signal s_2(t) meant for U_2, subtract the decoded signal corresponding to $s_1(t)$ from the received superimposed signal. Assuming that the near user's threshold data rate is R_2 and the far user's is R_1, the threshold signal-to-noise ratios (SNRs) that correlate to the distant and near users are as follows:

$$\theta_1 = 2^{2R_1} - 1 \tag{9.4}$$

and

$$\theta_2 = 2^{2R_2} - 1 \tag{9.5}$$

The threshold SNR at the far user for the relayed path is given as:

$$\theta_3 = 2^{\frac{2R_1}{1-\tau}} - 1 \tag{9.6}$$

Under TSR mode, the necessary SNRs for s_1(t) and s_2(t) are given by

$$\gamma_{SU_2}^{S_1} = \frac{\gamma_{in} a_1 |h_2|^2}{\gamma_{in} a_2 |h_2|^2 + 1} \tag{9.7}$$

$$\gamma_{SU_2}^{S_2} = \gamma_{in} a_2 |h_2|^2 \tag{9.8}$$

where $\gamma_{in} = \frac{P_s}{\sigma^2}$ denotes the input SNR.

The corresponding SNRs for s_1(t) and s_2(t) for PTS scheme are given by:

$$\gamma_{SU_2}^{S_1} = \frac{(1-\rho)\gamma_{in} a_1 |h_2|^2}{(1-\rho)\gamma_{in} a_2 |h_2|^2 + 1} \tag{9.9}$$

$$\gamma_{SU_2}^{S_2} = (1-\rho)\gamma_{in} a_2 |h_2|^2 \tag{9.10}$$

With TSR protocol, U_2 functions as an EH relay. Two halves comprise the overall time duration T. τT time is dedicated to energy harvesting and transmission of information from source to U_1, source to U_2 , and U_2 to U_1 takes up the residual $(1 - \tau)$T time. The harvested energy at U_2 for TSR and PTS scheme is respectively given by:

$$E_{U_2}^{TSR} = \eta P_s |h_2|^2 \tau T \tag{9.11}$$

$$E_{U_2}^{PTS} = \frac{\eta T P_s |h_2|^2 \left(2\tau + \rho(1-\tau)\right)}{2} \tag{9.12}$$

The variable η represents energy conversion efficiency, while P_s stands for total transmission power. In this instance, the near user U_2's transmit power is determined by the energy harvested at U_2, and it is represented as follows.

For TSR:

$$P_r^{TSR} = \frac{E_{U_2}^{TSR}}{\frac{(1-\tau)T}{2}} = \frac{2\eta P_s |h_2|^2 \tau}{(1-\tau)} \tag{9.13}$$

For PTS:

$$P_r^{PTS} = \frac{E_{U_2}^{PTS}}{\frac{(1-\tau)T}{2}} = \frac{\eta P_s |h_2|^2 \left(2\tau + \rho(1-\tau)\right)}{(1-\tau)} \tag{9.14}$$

It should be observed that U_2 is transmitting the signal of U_1 while preserving its own energy by using the energy it has acquired. As a result, U_1 gets two copies of the signals: one from U_2 and one straight from the base station. At time instant t, the instantaneous received signal at U_1 is represented as:

$$\gamma_{SU_1} = h_3\sqrt{P_r}S_2(t-1) + h_1\left(s_1(t)\sqrt{P_s a_1} + S_2(t)\sqrt{P_s a_2}\right) + n_1(t) \tag{9.15}$$

In order to decode s_1(t), the received SNR at U_1 from the base station is expressed as:

$$\gamma_{SU_1}^{S_1} = \frac{\gamma_{in} a_1 |h_1|^2}{\gamma_{in} a_2 |h_1|^2 + 1} \tag{9.16}$$

The SNR received at U_1 from EH relay through $U_2 - U_1$ path for TSR and PTS are, respectively:

$$\gamma_{U_2U_1}^{S_1} = \frac{P_r^{TSR} |h_3|^2}{\sigma^2} = \frac{2\gamma_{in}\eta\tau |h_2|^2 |h_3|^2}{(1-\tau)} \tag{9.17}$$

and

$$\gamma_{U_2U_1}^{S_1} = \frac{P_r^{PTS} |h_3|^2}{\sigma^2} = \frac{\eta\gamma_{in}\left(2\tau + \rho(1-\tau)\right)|h_2|^2 |h_3|^2}{(1-\tau)} \tag{9.18}$$

Next, to obtain an overall SNR at U_1, we take into account maximal ratio combining (MRC). Thus, following MRC's conversion, the total received SNR at U_1 becomes:

For TSR:

$$\gamma_{MRC} = \gamma_{SU_1}^{S_1} + \gamma_{U_2U_1}^{S_1} = \frac{\gamma_{in} a_1 |h_1|^2}{\gamma_{in} a_2 |h_1|^2 + 1} + \frac{2\gamma_{in}\eta\tau |h_2|^2 |h_3|^2}{(1-\tau)} \tag{9.19}$$

For PTS:

$$\gamma_{MRC} = \gamma_{SU_1}^{S_1} + \gamma_{U_2U_1}^{S_1} = \frac{\gamma_{in} a_1 |h_1|^2}{\gamma_{in} a_2 |h_1|^2 + 1} + \frac{\eta T P_s \left(2\tau + \rho(1-\tau)\right)|h_2|^2 |h_3|^2}{(1-\tau)} \tag{9.20}$$

9.4 OUTAGE ANALYSIS WITH TSR SCHEME AT THE RELAY

Outage probability of the near user (P_{out}^N) is defined as the outage when the SNR received at the near user falls under the target threshold SNR of the user. P_{out}^N is expressed as:

$$P_{out}^N = P(\gamma_{SU_2}^{S_1} < \theta_1 \, or \, \gamma_{SU_2}^{S_2} < \theta_2) \tag{9.21}$$

For TSR, putting the values from equations (7) and (8), the closed-form expression of equation (21) can be written as:

$$P_{out}^N = 1 - e^{-\frac{\theta_1 \lambda_2}{\gamma_{in}(a_1 - a_2\theta_1)}} + 1 - e^{-\frac{\theta_2 \lambda_2}{\gamma_{in} a_2}} - \left(1 - e^{-\frac{\theta_1 \lambda_2}{\gamma_{in}(a_1 - a_2\theta_1)}}\right)\left(1 - e^{-\frac{\theta_2 \lambda_2}{\gamma_{in} a_2}}\right) \tag{9.22}$$

For PTS scheme, putting the values from equations (9) and (10), the closed-form expression of equation (21) can be written as:

$$P_{out}^N = 1 - e^{-\frac{\theta_1 \lambda_2}{(1-\rho)\gamma_{in}(a_1 - a_2\theta_1)}} + 1 - e^{-\frac{\theta_2 \lambda_2}{(1-\rho)\gamma_{in} a_2}} - \left(1 - e^{-\frac{\theta_1 \lambda_2}{(1-\rho)\gamma_{in}(a_1 - a_2\theta_1)}}\right)\left(1 - e^{-\frac{\theta_2 \lambda_2}{(1-\rho)\gamma_{in} a_2}}\right) \tag{9.23}$$

If the far user's received SNR falls below the user's necessary threshold SNR, the far user experiences an outage. Consequently, the far user's outage probability (P_{out}^F) is given as:

$$P_{out}^F = P(\gamma_{MRC} < \theta_3) \tag{9.24}$$

For both TSR and PTS scheme, putting the values from equations (19) and (20), the closed-form expression of equation (24) can be written as

$$P_{out}^F = 1 - \lambda_3 \sqrt{\frac{4z\lambda_2}{\lambda_3}} K_1\left(\sqrt{4z\lambda_2\lambda_3}\right) \tag{9.25}$$

Where $z = \frac{(\theta_3 - \frac{a_1}{a_2})(1-\tau)}{\gamma_{in}\tau\eta}$ for TSR and $\frac{(\theta_3 - \frac{a_1}{a_2})(1-\tau)}{\gamma_{in}\eta\left(2\tau + \rho(1-\tau)\right)}$ for PTS scheme. K_1 is defined as the modified Bessel function of second kind.

The system throughput is determined by the total amount of information gathered from both the users. The expression appears as follows:

$$T_{total} = \left(1 - P_{out}^F\right)R_1 + \left(1 - P_{out}^N\right)R_2$$

9.5 RESULTS AND DISCUSSION

9.5.1 Results with TSR

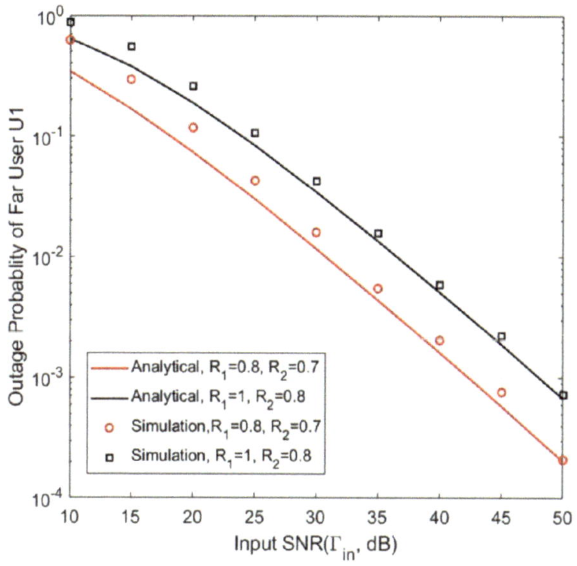

FIGURE 9.4 Outage performance analysis of far user (U_1) for several threshold data rate pairs.

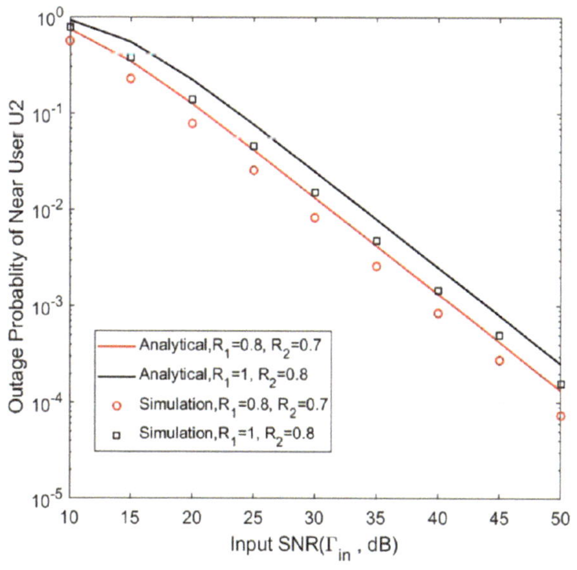

FIGURE 9.5 Outage performance analysis of near user (U_2) for several threshold data rate pairs.

The following parameter values have been taken into consideration: $\eta = 0.7$, $\lambda_1 = 0.5$, $\lambda_2 = 1$, and $\lambda_3 = 0.66$. Bits/sec is the unit used to express the threshold data rate. This scheme has power allocation coefficients set to $a_1 = 0.8$ and $a_2 = 0.2$, a time switching factor of $\tau = 0.4$, and a power splitting factor of $\rho = 0.4$.

Outage probability analysis of the far user utilizing the TSR protocol is shown in Figure 9.4 for various R_1, R_2 values. According to the figure, the probability of outage decreases as input signal strength increases, and an outage becomes more acceptable in situations where R_1, R_2 values are low (in this case, $R_1 = 0.8$, $R_2 = 0.7$).

The near user's outage probability analysis utilizing the TSR protocol for various R_1, R_2 values is displayed in Figure 9.5. This trend of the outage is similar to that of the far user.

For various target data rates, the system outage probability is displayed in Figure 9.6 in relation to input SNR. System outage displays the model's total outage based on each user's individual outage. Thus, the reliability of the two users' individual outage probability curves is shown by this figure.

The overall system throughput for various R_1, R_2 values is displayed in Figure 9.7 in relation to the input SNR. It is evident that as input signal-to-noise ratio (SNR) grows, corresponding throughput increases as outage decreases. We see a saturation in throughput at high-SNR region as outage <<1 at this region and hence the throughput ≈ R.

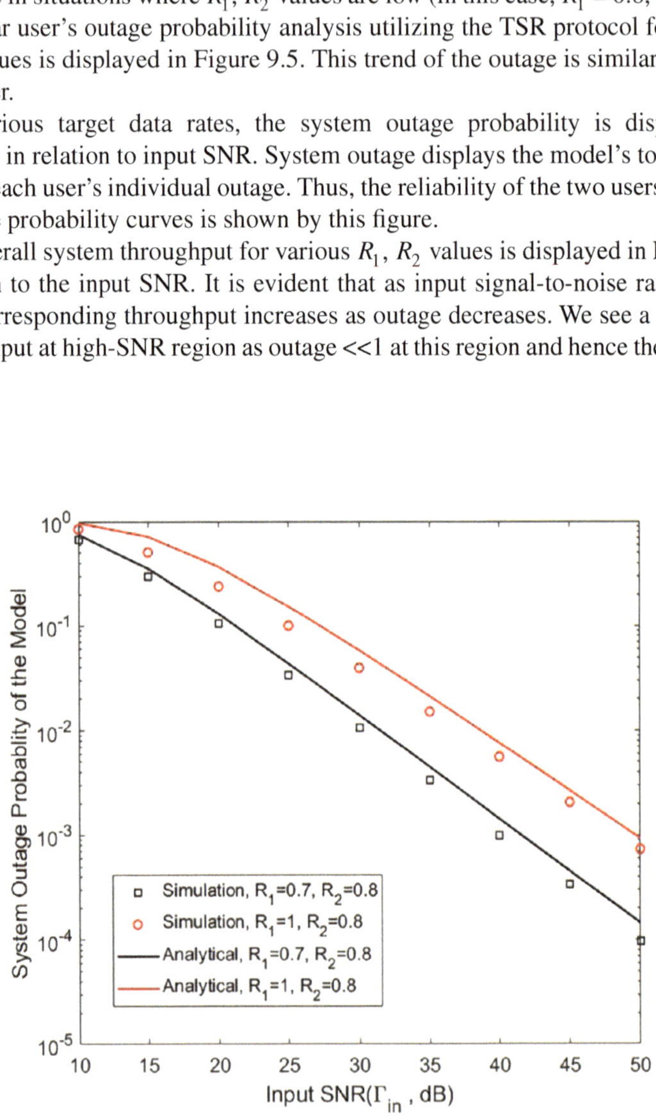

FIGURE 9.6 System outage analysis for several target data rate pairs.

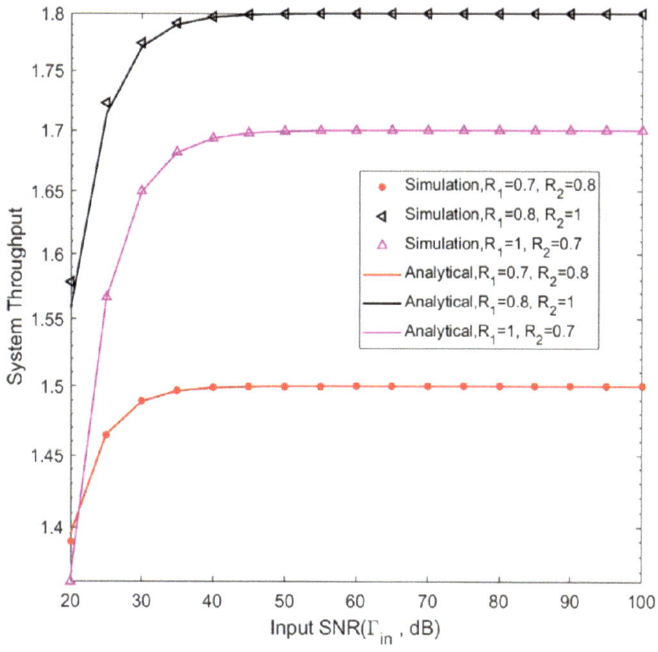

FIGURE 9.7 System throughput analysis for several target data rate pairs.

The system outage performance is plotted against the time switching factor (τ) in Figure 9.8. This figure shows that when τ increases, the outage increases as well. With the increase in τ, the harvesting time also increases, resulting in a reduction of the effective transmission time, capacity, and outage probability.

9.5.2 Results with PTS

The far user's outage probability analysis utilizing the PTS protocol is displayed in Figure 9.9 for various R_1, R_2 values. The figure shows that as input SNR increases, the outage probability decreases, and in the case of a low R_1 value (in this example, $R_1 = 0.7$), the outage becomes more acceptable.

Outage probability analysis of the nearby user utilizing the PTS protocol is displayed in Figure 9.10 for various target data rates. The similar pattern as the far user is also shown in this figure.

For a range of R_1, R_2 values, the model's system outage performance is displayed in Figure 9.11 in relation to the input SNR. As the input SNR increases from Figures 9.9 and 9.10, the outages experienced by two users also reduce, and this decrease is reflected in the total system outage.

The overall system throughput for various R_1, R_2 values is displayed in Figure 9.12 in relation to the input SNR. The trend in this picture is similar to that in the TSR mode with the same R_1 and R_2 values.

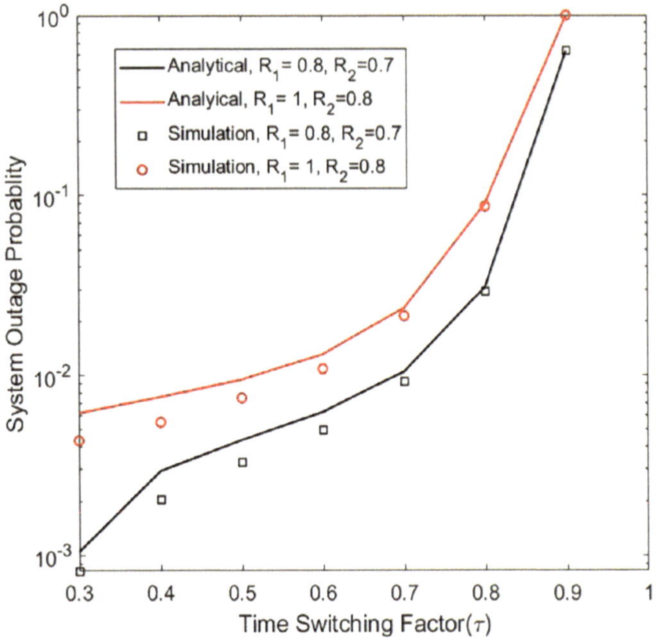

FIGURE 9.8 Outage probability of far user (U_1) versus time switching factor (τ) for several target data rate pairs.

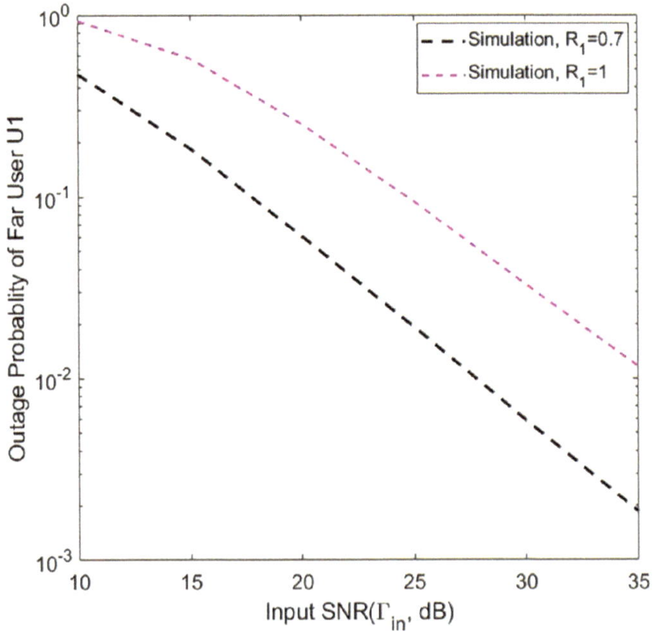

FIGURE 9.9 Outage performance analysis of far user (U_1) for several threshold data rate pairs.

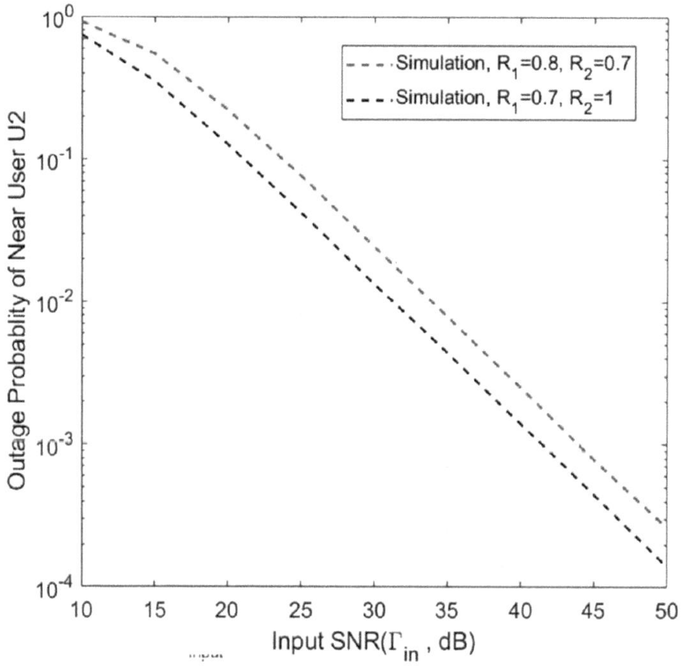

FIGURE 9.10 Outage performance analysis of near user (U_2) for several threshold data rate pairs.

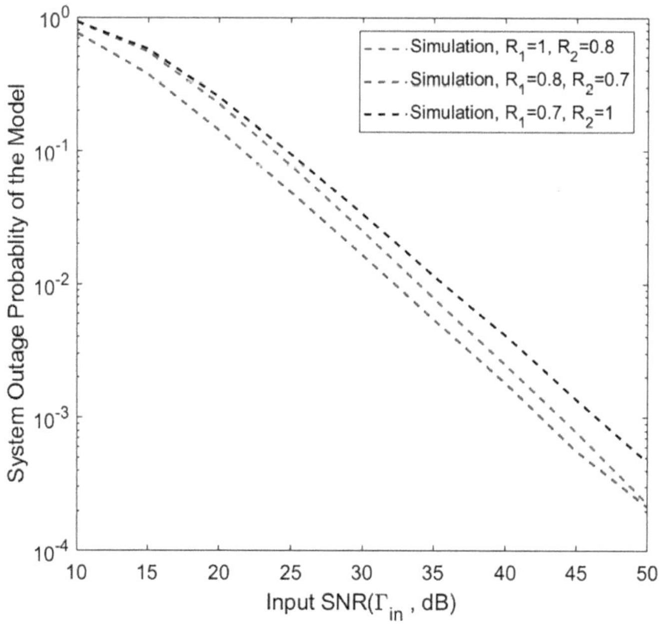

FIGURE 9.11 System outage analysis for several target data rate pairs.

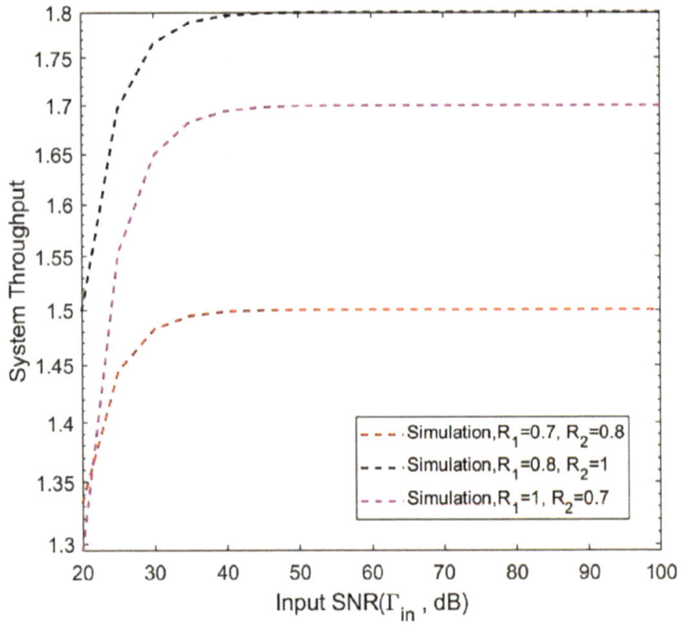

FIGURE 9.12 System throughput analysis for several target data rate pairs.

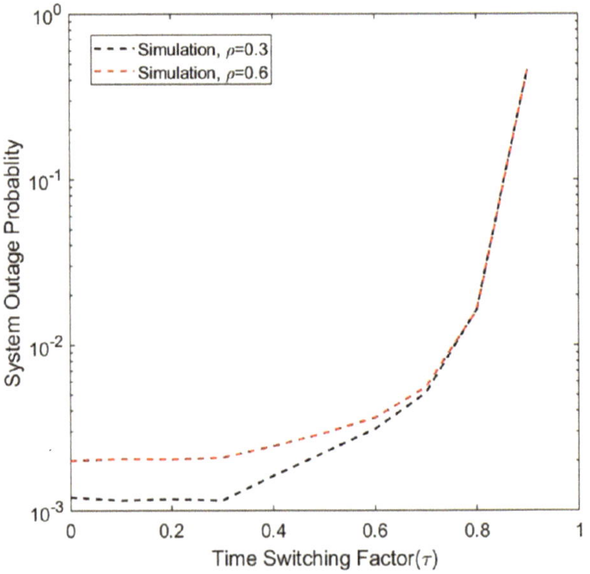

FIGURE 9.13 System outage probability verses time switching factor (τ) for several values of power splitting factor (ρ).

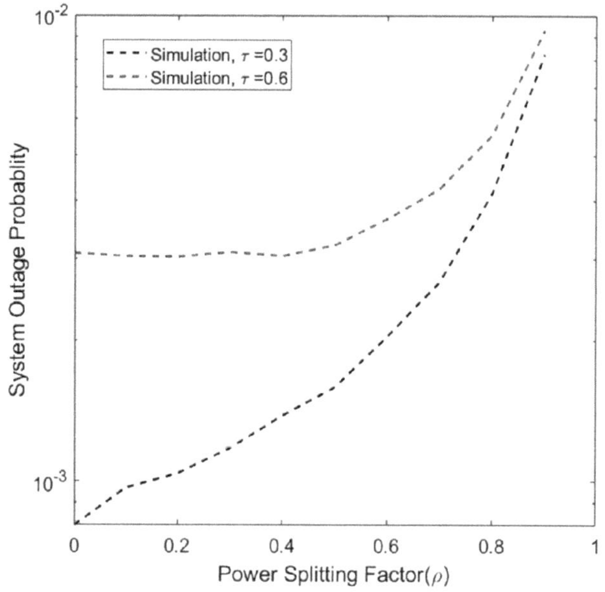

FIGURE 9.14 System outage probability verses power splitting factor (ρ) for several values of time switching factor (τ).

For two different choices of ρ, Figure 9.13 plots the system's outage probability against the time switching factor (τ). This figure also shows same character as in TSR mode. System outage is getting higher with the increase of the value of τ. We are getting comparatively better outage for lower value of power splitting factor (ρ).

The system outage probability against the power splitting factor (ρ) is displayed in Figure 9.14 for two distinct time switching factor (τ) values. It is evident that a lower value of τ corresponds to greater outage performance, or a reduced outage probability.

9.6 CONCLUDING REMARKS

This performance of a cooperative EH-assisted NOMA model based on SWIPT technology is presented. Energy harvesting based on TSR and hybrid PTS is considered following SWIPT NOMA at a near user, which acts as a relay for a far user. The probability of outage is examined for each user independently. The outage probability is found to improve significantly with the increase in input SNR for lower values of target data rate pairs for both TSR and hybrid PTS schemes. System outage also increases with the increase in time switching factor and power splitting factor in both the cases of hybrid PTS and TSR modes. The analysis presented helps to choose appropriate parameters for energy harvesting, like τ in TSR, ρ and τ in PTS, NOMA coefficients to achieve a desired level of outage under a given target data rate in the cooperative NOMA network. Our study provides valuable insights and practical guideline for the design of a cooperative NOMA network with appropriate choice of energy-harvesting scheme and associated parameters to achieve a desired level of outage and throughput performance.

REFERENCES

[1] Xingqin Lin, and Namyoon Lee, "5G and Beyond," Cham: Springer Nature, 2021.

[2] Linglong Dai, Bichai Wang, Yifei Yuan, Shuangfeng Han, I Chih-Lin, and Zhaocheng Wang, "Non-Orthogonal Multiple Access for 5G: Solutions, Challenges, Opportunities, and Future Research Trends," IEEE Commun. Mag., vol. 53, no. 9, pp. 74–81, Sept. 2015.

[3] M Zeng, W Hao, OA Dobre, and Z Ding, "Cooperative NOMA: State of the Art, Key Techniques, and Open Challenges," IEEE Netw., vol. 34, no. 5, pp. 205–211, 2020.

[4] M Xu, F Ji, M Wen, and W Duan, "Novel Receiver Design for the Cooperative Relaying System with Non-Orthogonal Multiple Access," IEEE Commun. Lett., vol. 20, no. 8, pp. 1679–1682, Aug. 2016.

[5] Abhinav Singh Parihar, Pragya Swami, Vimal Bhatia, and Zhiguo Ding, "Performance Analysis of SWIPT Enabled Cooperative-NOMA in Heterogeneous Networks Using Carrier Sensing," IEEE Trans. Veh. Technol., vol. 70, no. 10, pp. 10646–10656, Oct. 2021.

[6] O Abbasi, A Ebrahimi, and N Mokari, "NOMA Inspired Cooperative Relaying System Using an AF Relay," IEEE Wireless Commun. Lett., vol. 8, no. 1, pp. 261–264, Feb. 2019.

[7] M Bavethra, and U Palani. "Study of Simultaneous Wireless Information and Power Transfer (SWIPT) in 5G-IoT," 2023, doi: https://doi.org/10.21203/rs.3.rs-3419199/v1.

[8] Yafang Zhang, Suili Feng, and Weijun Tang, "Performance Analysis and Optimization for Power Beacon-Assisted Wireless Powered Cooperative NOMA Systems," IEEE Access, vol. 8, pp. 198436–198450, 2020.

[9] Priyanka Ghosh, Shashibhushan Sharma, S Dhar Roy, and S Kundu, "Energy Harvesting Assisted-Cooperative Non Orthogonal Multiple Access (NOMA) Using Time Switching Relaying (TSR) Protocol," In 2022 IEEE Calcutta Conference (CALCON). IEEE, pp. 113–116, 2022.

[10] Sayanti Ghosh, Abhijit Bhowmick, Sanjay Dhar Roy, and Sumit Kundu, "UAVs-Assisted Multi-Hop D2D Communication Using Hybrid PTS for Disaster Management," In 2023 5th International Conference on Recent Advances in Information Technology (RAIT). IEEE, pp. 1–6, 2023.

[11] Yulin Hu, Yao Zhu, M Cenk Gursoy, and Anke Schmeink, "SWIPT-Enabled Relaying in IoT Networks Operating with Finite Blocklength Codes," IEEE J. Sel. Areas Commun., vol. 37, no. 1, pp. 74–88, Jan. 2019, doi: 10.1109/JSAC.2018.2872361.

[12] Prasanna Raut, Keshav Singh, Chih-Peng Li, Mohamed-Slim Alouini, and Wan-Jen Huang, "Nonlinear EH-Based UAV-Assisted FD IoT Networks: Infinite and Finite Blocklength Analysis," IEEE Internet Things J., vol. 8, no. 24, Dec. 2021.

[13] Dinh-Thuan Do, "Time Power Switching Based Relaying Protocol in Energy Harvesting Mobile Node: Optimal Throughput Analysis," Mob. Inf. Syst., vol. 2015, p. 769286, Jun 2015, doi: 10.1155/2015/769286.

[14] Vaijayanti Panse, Prabhat Kumar Sharma, Tapan Kumar Jain, and Ashwin Kothari, "Relay Selection and Nonlinear Energy Harvesting in Full-Duplex Multi-Relay Cooperative Network," Phys. Commun. 56, p. 101946, 2023.

[15] Prabhat Kumar Sharma, Neha Sharma, Shivani Dhok, and Anamika Singh, "RIS-Assisted FD Short Packet Communication with Non-Linear EH," IEEE Commun. Lett., vol. 27, no. 2, pp. 522–526, Feb. 2023.

[16] Prasanna Raut, Prabhat Kumar Sharma, Theodoros A. Tsiftsis, and Yulong Zou, "Power-Time Splitting-Based Non-Linear Energy Harvesting in FD Short-Packet Communications," IEEE Trans. Veh. Technol., vol. 69, no. 8, Aug. 2020.
[17] AK Baranwal, SD Roy, and S Kundu, "Performance Analysis of Full-Duplex Cooperative NOMA with Power-Time Splitting Based Energy Harvesting," In 2023 4th International Conference on Computing and Communication Systems (I3CS), Shillong, pp. 1–6, 2023, doi: 10.1109/I3CS58314.2023.10127421.
[18] V Aswathi, and AV Babu, "Outage and Throughput Analysis of Full-Duplex Cooperative NOMA System with Energy Harvesting," IEEE Trans. Veh. Technol., vol. 70, no. 11, pp. 11648–11664, Nov. 2021.
[19] Yingting Liu, Haiyang Ding, Jianmei Shen, Ruyi Xiao, and Hongwu Yang, "Outage Performance Analysis for SWIPT Based Cooperative Non-Orthogonal Multiple Access Systems," IEEE Commun. Lett., vol. 23, no. 9, pp. 1501–1505, Sept. 2019.

10 Full-Duplex Energy Harvesting-Based Cooperative NOMA Network

Alok Baranwal, Sanjay Dhar Roy, and Sumit Kundu

10.1 INTRODUCTION: BACKGROUND AND MOTIVATION

The continuous emergence of various types of electronic devices, like wireless sensor networks and the Internet of Things (IoTs), has led researchers to think of some new kind of techniques that not only satisfy spectrum management but should also meet the energy management of all such devices. Most of these upcoming wireless devices are small and battery-powered, which need to be charged frequently for their longer operation and long life. Many times, such devices are placed in such a drastic condition that their battery recharging is very difficult or impossible. Recharging from renewable energy sources is one of the options to keep the long life of such devices; however, harvesting energy from renewable energy sources is quite unpredictable in nature and fails to provide a timely energy source for such data transmission [1, 2]. Energy harvesting (EH) from radio frequency (RF) is a more reliable and feasible solution for such small powered devices, where the transfer of information takes place and energy harvesting can be done simultaneously [3].

To serve the number of users in 5G and the next generation of communication, a new multiple access (MA) technique, named non-orthogonal multiple access (NOMA), is devised [4]. NOMA improves spectrum efficiency (SE) by transmitting multiple users' information over the same spectrum (time/frequency/code) simultaneously, allocating different powers, called power domain (PD) NOMA [5]. In PD-NOMA, user pairing on the basis of their channel status and transmission rates ensures more efficient NOMA transmission. In NOMA, at the transmitter side, superposition coding (SC) is used to superpose the signals from multiple users and transmit over a resource block, while on the receiver side, successive interference cancellation (SIC) is used to retrieve the message signals by each user [6]. Based on the channel status and transmission rates of users, users in NOMA can be labelled as near and far users. Cooperative NOMA (C-NOMA), with the help of relay (relay can be half-duplex [HD] or full-duplex [FD]) helps in the successful transmission of information from source to both users. In this scenario, the near user functions as a relay enhancing the communication robustness of the distant user [7, 8].

 DOI: 10.1201/9781032625034-10

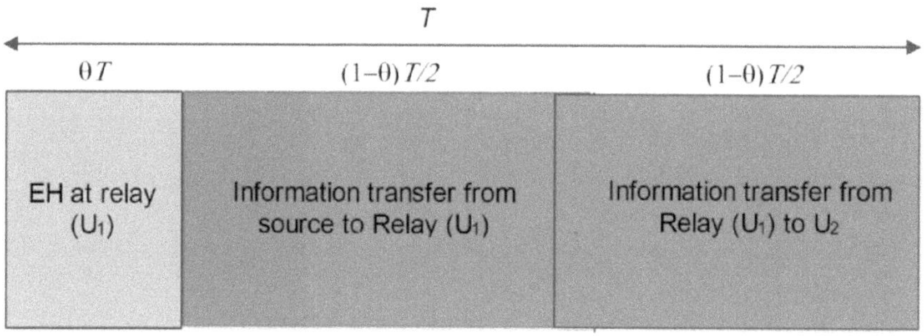

FIGURE 10.1 Time switching–based energy harvesting protocol.

The authors in [9] proposed an FD C-NOMA, where an FD relay helps signal transmission to a weak channel user. Several works of CNOMA are studied by authors in [10, 11], where far user signal information improves with the assistance of near user close to the source (S).

The use of a near user as a relay increases the energy demand of a cooperative NOMA network. In order to improve the energy efficiency in a cooperative NOMA network, the relay can harvest energy from the received RF signal of the source [12]. Further, in order to improve the multiplexing gain of a cooperative NOMA network, various relaying schemes have been developed, like two-way relay, decode-and-forward (DF) relay, and amplify-and-forward (AF) relay [8, 13, 14]. EH techniques can also be used by other users to store the harvested energy, which can be utilized later on. In NOMA, the near user harvests energy from the RF signal of source and stores it. In C-NOMA, the near user harvests energy, and depending on the availability of energy at near user, it uses the harvested energy either in forwarding source data to far user or storing the harvested energy. In order to harvest energy and process information separately and simultaneously, two receiver architectures are proposed. These receiver architectures are based on time switching (TS) techniques and power splitting techniques (PS), known as TS relay and PS relay, respectively [3]. *Simultaneous wireless information and power transfer* (SWIPT) is a term coined for this technique, in which energy harvesting and information processing are done simultaneously [15]. Numerous works based on either TS or PS were discussed in the literature [16, 17]. In TS relay, the total time T is divided in such a way that θT time is used in energy harvesting, while in each $^{(1-\theta)T}\!/_2$ time, the transmission of information occurs from source to relay (U_1) and relay to U_2, respectively, as shown in Figure 10.1, with θ as a time switching factor. Figure 10.2 shows energy harvesting using PS protocol, where total time frame T is divided into two parts with $^T\!/_2$ each, and source power P_s is divided with a power splitting factor ρ. In the first half of the time slot $^T\!/_2$, the relay performs energy harvesting using power ρP_s, and information is received from the source to the relay using power $(1-\rho)P_s$. In the subsequent time slot, the relay transmits source information to U_2 using power $(1-\rho)P_s$.

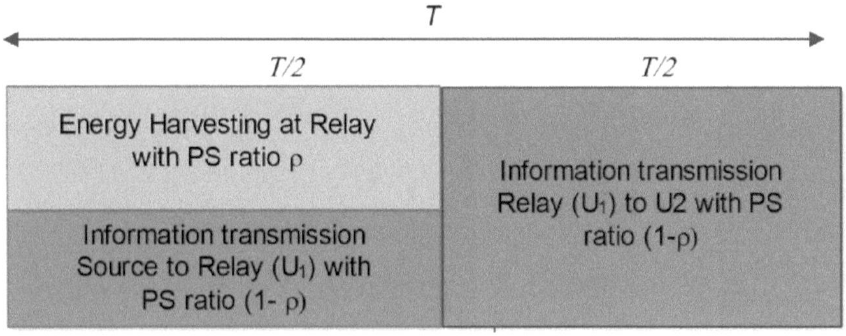

FIGURE 10.2 Power splitting–based energy harvesting protocol.

However, there are few works based on the hybrid mode which use power and time protocol to harvest energy and process information, called power–time splitting (PTS) protocol. The use of the PTS mode of energy harvesting increases the harvested energy at the relay, thus increasing the received power. Most of the work of energy harvesting is linear in nature. The harvested energy obeys a linear relation with transmit power. However, practical EH circuits are non-linear in nature, which means that harvested energy becomes constant after a saturation power is attained defined for the circuits [18, 19].

Motivated by the advantages of NOMA, FD, and EH, we showcase the outage performance and ergodic capacity of the C-NOMA network aided by an EH FD relay.

The main contributions of presenting the outage and ergodic performance of EH C-NOMA are presented in the following.

- Describing a cooperative NOMA network with the assistance of a near user acts as an FD relay for better utilization of available spectrum resources.
- Energy harvesting at near users adds another advantage of using such a system where far users can be served in a more efficient way.
- Using a hybrid mode of energy harvesting which uses TS and PS protocols at the relay.
- Recognizing the practicality of the energy harvesting receiver, we choose to employ the non-linear energy harvesting model and illustrate its comparison with the linear harvesting model.
- Analyzing outage performance and ergodic capacity of a C-NOMA network enabled by an EH relay.
- Novel analytical expressions for the outage probability of two users and system outage probability are developed considering PTS-based energy harvesting.
- We obtained novel closed-form expressions of the ergodic capacity of both users using a Gaussian–Laguerre approximation.

Further, we organize this chapter into three sections with detailed outlines as described next. Section 10.2 describes the system model of two-user NOMA network enabled by an EH relay and SINR equations. We derive the analytical expressions of the outage probability of both user and system outage probability in Section 10.3. Further, Section 10.4 derives the ergodic capacity of each user. We discuss numerical results in Section 10.5. Finally, Section 10.6 discusses the chapter conclusions.

10.2 A SYSTEM MODEL OF C-NOMA ENABLED BY AN FD RELAY

A C-NOMA network enabled by an EH FD relay is considered by the authors in [20], shown in Figure 10.3. In a C-NOMA network, a user U_2 with weak channel gain from the source receives the source signal via a near user U_1 having strong channel gain and is close to the source. Due to severe fading, we assume that there is no direct link that exists between the source and the far user U_2, while a direct link exists between the source and the near user U_1. The source transmits a composite signal, x_1 andx_2 of U_1 and U_2, respectively, in two phases. Here, U_1 serves as an FD relay, facilitating signal transmission from the source to U_1 and, subsequently, to U_2.

In the HD mode of operation, the relay needs two resource blocks for the continuous transmission of information between sources to U_2 via U_1. Full-duplex (FD) relay mode of operation ensures uninterrupted transmission of information between source to U_2 using the same resource blocks. Thus, the use of FD relay saves the spectrum bandwidth and doubles up the spectrum efficiency than does the HD mode of operation. Besides U_1 helping in signal transmission from the source to U_2 in the first phase, the additional power required by U_1 to transmit this signal is met by energy harvesting from the source signal power P_s. Here, U_1 uses TS-EH protocol to harvest the required power in forwarding source signal to U_2. Table 10.1 illustrates the key variables used in this chapter.

Without loss of generality, we assume the source and node U_2 work in HD mode with a single antenna, while the node U_1 is equipped with two antennas and operates in FD mode. Although the number of nodes can be extended to more than two in the present system model, it is, however, currently out of the scope of this work.

10.2.1 CHANNEL MODEL

Any wireless communication system considers the effect of large-scale fading and small-scale fading. We assume small-scale fading between the transmitter and the receiver, denoted by h_x, that follows an independent and identically distributed (i.i.d.)

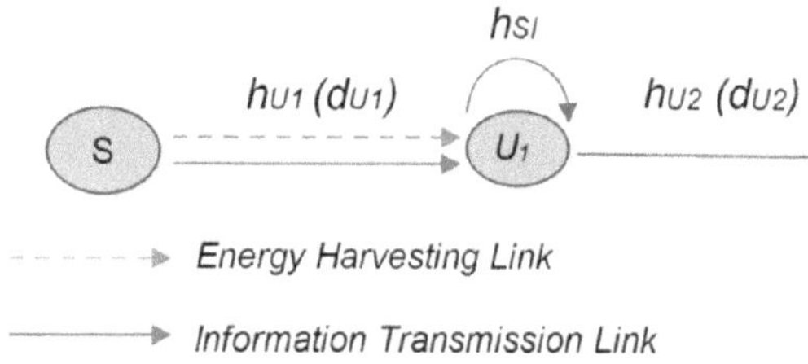

FIGURE 10.3 System model of a hybrid PTS-based NOMA network.

complex Gaussian random variable (Rayleigh envelope) and is modelled as Rayleigh fading channels. Thus, $g_x = |h_x|^2$ follows exponential PDF with average power gain $E\{g_x\} = \lambda_x$, while $E\{.\}$ represents the statistical expectation operation. The large-scale fading is contingent on the distance between the transmitter and the receiver, denoted as d_x, where $x \in \{U_1, U_2\}$.

The cumulative distribution function (CDF) and probability density function (PDF) of g_x is given by

$$F_{g_x}(z) = 1 - e^{\frac{-z}{\lambda_x}}$$

and

$$f_{g_x}(z) = \frac{1}{\lambda_x} e^{-\frac{z}{\lambda_x}}$$

10.2.2 Effect of Full-Duplex Operation on U_1

While the source transmits the composite signal of x_1 and x_2 to U_1, U_1 transmits earlier decoded signal $s(t)$ to U_2 because of FD operation of U_1, where $s(t)$ is the transmitted signal by U_1 to U_2, which is equal to $\widehat{x_2}(t-\tau)$, and $\widehat{x_2}$ is the decoded version of signal x_2. On successful detection of x_2 at U_1, $\widehat{x_2}$ becomes x_2. τ denotes the extra time required in processing the message signal during transmission and reception due to FD operation of U_1 [21]. However, the FD operation of the relay creates a self-interference (SI) channel with a fading coefficient h_{SI}. The advantage of a full-duplex relay over a half-duplex relay–based network depends on the full or partial removal of interference created by SI. The authors have extensively reviewed the literature in [22–24] concerning the partial removal of self-interference (SI) through both passive and active cancellation methods. Extensive research on ways of removing SI results in a partial reduction in the level of SI, and the final component left after removing SI is residual self-interference (RSI), denoted by I_s, which varies from –20 to –5 dB [24, 25].

10.2.3 Energy Harvesting and Information Transmission Phase

The near user U_1 harvests energy from the transmit power of the source using TS and PS together, called PTS protocol. The entire communication from energy harvesting to information transmission between source to U_2 via U_1 takes place in three phases, with θ and ρ as the TS factor and PS factor, illustrated in Figure 10.4. In $(1-\theta)T/2$, the relay performs EH using θT time in TS protocol, while $(1-\theta)T/2$ time is used by the relay to perform EH and information decoding (ID) with ρ and $(1-\rho)$, respectively. Next, in $(1-\theta)T/2$, information is transmitted between U_1 and U_2 using harvested energy by U_1. We assume the source with power P_s transmits a composite signal of $x_1(t)$ and $x_2(t)$ in time 't', denoted by $x(t) = \sqrt{a_1 P_s}\, x_1(t) + \sqrt{a_2 P_s}\, x_2(t)$. Here, a_1 and a_2 denote the power coefficient allocated to U_1 and U_2, respectively, such that $a_1 > a_2$.

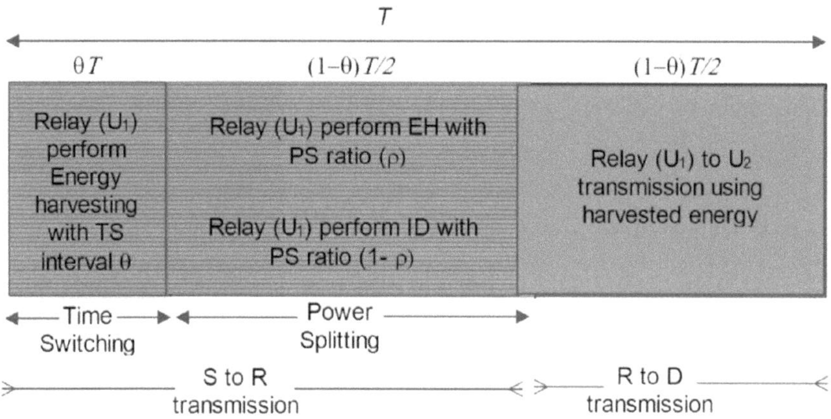

FIGURE 10.4 Frame structure of PTS-based EH protocol.

Thus, the signal received by U_1 in time 't' is given by:

$$Y_{U_1}(t) = \frac{h_{U_1}}{\sqrt{d_1^m}} x(t) + I_S + n_{U_1}(t) \tag{10.1}$$

We assume negligible power is harvested from noise. Therefore, the energy harvested at U_1 using TS and PS protocols are respectively given as:

$$E_{U_1}^{TS} = \frac{\eta \left|h_{U_1}\right|^2 P_s}{d_1^m} \theta T \tag{10.2}$$

$$E_{U_1}^{PS} = \frac{\eta \rho P_s \left|h_{U_1}\right|^2 P_s}{2d_1^m} (1-\theta)T \tag{10.3}$$

Thus, the energy harvested using PTS protocol is given by:

$$E_{U_1}^{PTS} = \frac{\eta P_s \left|h_{U_1}\right|^2 T}{d_1^m} \left(\theta + \frac{\rho(1-\theta)}{2}\right) \tag{10.4}$$

The average power harvested is given by:

$$P_{U_1}^{PTS} = \frac{\eta P_s \left|h_{U_1}\right|^2}{d_1^m} \left(\rho + \frac{2\theta}{(1-\theta)}\right) \tag{10.5}$$

This is the power used by U_1 in forwarding signals to U_2.

Although following linear relation in energy harvesting is rampant in research, practical EH circuits are, however, non-linear (NL) in nature, which works on two different points – sensitivity and saturation [19]. So in accordance with the practical EH circuits, which are non-linear, this work uses the NLEH way of harvesting energy. The working principle of this NLEH circuit depends on two values; one is called minimum threshold (*i.e.*, P_{min}), and the other is called maximum allowable saturation (*i.e.*, $P_{\max}$). The harvested power (here, $P_{U_1}^{PTS}$) must be greater than P_{in} to trigger the circuit; however, if $P_{U_1}^{PTS}$ becomes greater than $P_{\max}$, then harvested power becomes constant. Due to this unique working of NLEH circuits, this is also termed as constant-linear-constant.

Following NLEH, the transmit power is given by:

$$P_{U_1} = \begin{cases} 0, & P_{U_1}^{PTS} < P_{\min} \\ \dfrac{\eta P_s \left|h_{U_1}\right|^2}{d_1^m}\left(\rho + \dfrac{2\theta}{(1-\theta)}\right), & P_{\min} \le P_{U_1}^{PTS} \le P_{\max} \\ P_{\max}\left(\eta\rho + \dfrac{2\eta\theta}{(1-\theta)}\right), & P_{U_1}^{PTS} > P_{\max} \end{cases} \tag{10.6}$$

10.2.4 SINR Equation

Following the NOMA principle of decoding, the near user U_1, after receiving the signal from the source, first detects the message of U_2, performs SIC, and then decodes its own signal. The SINR equation of decoding signal x_2 and x_1 at U_1 is written in the following.

$$\gamma_{x_2}^{U_1} = \frac{a_2 P_s \left|h_{U_1}\right|^2}{a_1 P_s \left|h_{U_1}\right|^2 + (I_s + N_0)d_1^m} \tag{10.7}$$

$$\gamma_{x_1}^{U_1(pSIC)} = \frac{a_1 P_s \left|h_{U_1}\right|^2}{(I_s + N_0)d_1^m} \tag{10.8}$$

However, due to some imperfection in the SIC operation, achieving perfect SIC (pSIC) is not possible; there are some residual left of the signal x_2. Incorporating imperfect SIC (ipSIC), the SINR equation of decoding x_1 at U_1 is written as

$$\gamma_{x_1}^{U_1(ipSIC)} = \frac{a_1 P_s \left|h_{U_1}\right|^2}{a_2 P_s \left|h_{U_1}\right|^2 \beta + (I_s + N_0)d_1^m} \tag{10.9}$$

Where β denote an ipSIC factor whose value lies between 0 and 1.

Due to FD operation at U_1, while U_1 receives the signal from the source and harvests energy, it also forwards the decoded signal to U_2. Thus, the SINR equation of decoding x_2 at U_2 is written as

$$\gamma_{x_2}^{U_2} = \frac{P_{U_1} \left| h_{U_2} \right|^2}{N_0 d_2^m} \tag{10.10}$$

10.3 OUTAGE PROBABILITY FORMULATION

Here we calculate the outage probability of each user and system outage probability. The *outage probability* is defined as the probability of an event when data capacity falls below a required data rate. We assume r_1 and r_2 as the required data rate to decode x_1 and x_2, respectively.

10.3.1 OUTAGE PROBABILITY OF U_1

The outage probability of U_1 is defined as

$$P_{out}^{U_1} = \Pr\left(C_{21} < r_2 \cup C_{11} < r_1\right) \tag{10.11}$$

where C_{11}, C_{21} denote the data capacity to decode the message x_1 at U_1, x_2 at U_1, and given as

$$C_{11} = \frac{(1-\theta)}{2} \log_2(1 + \gamma_{x_1}^{U_1(ipSIC)}) \tag{10.12}$$

$$C_{21} = \frac{(1-\theta)}{2} \log_2(1 + \gamma_{x_2}^{U_1}) \tag{10.13}$$

Substituting C_{11} and C_{21} in equation (10.11), the outage probability of U_1 is written as

$$P_{out}^{U_1} = \Pr\left(\frac{(1-\theta)}{2} \log_2(1 + \gamma_{x_1}^{U_1(ipSIC)}) < r_1 \cup \frac{(1-\theta)}{2} \log_2(1 + \gamma_{x_2}^{U_1}) < r_2\right) \tag{10.14}$$

Equation (10.14) can be simplified as

$$P_{out}^{U_1} = \Pr\left(\gamma_{x_1}^{U_1(ipSIC)} < \omega_1 \cup \gamma_{x_2}^{U_1} < \omega_2\right) \tag{10.15}$$

where ω_1 and ω_2 are defined as SINR threshold and given by $\omega_1 = \left(2^{2r_1/(1-\theta)} - 1\right)$ and $\omega_2 = \left(2^{2r_2/(1-\theta)} - 1\right)$.

Here, we have not calculated the outage probability of U_1, taking into account equation (10.8), $\gamma_{x_1}^{(pSIC)}$. This is because the mathematical derivation would be very simple in this case. In fact, the mathematics involved using $\gamma_{x_1}^{(ipSIC)}$ in calculating outage probability of U_1 is more complex.

Substituting equations (10.7) and (10.9), the $P_{out}^{U_1}$ of equation (10.15) can further be solved as

$$\begin{aligned} P_{out}^{U_1} &= 1 - \Pr\left(\gamma_{x_1}^{U_1(ipSIC)} > \omega_1, \gamma_{x_2}^{U_1} > \omega_2\right) \\ &= 1 - \Pr\left(\left|h_{U_1}\right|^2 > \phi_1, \left|h_{U_1}\right|^2 > \phi_2\right) \\ &= 1 - e^{-\frac{\phi}{\lambda_1}} \end{aligned} \tag{10.16}$$

where $\phi = \max(\phi_1, \phi_2)$, $\phi_1 = \frac{\omega_1\left(1+\left(I_s/N_o\right)\right)}{P_s/N_o\left(a_2 - a_1\omega_1\right)}$, and $\phi_2 = \frac{\omega_2\left(1+\left(I_s/N_o\right)\right)}{P_s/N_o\left(a_1 - \beta a_2\omega_2\right)}$.

10.3.2 Outage Probability of U_2

The outage probability of U_2 is defined as

$$P_{out}^{U_2} = \Pr\left(C_{21} < r_2 \cup C_{22} < r_2\right) \tag{10.17}$$

Where C_{22} denotes the data capacity of x_2 at U_2 and is given by

$$C_{22} = \frac{(1-\theta)}{2}\log_2(1+\gamma_{x_2}^{U_2}) \tag{10.18}$$

Similarly to the steps in calculating $P_{out}^{U_1}$, the outage probability of U_2 is solved as

$$\begin{aligned} P_{out}^{U_2} &= \Pr\left(C_{21} < r_2 \cup C_{22} < r_2\right) \\ &= \Pr\left(\gamma_{x_2}^{U_1} < \omega_2 \cup \gamma_{x_2}^{U_2} < \omega_2\right) \\ &= 1 - \Pr\left(\gamma_{x_2}^{U_1} > \omega_2, \gamma_{x_2}^{U_2} > \omega_2\right) \\ &= 1 - \Pr\left(\left|h_{U_1}\right|^2 > \phi_1, \left|h_{U_1}\right|^2 \left|h_{U_2}\right|^2 > \phi_3\right) \\ &= 1 - \int_{\phi_1}^{\infty} \frac{1}{\lambda_1} e^{-\left(\frac{x}{\lambda_1} + \frac{\phi_3}{x\lambda_2}\right)} dx \\ &= 1 - \psi K_1(\psi) + \int_0^{\phi_1} \frac{1}{\lambda_1} e^{-\left(\frac{x}{\lambda_1} + \frac{\phi_3}{x\lambda_2}\right)} dx \end{aligned} \tag{10.19}$$

where $\psi = \sqrt{\frac{4\phi_3}{\lambda_1\lambda_2}}$, $\phi_3 = \frac{\omega_2 d_1^m d_2^m N_o}{P_s(\eta\rho + \frac{2\eta\theta}{(1-\theta)})}$, and $K_1(.)$ denote the first-order and second-kind Bessel function.

10.3.3 Throughput of the System

In a delay-limited mode of transmission, the average throughput of the system with rates r_1 and r_2 of user U_1, U_2, respectively, is given as [26]:

$$r(\theta) = \frac{1}{2}\left[(1-P_{out}^{U_1})r_1(1-\theta) + (1-P_{out}^{U_2})r_2(1-\theta)\right] \tag{10.20}$$

10.4 ERGODIC CAPACITY FORMULATION

The *ergodic capacity* is defined as the upper bound of the rate based on the statistical characterization of the channel.

10.4.1 Ergodic Rate of U_1

The SINR of decoding x_1 at U_1 is given by equation (10.9), considering the impact of ipSIC.

Therefore, the ergodic rate of U_1 is given by

$$R_e^{U_1} = \frac{(1-\theta)}{2}\mathrm{E}\{\left[\log_2(1+W)\right]\} \tag{10.21}$$

where $W = \gamma_{x_1}^{U_1(ipSIC)} = \frac{a_1 P_s |h_{U_1}|^2}{a_2 P_s |h_{U_1}|^2 \beta + (I_s + N_0)d_1^m}$.

In order to solve equation (10.19), first, we have to find the CDF of W.

The CDF of W is calculated as

$$\begin{aligned} F_W(w) &= \Pr(W < w) = 1 - \Pr(W > w) \\ &= 1 - \Pr\left(|h_{U_1}|^2 > \frac{w(I_s + N_o)d_1^m}{(a_1 P_s - a_2 P_s \beta w)}\right) \\ &\overset{\Delta S1}{=} 1 - e^{-\frac{m_2 w}{(a_1 - a_2 w\beta)}} \end{aligned} \tag{10.22}$$

where $m_2 = \frac{(I_s + N_o)d_1^m}{P_s\lambda_1}$.

And equation (10.20) holds on the condition that $(a_1 - a_2\beta w) > 0$.
Thus, the ergodic capacity of U_1 is solved as

$$R_e^{U_1} = \frac{(1-\theta)}{2\ln 2} \int_0^{a_1/a_2\beta} \frac{1 - F_W(w)}{1+w} dw \tag{10.23}$$

Substituting the CDF of W in equation (10.21), we have the expression of $R_e^{U_1}$ as:

$$R_e^{U_1} = \frac{(1-\theta)}{2\ln 2} \int_0^{a_1/a_2\beta} \frac{e^{-\frac{m_2 w}{(a_1 - a_2\beta w)}}}{1+w} dw \tag{10.24}$$

The preceding equation (10.24) is intractable to solve. Using a few substitutions and mathematical manipulation, we have obtained the value of $R_e^{U_1}$ as

$$R_e^{U_1} = \frac{(1-\theta)m_5}{2\ln 2} \int_{-1}^{1} \frac{f(x)}{\sqrt{1-x^2}} dx \tag{10.25}$$

Now, applying Gaussian–Chebyshev quadrature, the final closed-form expression is derived as

$$R_e^{U_1} = \frac{(1-\theta)m_5}{2\ln 2} \sum_{i=1}^{N_1} w_i f(x_i) \tag{10.26}$$

where $m_5 = \frac{a_1}{2a_2\beta}$, $f(x) = \frac{e^{\frac{m_2 m_5 (1+x)}{a_1 x}} \sqrt{1-x^2}}{(1+m_5(1+x))}$, $w_i = \frac{\pi}{N_1}$, and $x_i = \cos\left[\frac{(2i-1)\pi}{2N_1}\right]$.

w_i are weights, so $w(x)$ is a weight function.

10.4.2 Ergodic Rate of U_2

The SINRs of decoding x_2 at U_1 and U_2 are given by equations (10.7) and (10.10), respectively.

Therefore, the SINR of decoding x_2 is given by $\gamma_{x_2} = \min(\gamma_{x_2}^{U_1}, \gamma_{x_2}^{U_2})$.

Therefore, the ergodic rate of U_2 is given by

$$R_e^{U_2} = \frac{(1-\theta)}{2} \mathrm{E}\{[\log_2(1+Z)]\} \tag{10.27}$$

$$R_e^{U_2} = \frac{(1-\theta)}{2\ln 2}\int_0^{\infty}\frac{1-F_Z(z)}{1+z}dz \tag{10.28}$$

where $Z = \min(\gamma_{x_2}^{U_1}, \gamma_{x_2}^{U_2})$, $X = \gamma_{x_2}^{U_1}$, and $Y = \gamma_{x_2}^{U_2}$.

First, we find the CDF of Z.

$$\begin{aligned} F_Z(z) &= \Pr(Z < z) = 1 - \Pr(Z > z) \\ &= 1 - \Pr\left[\min(\gamma_{x_2}^{U_1}, \gamma_{x_2}^{U_2}) > z\right] = 1 - \left[\Pr(\gamma_{x_2}^{U_1} > z).\Pr(\gamma_{x_2}^{U_2} > z)\right] \\ &= 1 - \left[1 - F_X(z)\right]\left[1 - F_Y(z)\right] \end{aligned} \tag{10.29}$$

Following the similar steps used in finding the CDF of W, the CDF of X is calculated as

$$F_X(x) \overset{\Delta S2}{=} 1 - e^{-\frac{m_2 x}{(a_2 - a_1 x)}} \tag{10.30}$$

The equation $S2$ holds on condition $(a_2 - a_1 x) > 0$.

Now, we find the CDF of Y:

$$\begin{aligned} F_Y(y) &= \Pr(Y < y) = \Pr\left(\left|h_{U_1}\right|^2 \left|h_{U_2}\right|^2 < y m_3\right) \\ &= \int_0^{\infty}(1 - e^{-\frac{y m_3}{x\lambda_1}}).\left(\frac{1}{\lambda_2}e^{-\frac{x}{\lambda_2}}\right)dx \\ &= \int_{x=0}^{\infty}\frac{1}{\lambda_2}e^{-\frac{x}{\lambda_2}}dx - \int_{x=0}^{\infty}\frac{1}{\lambda_2}e^{-\left(\frac{y m_3}{x\lambda_1}+\frac{x}{\lambda_2}\right)}dx \\ &= 1 - \frac{1}{\lambda_2}\int_{x=0}^{\infty}e^{-\left(\frac{x}{\lambda_2}+\frac{y m_3}{x\lambda_1}\right)}dx \end{aligned} \tag{10.31}$$

Further, this integration is intractable. So using $\int_0^{\infty} e^{-\left(\nu x + \frac{\kappa}{4x}\right)}dx = \sqrt{\frac{\kappa}{\nu}}K_1(\sqrt{\kappa\nu})$, equation (10.31) can be further simplified as

$$F_Y(y) = 1 - 2\sqrt{m_4 y}K_1(\sqrt{m_4 y}) \tag{10.32}$$

where $m_3 = \frac{d_1^m d_2^m}{P_s(\eta\rho + \frac{2\eta\theta}{(1-\theta)})}$ and $m_4 = \frac{m_3}{\lambda_1\lambda_2}$.

Combining equations (10.27) and (10.29), under the condition $z < \frac{a_2}{a_1}$, $F_Z(z)$ is calculated as:

$$F_Z(z) = 1 - \left[e^{-\frac{m_2 z}{(a_2 - a_1 z)}}\right]\left[\sqrt{m_4 z}K_1(\sqrt{m_4 z})\right] \tag{10.33}$$

Finally, substituting $F_Z(z)$ in equation (10.26), the ergodic rate of U_2 is written as

$$R_e^{U_2} = \frac{(1-\theta)}{2\ln 2}\int_0^{\frac{a_2}{a_1}} \frac{2e^{-\frac{m_2 z}{(a_2 - a_1 z)}}\sqrt{m_4 z}K_1(\sqrt{m_4 z})}{1+z}dz \tag{10.34}$$

After performing a few mathematical manipulations and substitutions, we obtain

$$R_e^{U_2} = \frac{(1-\theta)m_5}{2\ln 2}\int_{-1}^{1} \frac{f(x)}{\sqrt{1-x^2}}dx \tag{10.35}$$

where $f(x) = \frac{e^{-\frac{m_2(1+x)}{a_1(1-x)}} 2\sqrt{m_4 m_5(1+x)(1-x^2)}K_1(\sqrt{m_4 m_5(1+x)})}{1+m_5(1+x)}$.

Equation (10.33) is intractable to solve. Therefore, we use Gaussian–Chebyshev quadrature to solve the ergodic rate obtained in equation (10.33):

$$R_e^{U_2} = \frac{(1-\theta)m_5}{2\ln 2}\sum_{i=1}^{N_1} w_i f(x_i) \tag{10.36}$$

where $m_5 = \frac{a_2}{2a_1}$, $w_i = \frac{\pi}{N_1}$, and $x_i = \cos\left[\frac{(2i-1)\pi}{2N_1}\right]$.

10.5 NUMERICAL RESULTS BASED ON THE OUTAGE PROBABILITY AND ERGODIC RATE

The numerical values for the NOMA-based network are given in the following table.

Figure 10.5 illustrates the outage probability against the source transmit SNR of user 1, considering various values of the target data rate. Based on the outage

TABLE 10.1
Numerical Parameters with Their Values

Numerical parameters	Numerical values
Power allocation coefficients to U_1 and U_2	0.3, 0.7
Target data rate for U_1 (r_1)	0.2, 0.4, 0.6 bps/Hz
Target data rate for U_2 (r_2)	0.1, 0.3, 0.4 bps/Hz
Distance of U_1 from source (d_{U1})	1 m
Distance of U_2 from source (d_{U2})	1.5 m
Value of self-interference (I_s)	1, 10
Value of ipSIC factor (β)	0.1
Value of energy harvesting efficiency (η)	0.7

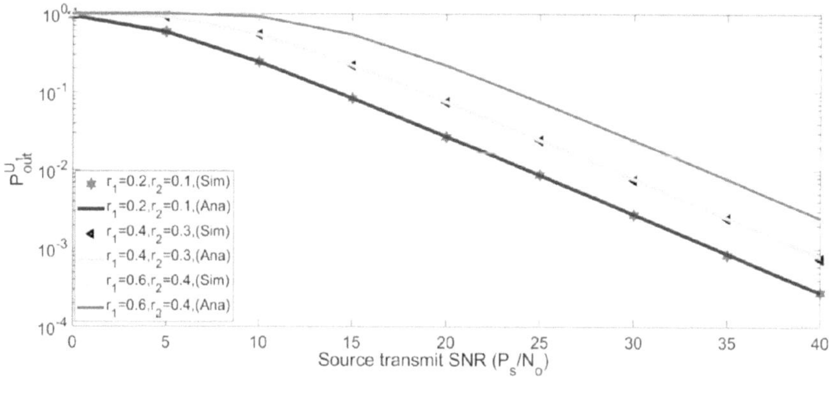

FIGURE 10.5 Outage probability of U_1 versus source transmit SNR for various target rate values.

probability of user 1 as defined by equation (10.11), the outage probability of user 1 is influenced by the target data rate, r_1 and r_2. As the source transmit SNR increases, the SINR in equation (10.9) increases, which helps in a decrease in outage probability. However, the increase in the target data rate has an adverse effect, which increases the outage probability of user 1. In order to find the accuracy in the derived analytical result, we have performed a Monte Carlo simulation, which perfectly matches with the analytical results.

Figure 10.6 shows the outage probability versus transmit SNR for U_2 under different values of the target data rate r_2. As the symbol x_2 is decoded and removed at U_1 and then decoded at U_2, the outage probability of user 2 is defined in that way, as shown by equation (10.17). Parallel to the outage probability trend of U_1, as the transmit SNR rises, the outage probability of U_2 diminishes. The impact of the target rate r_2 on the outage probability of user 2 is also noticeable in Figure 10.6. As the target data rate increases, the outage probability also increases.

Figure 10.7 shows the effect of energy harvesting using linear and non-linear ways on the outage probability of user 2. According to the definition of *harvesting energy* as defined in equation (10.6), the harvesting energy saturates after P_{max}. In this work, we have chosen P_{min} as 10% of source transmit SNR, and P_{max} as 80% of source transmit SNR. Therefore, as amount of harvesting energy crosses P_{max}, the transmit power of U_1 becomes constant, which clearly shows the effect on the outage probability of user 2. As the transmit SNR goes beyond 15 dB, the outage probability of

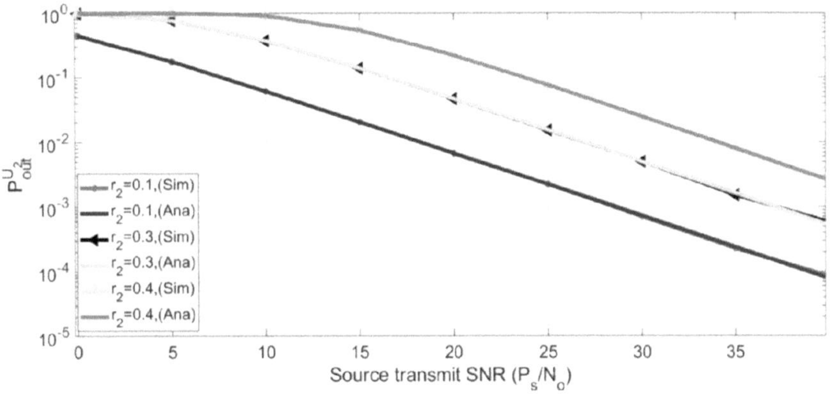

FIGURE 10.6 Outage probability of U_2 versus source transmit SNR for various target rate values.

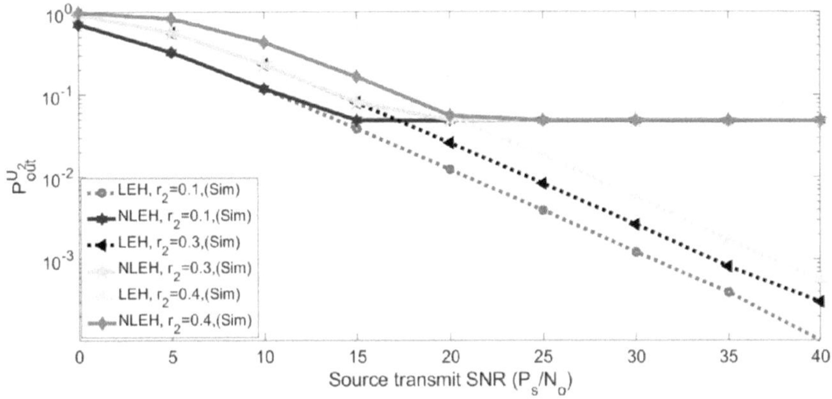

FIGURE 10.7 Outage probability of U_2 versus source transmit SNR for different values of target rate in linear energy harvesting and non-linear harvesting mode.

U_2 continues to decrease in the linear mode of energy harvesting, while in a non-linear mode of energy harvesting, harvesting energy saturates after 15 dB SNR, so the outage probability curve becomes flat. The analytical result perfectly matches with the Monte Carlo result for various values of the target data rate.

In Figure 10.8, the simulation and analytical results of system throughput are plotted against the source transmit SNR. Again, it is apparent that our analytical result in (10.19) matches precisely with the Monte Carlo simulation for the various number of TSR factor. As anticipated, the increase in source transmit SNR increases system throughput. However, after 25 dB SNR, the system throughput remains flat; the reason for this is the counter-effect of self-interference and ipSIC. Due to SI and ipSIC, the SNR starts decreasing after 25 dB SNR.

In Figures 10.9 and 10.10, the ergodic rate curves for user 1 and user 2 of an FD cooperative NOMA network are depicted. For both figures, we observe that

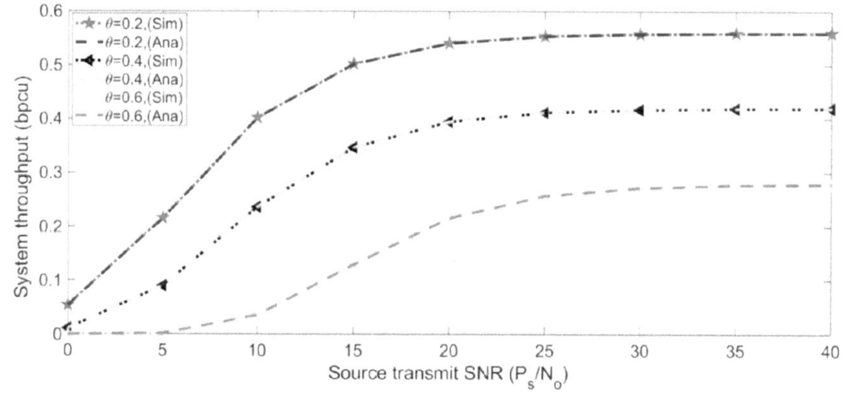

FIGURE 10.8 System throughput vs. source transmit SNR for different values of θ.

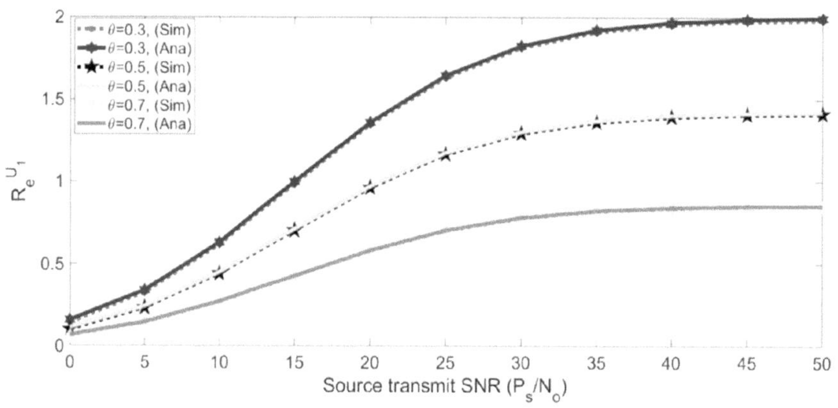

FIGURE 10.9 Ergodic rate of U_1 vs. source transmit SNR for different values of θ.

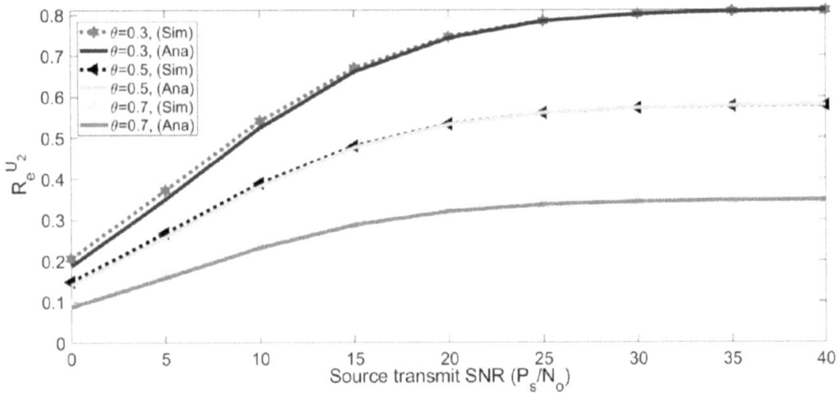

FIGURE 10.10 Ergodic rate of U_2 vs. source transmit SNR for different values of θ and $\rho = 0.5$.

the analytical result perfectly matches with the simulation result. As predicted, the increase in transmit SNR increases the ergodic rate; however, the rate of increase is higher in user 1 in comparison to user 2. Achieving a higher ergodic rate in user 1 is due to the higher SNR compared to the SNR achieved in user 2, since the transmit power for user 1 is higher than that of user 2 because of loss in harvesting efficiency.

10.6 CONCLUDING REMARKS

The outage performance of a full-duplex decode-and-forward cooperative NOMA network with energy harvesting is examined. The energy harvesting protocol used is of hybrid in nature, which means near user harvests energy using TS and PS relays both. The near user, which also acts as a DF relay, operates in full-duplex. We have shown the effect of different target data rate and TS relay factors on the outage performance of both users. The outage probability of both users decreases as SNR increases. Likewise, the system throughput and ergodic capacity of both users 1 and 2 are depicted. Regarding the system throughput, there is an increase that plateaus at high SNR, specifically beyond 25 dB SNR. Similarly, the ergodic rate of user 1 has a higher value than the ergodic rate of user 2 as SNR increases. We have shown the effect of non-linear energy harvesting and compared it with harvesting energy in linear mode. The said model and study are important in designing the NOMA network based on FD.

REFERENCES

[1] K. Huang and V. K. N. Lau, "Enabling Wireless Power Transfer in Cellular Networks: Architecture, Modeling and Deployment," in *IEEE Transactions on Wireless Communications*, vol. 13, no. 2, pp. 902–912, Feb. 2014, doi: 10.1109/TWC.2013.122313.130727.

[2] L. R. Varshney, "Transporting Information and Energy Simultaneously," in 2008 IEEE International Symposium on Information Theory, pp. 1612–1616, 2008.

[3] X. Zhou, R. Zhang and C.-K. Ho, "Wireless Information and Power Transfer: Architecture Design and Rate-Energy Trade-Off," in IEEE Transactions on Communications, vol. 61, pp. 4754–4767, Nov. 2013.
[4] L. Dai, B. Wang, Z. Ding, Z. Wang, S. Chen and L. Hanzo, "A Survey of Non-Orthogonal Multiple Access for 5G," in IEEE Communications Surveys Tutorials, vol. 20, no. 3, pp. 2294–2323, 2018.
[5] G. Wunder, P. Jung, M. Kasparick, T. Wild, F. Schaich, Y. Chen, S. T. Brink, I. Gaspar, N. Michailow, A. Festag, L. Mendes, N. Cassiau, D. Ktenas, M. Dryjanski, S. Pietrzyk, B. Eged, P. Vago, and F. Wiedmann, "5GNOW: Non-Orthogonal, Asynchronous Waveforms for Future Mobile Applications," in IEEE Communications Magazine, vol. 52, no. 2, pp. 97–105, Feb. 2014.
[6] Z. Ding, X. Lei, G. K. Karagiannidis, R. Schober, J. Yuan and V. K. Bhargava, "A Survey on Non-Orthogonal Multiple Access for 5G Networks: Research Challenges and Future Trends," in IEEE Journal on Selected Areas in Communications, vol. 35, no. 10, pp. 2181–2195, Oct. 2017.
[7] Z. Ding, M. Peng and H. V. Poor, "Cooperative Non-Orthogonal Multiple Access in 5G Systems," in IEEE Communications Letters, vol. 19, no. 8, pp. 1462–1465, Aug. 2015.
[8] H. Liu, Z. Ding, K. J. Kim, K. S. Kwak and H. V. Poor, "Decode-and-Forward Relaying for Cooperative NOMA Systems with Direct Links," in IEEE Transactions on Wireless Communications, vol. 17, no. 12, pp. 8077–8093, Dec. 2018.
[9] C. Zhong and Z. Zhang, "Non-Orthogonal Multiple Access with Cooperative Full-Duplex Relaying," in IEEE Communications Letters, vol. 20, no. 12, pp. 2478–2481, Dec. 2016.
[10] Z. Yu, C. Zhai, J. Liu and H. Xu, "Cooperative Relaying Based Non-Orthogonal Multiple Access (NOMA) with Relay Selection," in IEEE Transactions on Vehicular Technology, vol. 67, no. 12, pp. 11606–11618, Dec. 2018.
[11] G. Im and J. H. Lee, "Outage Probability for Cooperative NOMA Systems with Imperfect SIC in Cognitive Radio Networks," in IEEE Communications Letters, vol. 23, pp. 692–695, 2019.
[12] A. A. Nasir, X. Zhou, S. Durrani and R. A. Kennedy, "Relaying Protocols for Wireless Energy Harvesting and Information Processing," in IEEE Transactions on Wireless Communications, vol. 12, no. 7, pp. 3622–3636, Jul. 2013.
[13] M. Ju and H.-C. Yang, "Optimum Design of Energy Harvesting Relay for Two-Way Decode-and-Forward Relay Networks Under Max–Min and Max-Sum Criterions," in IEEE Transactions on Communications, vol. 67, no. 10, pp. 6682–6697, Oct. 2019, doi: 10.1109/TCOMM.2019.2927210.
[14] Y. Liu, G. Pan, H. Zhang and M. Song, "Hybrid Decode-and-Forward $&$ Amplify-and-Forward Relaying with Non-Orthogonal Multiple Access," in IEEE Access, vol. 4, pp. 4912–4921, 2016.
[15] G. Singhal, S. B. Kotwal, S. Modem and S. Prakriya, "Performance of Cooperative Relayed NOMA with Energy Harvesting Nodes in Underlay Networks," in 2020 International Conferences on Internet of Things (iThings) and IEEE Green Computing and Communications (GreenCom) and IEEE Cyber, Physical and Social Computing (CPSCom) and IEEE Smart Data (SmartData) and IEEE Congress on Cybermatics (Cybermatics), Rhodes, pp. 385–391, 2020.
[16] Y. Liu, H. Ding, J. Shen, R. Xiao and H. Yang, "Outage Performance Analysis for SWIPT-Based Cooperative Non-Orthogonal Multiple Access Systems," in IEEE Communications Letters, vol. 23, no. 9, pp. 1501–1505, Sept. 2019.
[17] A. S. Parihar, P. Swami, V. Bhatia and Z. Ding, "Performance Analysis of SWIPT Enabled Cooperative-NOMA in Heterogeneous Networks Using Carrier Sensing," in IEEE Transactions on Vehicular Technology, vol. 70, no. 10, pp. 10646–10656, Oct. 2021.

[18] X. Xie, J. Chen and Y. Fu, “Outage Performance and QoS Optimization in Full-Duplex System with Non-Linear Energy Harvesting Model,” in IEEE Access, vol. 6, pp. 44281–44290, 2018.
[19] J. Zhang and G. Pan, “Outage Analysis of Wireless-Powered Relaying MIMO Systems with Non-Linear Energy Harvesters and Imperfect CSI,” in IEEE Access, vol. 4, pp. 7046–7053, 2016.
[20] A. K. Baranwal, S. D. Roy and S. Kundu, “Performance Analysis of Full-Duplex Cooperative NOMA with Power-Time Splitting Based Energy Harvesting,” in 2023 4th International Conference on Computing and Communication Systems (I3CS), Shillong, pp. 1–6, 2023.
[21] Z. Ding, P. Fan and H. V. Poor, “On the Coexistence Between Full-Duplex and NOMA,” in IEEE Wireless Communications Letters, vol. 7, no. 5, pp. 692–695, Oct. 2018.
[22] K. Chang and Y. Choi, “Performance Evaluation of In-Band Full-Duplex System Using One-Time-Slot Two-Way Relay,” in IEEE Systems Journal, vol. 14, no. 1, pp. 510–519, Mar. 2020.
[23] B. C. Nguyen, X. N. Tran, D. T. Tran, X. N. Pham and L. T. Dung, “Impact of Hardware Impairments on the Outage Probability and Ergodic Capacity of One-Way and Two-Way Full-Duplex Relaying Systems,” in IEEE Transactions on Vehicular Technology, vol. 69, no. 8, pp. 8555–8567, Aug. 2020.
[24] G. Liu, W. Feng, Z. Han and W. Jiang, “Performance Analysis and Optimization of Cooperative Full-Duplex D2D Communication Underlying Cellular Networks,” in IEEE Transactions on Wireless Communications, vol. 18, no. 11, pp. 5113–5127, Nov. 2019.
[25] V. Aswathi and A. V. Babu, “Outage and Throughput Analysis of Full-Duplex Cooperative NOMA System with Energy Harvesting,” in IEEE Transactions on Vehicular Technology, vol. 70, no. 11, pp. 11648–11664, Nov. 2021.
[26] A. Hakimi, M. Mohammadi, Z. Mobini and Z. Ding, “Full-Duplex Non-Orthogonal Multiple Access Cooperative Spectrum-Sharing Networks with Non-Linear Energy Harvesting,” in IEEE Transactions on Vehicular Technology, vol. 69, no. 10, pp. 10925–10936, Oct. 2020.

11 Design of Efficient Signal Detection Schemes for Large-Scale MU-MIMO System

Naga Raju Challa, Kalapraveen Bagadi, Visalakshi Annepu, Kamireddy Vijay Chandra, and Deepika Rani Sona

11.1 INTRODUCTION

A large-scale MU-MIMO system is an advanced wireless communication architecture designed to handle a substantial number of antennas and users simultaneously. In traditional MIMO systems, multiple antennas are employed at both the transmitter and receiver, enabling improved data throughput and reliability. In large-scale MU-MIMO, this concept is extended to accommodate an even greater number of antennas, typically in the order of tens or hundreds, at both ends of the communication link. Key features and considerations in large-scale MU-MIMO systems include:

Increased capacity. One of the primary advantages of large-scale MU-MIMO is its ability to support a higher number of concurrent users and devices. The system can serve multiple users simultaneously, enhancing spectral efficiency and overall capacity.

Spatial multiplexing. Large-scale MU-MIMO exploits spatial diversity by transmitting multiple data streams concurrently to different users on the same frequency. This is achieved by using the spatial dimension provided by the multitude of antennas.

Channel estimation and precoding. Managing the large number of antennas requires sophisticated channel estimation techniques to understand the complex propagation environment. Precoding algorithms are then applied to optimize signal transmission based on the estimated channel conditions.

Interference management. As the number of antennas and users increases, so does the potential for interference. Large-scale MU-MIMO systems often incorporate advanced interference management techniques to maintain reliable communication in dense network scenarios.

Beamforming. *Beamforming* is a crucial aspect of large-scale MU-MIMO, where the system dynamically adjusts the direction of the transmitted signal to enhance signal strength and quality at the intended user while minimizing interference.

DOI: 10.1201/9781032625034-11

Challenges in implementation. Implementing large-scale MU-MIMO poses computational challenges due to the increased complexity in signal processing and the need for real-time adaptation. Efficient algorithms and hardware are essential for practical deployment.

Applications. Large-scale MU-MIMO finds applications in various sectors, including telecommunications, 5G and beyond, wireless broadband, and IoT connectivity, where the demand for high data rates and connectivity for numerous devices is critical.

Overall, large-scale MU-MIMO systems play a crucial role in meeting the escalating demands for data capacity and connectivity in modern wireless communication networks. Advances in this technology contribute to the evolution of telecommunications infrastructure, particularly in the context of emerging technologies like 5G and beyond.

In the ever-evolving landscape of wireless communication technology, large-scale MU-MIMO systems have emerged as a cornerstone of progress. These systems hold the promise of substantial improvements in both capacity and spectral efficiency, marking a significant leap in the quest for faster and more reliable wireless communications. At the heart of realizing this promise lies the development of efficient multiuser detection techniques. To fully harness the potential of large-scale MU-MIMO systems, it is imperative to design signal detection schemes that are not only accurate but also computationally efficient.

This work is devoted to addressing this pivotal challenge through the design and evaluation of innovative signal detection approaches tailored for large-scale MU-MIMO systems. Specifically, we delve into the realm of the MZF-LAS algorithm, which has been engineered to overcome the obstacles posed by the increasing number of antennas and users in these systems. By systematically comparing the performance of the MZF-LAS algorithm with that of conventional techniques such as zero forcing (ZF), likelihood ascent search (LAS), and zero forcing–assisted likelihood ascent search (ZF-LAS), we aim to assess its effectiveness in terms of detection accuracy and computational complexity. The heart of our research centres on a comprehensive exploration that involves rigorous simulations and meticulous performance analyses. Critical metrics, including signal detection accuracy, computational complexity, and bit error rate (BER), will be scrutinized to provide a detailed assessment of the MZF-LAS algorithm. Preliminary insights into the findings already hint at the superior performance of this novel approach.

In this ever-evolving field, the potential of the MZF-LAS algorithm to improve the overall performance and effectiveness of wireless communications carries great significance. As we confront the demands of large-scale MU-MIMO systems, it is paramount to develop innovative solutions that not only meet but also exceed the expectations of the industry. The outcome of this research has the potential to contribute significantly to the advancement of this critical technology, shaping the future of wireless communication in profound ways. This work represents a promising step forward in the pursuit of realizing the full potential of large-scale MU-MIMO systems and their role in shaping the telecommunications industry.

11.1.1 Related Work

Over the past decades, revolutionary improvements have been developed in the field of wireless networks by the reason of 5G technologies [1]. The novel approaches are also demonstrated in fewer areas, like artificial intelligence (AI), contactless payments, e-commerce, full high-definition (HD) video transmissions for connecting drones and robots, Internet of Everything (IOE), Internet of Things (IoT), smart traffic, telemedicine, weather monitoring, and control and virtual and augmented reality (VAR) [2–5]. Already, the third-generation partnership project (3GPP) is working on the advanced upgrades in 5G technologies (beyond 5G and 6G), which include massive MU-MIMO, millimetre wave, enhanced mobile broadband (e-MBB), software-defined radio (SDR), ultra-densification, and massive machine-type communication (m-MTC), that significantly benefited a variety of stakeholders, commercial solution providers, and end users [6–8].

Large-scale MU-MIMO, also referred to as large-scale space division multiple access, plays a pivotal role in enabling the practical realization of 5G and future 5G wireless communication systems [9]. In this system, a significant number of antennas, often exceeding 100, are integrated into e-MBB stations, while user equipment (UE) incorporates a large quantity of antennas, typically more than one [10]. The utilization of this massive array of antennas in large-scale MU-MIMO systems brings about several benefits, including increased data rates, enhanced link reliability, improved spectral and energy efficiency, and expanded network capacity, particularly in densely populated urban areas [11–13]. However, this impressive array of advantages is not without its challenges. Large-scale MU-MIMO systems encounter various hurdles, including computational complexity, channel estimation difficulties, antenna synchronization issues, and interference management complexities [14, 15]. Of these challenges, complexity and interference management stand out as major factors that can hinder the performance of massive MU-MIMO systems [16–18]. While the maximum likelihood (ML) detector is capable of delivering optimal BER performance in traditional MIMO systems, its computational demands become undesirable when dealing with the redundant antennas present in e-MBB stations within massive MU-MIMO systems [19]. The ZF and minimum mean squared error (MMSE) can achieve error rate performance that approaches the ideal. However, these methods necessitate matrix inversions, a computationally intensive operation that becomes impractical when dealing with the vast number of antennas found in large MU-MIMO systems [20–22]. In light of these challenges, the development of low-complexity methods based on matrix inverse approximations becomes a critical endeavour, as they hold the potential to enable more efficient and ideal detection in large MU-MIMO [23–25].

In the modern, internet-connected world, wireless networks play a crucial role in enabling communication among various devices [26]. To accommodate a growing number of users within the limited available spectral resources, advanced techniques like large-scale massive MU-MIMO systems are essential [27]. It offers high spectral efficiency by allowing multiple users to transmit their signals simultaneously over the same channel [28]. Within these systems, the channel impulse response (CIR) of each user serves as their spatial signature, aiding in the detection and differentiation of users at the receiver's end [29]. However, large-scale MU-MIMO systems often face challenges related to multiuser interference (MUI), which occurs when undesired users are in close proximity to the desired users [30]. This close proximity can result in the CIRs of different users becoming almost identical, leading to MUI. To address and mitigate MUI effectively, advanced multiuser detection (MUD) schemes are required, as mentioned in references [31]. The primary goal of MUD is to detect the desired user's signal while minimizing or eliminating interference from undesired users. It is worth noting that linear detectors, which are commonly used, have limitations. They can amplify noise, which can degrade system performance, and they often have suboptimal performance with polynomial-time detection complexity [32]. Therefore, researchers and engineers are exploring and developing more advanced signal processing MUD techniques to balance the performance–complexity trade-off of large-scale massive MIMO systems and to ensure efficient use of limited spectral resources in the future generation wireless communication landscape [33].

Recently, several efficient signal processing algorithms have been proposed based on the iterative search to avoid the MUI at e-MBB. In the LAS detector, the search for a solution is conducted in the vicinity of an initial solution vector, and the final solution is determined by selecting the best option from this neighbourhood, as indicated in references [34, 35]. The LAS detector requires an initial solution vector that is usually derived from the ZF/MMSE method. However, relying on the ZF/MMSE solution as the initial point for LAS can result in suboptimal performance because the ZF/MMSE solution is often not ideal for the specific problem being addressed. However, in this work, a modified ZF (M-ZF) scheme is considered as an initial solution for a LAS detector that achieves near-optimal performance in large-scale massive MU-MIMO systems.

11.1.2 Contributions

Contributions of the work are outlined as follows:

- **Innovative signal detection scheme.** One of the key contributions of this work is the development of the MZF-LAS algorithm. This innovative signal detection scheme is designed specifically for large-scale MU-MIMO systems, addressing the unique challenges posed by these systems. The introduction of MZF-LAS represents a novel approach to signal detection in the context of MU-MIMO.
- **Performance evaluation.** The study systematically compares the MZF-LAS algorithm with traditional signal detection techniques, including ZF, LAS, and ZF-LAS. By conducting extensive simulations and performance analyses, the work provides a comprehensive evaluation of MZF-LAS in terms of its detection accuracy and computational complexity. This comparative analysis provides valuable insights into the merits and limitations of the novel algorithm.
- **Enhanced detection accuracy.** The results of the performance evaluations clearly indicate that the MZF-LAS algorithm outperforms conventional methods in terms of detection accuracy. This improvement is a crucial contribution, as accurate signal detection is fundamental to the reliability and efficiency of wireless networks. The MZF-LAS algorithm's superior performance suggests its potential to enhance the quality of signal reception in large-scale MU-MIMO systems.

- **Reduced computational complexity.** Another significant contribution of this work is the reduction in computational complexity achieved by the MZF-LAS algorithm. The study demonstrates that MZF-LAS offers computational efficiency, which is essential for practical implementation in real-world large-scale MU-MIMO systems. This reduction in complexity has the potential to lead to more cost-effective and power-efficient solutions.
- **Advancement of wireless communication technology.** Overall, this research contributes to the advancement of wireless communication technology, particularly in the context of large-scale MU-MIMO systems. The development and evaluation of the MZF-LAS algorithm pave the way for more efficient and reliable wireless communication, thereby supporting the growth and evolution of the telecommunications industry. This work has the potential to significantly impact the practical implementation of large-scale MU-MIMO systems, making them more accessible and effective in real-world applications.

The structure of the document is as follows: Section 11.2 delineates the mathematical model for a large-scale MU-MIMO system. Section 11.3 delves into classical multiuser detection (MUD) techniques, encompassing ZF, ML, M-ZF, and LAS. Section 11.4 introduces the innovative M-ZF LAS detection approach. Section 11.5 showcases the simulation results, and the conclusion is provided in Section 11.6.

11.2 MATHEMATICAL MODEL

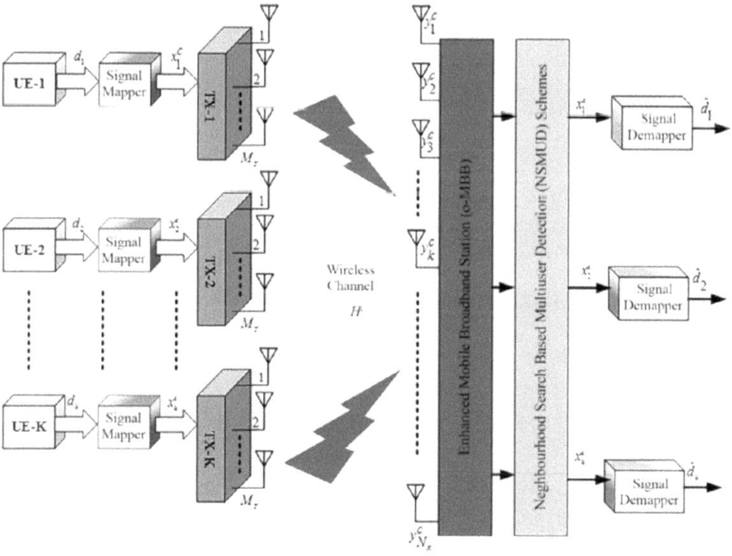

FIGURE 11.1 The uplink structure of a large-scale MU-MIMO system, along with an e-MBB station that has M_T transmitting antennas placed for each user. As a result, $N_T = K \times M_T$ and N_R receiving antennas and K users, and $N_T \leq N_R$, where N_T is the total number of transmitting antennas installed for all users. It is assumed that the transmitting and receiving antennas will both be large-scale or a few hundred in number.

Additionally, it is assumed that the e-MBB receiver has access to the instantaneous CIR and that the kth user is the only one with access to the CIR between the kth user and the e-MBB. The CIR between other $(K - 1)^{th}$ UEs and e-MBB is unknown at the kth UE because there is no user cooperation. In the schematic of a large-scale MU-MIMO system, a signal mapper modulates the $(M_T \times 1)$ dimensional data vector $\boldsymbol{d}_k = [d_{k,1}, d_{k,2}, ..., d_{k,M_T}]^T$ of the kth user to produce a complex data symbol vector, $\boldsymbol{s}_k^c = [s_{k,1}, s_{k,2}, ..., s_{k,M_T}]^T \in \mathbb{C}^{M_T}$, where $(.)^T$ denotes a transposition operation. The complex data vector is communicated through a complex channel matrix $\boldsymbol{G}_k^c \in \mathbb{C}^{N_R \times M_T}$. The signal vector which is received at the e-MBB from K simultaneous UEs is mathematically simulated as

$$\boldsymbol{x}^c = \sum_{k=1}^{K} \boldsymbol{G}_k^c \boldsymbol{s}_k^c + \boldsymbol{w}^c \tag{11.1}$$

where $\boldsymbol{w}^c = [w_1^c, w_2^c, ..., w_{N_R}^c]^T$ is the additive white Gaussian noise (AWGN) vector with zero mean and the covariance matrix $\sigma_n^2 I_{N_R}$. The matrix of dimensions $(N_R \times M_T)$ denoted by $\boldsymbol{G}_k^c$ represents the complex channel gains between the e-MBB receiver and the kth UE. Moreover, for the sake of simplicity, the complex MIMO system is equivalently represented as an identical real-valued system, adhering to the following model:

$$\boldsymbol{x} = \sum_{k=1}^{K} \boldsymbol{G}_k \boldsymbol{s}_k + \boldsymbol{w} \tag{11.2}$$

where $\boldsymbol{s}_k = \left[\Re(\boldsymbol{s}_k^c)^T \ \Im(\boldsymbol{s}_k^c)^T \right]^T_{(2M_T \times 1)}$, $\boldsymbol{x} = \left[\Re(\boldsymbol{x}^c)^T \ \Im(\boldsymbol{x}^c)^T \right]^T_{(2N_R \times 1)}$,

$\boldsymbol{w} = \left[\Re(\boldsymbol{w}^c)^T \ \Im(\boldsymbol{w}^c)^T \right]^T_{(2N_R \times 1)}$, and $\boldsymbol{G}_k = \begin{bmatrix} \Re\left(\boldsymbol{G}_k^c\right) & -\Im\left(\boldsymbol{G}_k^c\right) \\ \Im\left(\boldsymbol{G}_k^c\right) & \Re\left(\boldsymbol{G}_k^c\right) \end{bmatrix}_{(2N_R \times 2M_T)}$.

The output likelihood function of the massive MU-MIMO system model in (2) is given by

$$p(\boldsymbol{x}/\boldsymbol{G}, \boldsymbol{s}) = \frac{1}{\left(\pi\sigma^2\right)^{N_R}} \exp\left(\frac{1}{\sigma^2} \left\| \boldsymbol{x} - \boldsymbol{G}\boldsymbol{s} \right\|^2 \right) \tag{11.3}$$

To reconstruct the original signal at e-MBB, efficient multiuser detection (MUD) schemes are performed at the received terminal.

11.3 CLASSICAL MUD SCHEMES

Deploying an array of antennas for signal reception is a prominent uplink strategy in the development of future wireless communication systems. In practice, a significant challenge when implementing massive MU-MIMO systems lies in designing a signal detector that is both computationally efficient and reliable, particularly when dealing with higher modulation orders and large numbers of antennas. The primary objective of multiuser signal detection schemes is to reconstruct the transmitted input vector using the received vector and the channel matrix. For traditional MIMO systems, various classical MUD schemes have been proposed in prior research. However, these methods become unfeasible as the number of antennas increases when using linear detection techniques. In non-linear systems, the computational complexity of large-scale MU-MIMO systems becomes exponential. Consequently, there exists a clear trade-off between system performance and computational complexity. To mitigate this trade-off, search-based detection schemes have been introduced. In this section, various linear, non-linear, optimal, and search-based detection schemes are discussed.

11.3.1 Zero Forcing Detector

The ZF detector is a signal detection technique widely employed in large-scale MU-MIMO systems. In the context of MU-MIMO, signal detection becomes a complex task due to simultaneous transmissions from multiple users. The primary objective of the ZF detector is to eliminate or 'zero out' the interference caused by signals from other users, allowing the system to detect the desired signal accurately. This is achieved by employing a linear filtering technique that nullifies the interference introduced by other users' signals. The ZF detector operates based on the principle of interference cancellation, aiming to create orthogonal channels for each user. In a large-scale MU-MIMO system, the ZF detector becomes particularly relevant. By addressing interference issues, the ZF detector enhances the system's capacity to handle multiple users simultaneously, contributing to improved spectral efficiency and overall system performance. However, it is essential to note that the ZF detector assumes perfect channel state information, and its performance can degrade in practical scenarios with channel estimation errors or limited feedback. In summary, the ZF detector is a valuable tool in large-scale MU-MIMO systems, providing a means to mitigate interference and enhance the accuracy of signal detection in scenarios with numerous antennas and users.

The ZF detector is a linear detection method that eliminates interference by utilizing the pseudoinverse of the channel matrix.

$$\boldsymbol{G} = \left(\boldsymbol{G}^{H}\boldsymbol{G}\right)^{-1}\boldsymbol{G}^{H} \tag{11.4}$$

ZF detector estimates the signal vector as:

$$\hat{\boldsymbol{s}}_{ZF} = \left(\boldsymbol{G}^{T}\boldsymbol{G}\right)^{-1}\boldsymbol{G}^{T}\boldsymbol{x} \tag{11.5}$$

Here, $\boldsymbol{G}^{\dagger}\boldsymbol{G} = \boldsymbol{I}_{N_T}$. The ZF detector effectively eliminates spatial interference from other data streams. However, it does so at the cost of increasing the noise variance, which can negatively impact system performance, especially in scenarios with low signal-to-noise ratios (SNR). In addition to its performance characteristics, the ZF detector involves an inversion operation and requires a $o\left(N_T^3\right)+o\left(N_T^2 N_R\right)+o\left(N_T N_R^2\right)$ number of multiplications and additions. As a result, it exhibits quadratic per-symbol complexity. In order to suppress noise enhancement and achieve better performance in the large-scale massive MU-MIMO system, a modified zero forcing detector is introduced.

11.3.2 Modified Zero Forcing Detector

The modified zero forcing detector, with a focus on mitigating inherent noise amplification, achieves noteworthy performance gains. This algorithm involves a straightforward adjustment to the QR factorization of the channel matrix $\boldsymbol{G}$, that is,

$$\boldsymbol{G} = QR \tag{11.6}$$

where R is diagonal components of the upper triangular matrix, and Q is the orthogonal matrix. The R is remodified as $R = \hat{R}R_D$, where $\hat{R}$ is the unit upper triangular matrix and R_D is the diagonal matrix. It should be noted that the $cond\left(\boldsymbol{G}\right) = cond\left(R\right)$. Let us consider $r_{i,j}$ to be the element of R, and $R = \hat{R}R_D\ for\ j \geq i$, where $\hat{R}$ is the unit upper triangular matrix with elements $\hat{r}_{ij} = r_{ij} / r_{jj}$, $\hat{r}_{ii} = 1$, and R_D the diagonal matrix, whose diagonal elements are r_{ii}. The $\hat{r}_{ij}$ is the upper triangular channel correlation matrix, which is responsible for the channel ill-condition. The received signal vector of the MZF detector at e-MBB is

$$\hat{\boldsymbol{s}}_{MZFD} = \left[\boldsymbol{R}_D^{-1}\boldsymbol{Q}^H\boldsymbol{x}\right] \tag{11.7}$$

11.3.3 Maximum Likelihood Detector

The application of the ML detector is pivotal in the realm of large-scale MU-MIMO systems. Expressed mathematically, ML detection entails evaluating the joint probability distribution of the received signals conditioned on every possible transmitted symbol vector. The symbol vector that maximizes this probability becomes the chosen estimate for the transmitted signal. However, in large-scale MU-MIMO systems with a substantial number of antennas at both ends, the ML detector grapples with heightened computational complexity. The exhaustive search across all conceivable symbol vectors becomes computationally demanding as the system scales up. Despite these challenges, the ML detector is renowned for delivering optimal detection performance by minimizing error rates. To make ML detection viable in large-scale MU-MIMO systems, researchers often delve into algorithmic optimizations or employ parallel processing techniques. While computationally intensive, the ML detector persists as a benchmark for performance evaluation in scenarios where achieving the utmost accuracy is paramount.

The ML detector is considered an optimal method for estimating the transmitted symbol $\hat{s}$ from the received vector $\hat{x}$ and channel matrix G. It employs a comprehensive search strategy involving all potential symbol combinations across all transmitting antennas. This technique is particularly effective when the symbol x is uniformly selected from a constellation set $\mathcal{A}^{2N_T}$, where $\mathcal{A} = \left[a_1, a_2, \ldots, a_L\right]$, with a size denoted by L. The primary aim of the ML detector is to minimize the probability of error in symbol estimation $p\left(\hat{x} \neq x\right)$. Based on the probability density function given in (3), the estimated maximum likelihood for s given x is

$$\begin{aligned} \hat{x}_{ML} &= \arg\max_{s \in \mathcal{A}^{N_T}} p\left(x/G, s\right) \\ &= \arg\max_{s \in \mathcal{A}^{N_T}} \frac{1}{\left(\pi\sigma^2\right)^{N_R}} \exp\left(\frac{1}{\sigma^2}\left\|x - Gs\right\|^2\right) \\ &= \arg\min_{s \in \mathcal{A}^{N_T}} \left\|x - Gs\right\|^2 \end{aligned} \tag{11.8}$$

The output solution of ML detector is defined as

$$\hat{x}_{ML} = \arg\min_{s \in \mathcal{A}^{N_T}} \left\|x - Gs\right\|^2 \tag{11.9}$$

Here, $\hat{x}$ represents the estimated symbol vector, and $\mathcal{A}^{N_T}$ is the collection of all potential candidates for the $2N_T$ real-valued transmitted data vectors. The cardinality of $\mathcal{A}^{N_T}$ is determined by the particular modulation scheme in use.

The ML detector, while capable of achieving maximum diversity gain by exploring all possible symbol combinations across transmitting antennas, faces a significant challenge due to its exponential computational complexity. This complexity grows rapidly, making it impractical for even a small number of antennas. When dealing with massive structures and high-order constellation schemes, the ML detector's complexity becomes so overwhelming that it is deemed unfeasible. Consequently, for large-scale MU-MIMO systems, the ML detector is not a viable choice. In recent times, there has been a shift towards the development of neighbourhood search-based detection techniques aimed at narrowing the gap between computational complexity and performance.

11.3.4 Likelihood Ascent Search Detector

The ML detector offers optimal performance; however, it is impractical for implementing large-scale MU-MIMO systems due to its exponential complexity. Low-complexity detectors like the zero forcing (ZF) detector can achieve acceptable bit error rate (BER) performance when the signalling dimensions are significantly high. This leads to a trade-off between complexity and performance. In addressing this trade-off within extensive MIMO systems, search-based detection approaches offer a compelling solution. They can deliver performance that approaches optimality while requiring a manageable computational load. Among these search-based detectors, the LAS detector stands out as a noteworthy example.

The LAS detector is a signal detection technique employed in large-scale MU-MIMO systems. It is designed to enhance the accuracy of signal detection in scenarios with numerous antennas and users. The LAS detector operates by iteratively maximizing the likelihood function to improve the estimation of transmitted signals. It searches for the optimal solution in the likelihood space, gradually ascending towards the maximum likelihood estimate. This iterative process allows the detector to navigate through the complex signal space efficiently, making it well-suited for large-scale MU-MIMO environments. One notable application of the LAS detector is in the context of massive MU-MIMO systems, where traditional detection methods may face challenges due to increased dimensions and potential interference. The LAS detector's iterative nature enables it to adapt to the dynamic and complex nature of the channel, providing improved signal detection performance. Thus, the LAS detector is a valuable tool for signal detection, offering enhanced accuracy and adaptability to the challenges posed by the scale and complexity of such communication environments.

Within the family of neighbourhood search-based detectors, LAS stands out as the first reported detector primarily employing a local search strategy through an iterative process. The primary strength of the LAS detector resides in its linear average per-bit complexity, providing competitive performance, particularly within the domain of large-scale MU-MIMO systems. The LAS detector navigates through a series of bit vectors, structured according to LAS update rules. It initiates the process with an output vector from the zero forcing (ZF) detector, known as the initial solution vector. Subsequently, the detector explores the neighbourhood to identify the optimal solution vector closest to the one determined by the maximum likelihood (ML) detector. Let $\boldsymbol{A} = [a_1, a_2, \ldots, a_L]$ represent a real-valued constellation signal set, where L is the size of the signal set $\boldsymbol{A}$. Consider $\hat{\boldsymbol{s}}_{ZF} = \left[(\hat{\boldsymbol{s}}_1)^T, (\hat{\boldsymbol{s}}_2)^T, \ldots, (\hat{\boldsymbol{s}}_K)^T\right]^T = \left[\hat{s}_1, \hat{s}_2, \ldots, \hat{s}_{2N_T}\right]^T$ as the output vector solution derived from the ZF detector, chosen as the initial solution for the LAS algorithm. The LAS algorithm utilized for detecting the nth data symbol in the MU-MIMO system is formulated as follows:

$$\hat{\boldsymbol{s}}_{n,LAS} = \arg \min_{v_n = A_l, l=1,2,\ldots,L} \left\| \boldsymbol{x} - \boldsymbol{G}\boldsymbol{v} \right\|^2, \quad n = 1,2,\ldots,2N_T \tag{11.10}$$

where $\boldsymbol{v} = [v_1, v_2, \ldots, v_{2N_T}]^T$ is a trail vector of size $(2N_T \times 1)$, and the trail vector $\boldsymbol{v}$ is defined as

$$\boldsymbol{v}_i = \begin{cases} \boldsymbol{v}_n \in A & if \ i = n \\ \hat{s}_i & if \ i \neq n \end{cases}, \quad \text{where } i = 1,2,\ldots,2N_T \tag{11.11}$$

11.4 PROPOSED MODIFIED ZERO FORCING–ASSISTED LIKELIHOOD ASCENT SEARCH DETECTOR

LAS detector performance is predominantly influenced by two key factors: the initial solution vector and the selection of the search space. When an excellent initial solution is chosen, LAS can achieve superior BER performance with reduced computational complexity. Nevertheless, LAS faces a significant obstacle related to early termination, particularly when dealing with a limited search space, which becomes impractical for larger constellations and massive MU-MIMO systems. To terminate this issue and achieve near-optimal performance with fewer number of computations, a modified zero forcing–assisted likelihood ascent search (MZF-LAS) detector is here proposed. Let $\boldsymbol{A} = [a_1, a_2, \ldots, a_L]$ represent a real-valued constellation signal set, where L is the size of the signal set $\boldsymbol{A}$. Let $\hat{\boldsymbol{s}}_{MZFD} = \left[(\hat{\boldsymbol{s}}_1)^T, (\hat{\boldsymbol{s}}_2)^T, \ldots, (\hat{\boldsymbol{s}}_K)^T\right]^T = \left[\hat{s}_1, \hat{s}_2, \ldots, \hat{s}_{2N_T}\right]^T$ be the output vector solution obtained from MZF detector and taken to be the initial solution of the LAS algorithm. The MZF-LAS algorithm engaged for detecting n^{th} data symbol of large-scale MU–MIMO system is conveyed as

$$\hat{\boldsymbol{s}}_{n,MZF\text{-}LAS} = \arg \min_{z_n = A_l, l=1,2,\ldots,L} \|\boldsymbol{x} - \boldsymbol{G}\boldsymbol{z}\|^2, \quad n = 1,2,\ldots,2N_T \tag{11.12}$$

where $\boldsymbol{z} = [z_1, z_2, \ldots, z_{2N_T}]^T$ is a trail vector of size $(2N_T \times 1)$, and the trail vector $\boldsymbol{z}$ is defined as

$$z_i = \begin{cases} z_n \in A & if\ \ i = n \\ \hat{s}_i & if\ \ i \neq n \end{cases}, \quad \text{where } i = 1,2,\ldots,2N_T \tag{11.13}$$

The procedure for the suggested MZF-LAS detector in a large-scale MU-MIMO system is outlined next.

Algorithm 1: The MZF-assisted LAS

1. $INPUTS:$ x, G, s, z.
2. OUTPUT: $\hat{\boldsymbol{s}}_{MZF\text{-}LAS}$.
3. Initialization n=1& i=1.
4. Compute $\boldsymbol{G}$ and $\boldsymbol{x}$.
5. $\hat{\boldsymbol{s}}_{ZF} \leftarrow \left(\boldsymbol{H}^H \boldsymbol{H}\right)^{-1} \boldsymbol{H}^H \boldsymbol{y}$; zero forcing output.
6. $\hat{\boldsymbol{s}}_{MZFD} \leftarrow \left[\boldsymbol{R}_D^{-1} \boldsymbol{Q}^H \boldsymbol{x}\right]$; modified zero forcing output.
7. Compute the condition number of channel matrix $cond(\boldsymbol{G})$.
8. Fix a threshold value $\beta > 1$.
9. $if\ \beta > cond(\boldsymbol{G})$,
   $\hat{\boldsymbol{s}}_{MZFD} \leftarrow \hat{\boldsymbol{s}}$;
   else,
   $\hat{\boldsymbol{s}}_{ZF} \leftarrow \hat{\boldsymbol{s}}$.
   End.
10. $\boldsymbol{z} \leftarrow \hat{\boldsymbol{s}}_{MZFD}$; initial solution for MZF-LAS detector.
    where, $\boldsymbol{z} = [z_1, z_2, \ldots, z_{2N_T}]^T$.
11. *for* n=1 to $2N_T$; do.
12. $\text{Cost}_{prev} \leftarrow \|\boldsymbol{x} - \boldsymbol{G}\boldsymbol{z}\|^2$.

13. *for* i=1 to $2N_T$; do.
14. $z_n = A$.
15. $\text{Cost}_{next} \leftarrow \|x - Gz\|^2$.
16. *if* $\boldsymbol{Cost}_{\text{next}} \leq \boldsymbol{Cost}_{\text{Prev}}$.
17. $z_i = \hat{s}_i$.
18. *End.*
19. $\boldsymbol{Cost}_{\text{prev}} \leftarrow \boldsymbol{Cost}_{\text{next}}$.
20. *End.*
21. *End.*
22. Choose 'z' as solution of MZF-LAS MUD.

11.5 SIMULATION RESULTS

The simulation analysis is centred on an uplink model within a large-scale MU-MIMO system, where UEs operate as transmitters while e-MBBS serve as receivers. The study assesses the performance of the proposed MZF-LAS detector by comparing the average BER of each UE with that obtained through traditional detection techniques. These techniques encompass ZF, MZF, ZF-LAS, and single-input, single-output with AWGN (SISO-AWGN) detectors. Moreover, the results from these comparisons are derived from averaging the performance metrics over 100 data frames, with each frame containing 1,000 data symbols. This averaging procedure ensures result accuracy and minimizes the potential for misinterpretation. The simulation model assumes that multiple UEs transmit their data over a block fading channel, which remains constant during the transmission of an entire data frame. The elements within the channel matrix are represented as complex random variables that are statistically independent, a characteristic commonly known as a Rayleigh flat fading channel. Table 11.1 presents a comprehensive list of the parameters employed in the simulation evaluations, providing a detailed reference for the analysis conducted.

TABLE 11.1
Simulation Set-Up

Parameters	Value
Number of antennas at e-MBB station (N_R)	128
Number of antennas per each UE (M_T)	8
Number of UEs (K)	16
Number of data frames	100
Number of symbols per data frames	1,000
Type of modulations	4-QAM, 16–QAM, and 64 QAM
Channel	Rayleigh flat fading
MZF threshold value β	1.2
Signal-to-noise ratio range (SNR)	−10 to 15 dB

11.5.1 Performance Analysis

Within this section, we evaluate the BER performance of the recently introduced MZF-LAS scheme, contrasting it with classical ZF, MZF, and ZF-LAS schemes. BER comparisons across various schemes are illustrated in Figures 11.2 and 11.3, utilizing both BPSK and 16-QAM. The outcomes presented in these figures indicate that the performance achieved with BPSK modulation exceeds that attained with 16-QAM modulation. This observation aligns with the established notion that lower-order modulation schemes generally exhibit superior BER performance. From Figures 11.2 and 11.3, it becomes evident that the traditional ZF detector exhibits a relatively high BER due to its limitations in completely eliminating MUI at the e-MBB receiver. In contrast, the MZF detector proves to be more proficient, as it effectively suppresses noise enhancement and results in a better-conditioned channel matrix, thereby achieving a more favourable BER performance compared to the ZF detector. The conventional LAS detector predominantly mitigates MUI at the e-MBB station and achieves suboptimal BER compared to ZF and MZF detectors. The newly proposed MZF-assisted LAS aims to reduce MUI, albeit with a slight increase in complexity, and it exhibits superior performance in comparison to traditional detection methods.

As shown in Figure 11.2, when targeting a BER of 0.1, the necessary signal-to-noise ratio (SNR) values for ZF, MZF, ZF-LAS, and SISO-AWGN systems are 9.5 dB, 8 dB, 6 dB, and 2 dB, respectively. On the other hand, the newly proposed MZF-LAS system demonstrates remarkable efficiency, requiring only a 4 dB SNR when each user transmits a 16-QAM signal. To further assess the performance of both the proposed MZF-LAS and conventional detectors regarding BER, a comparison is

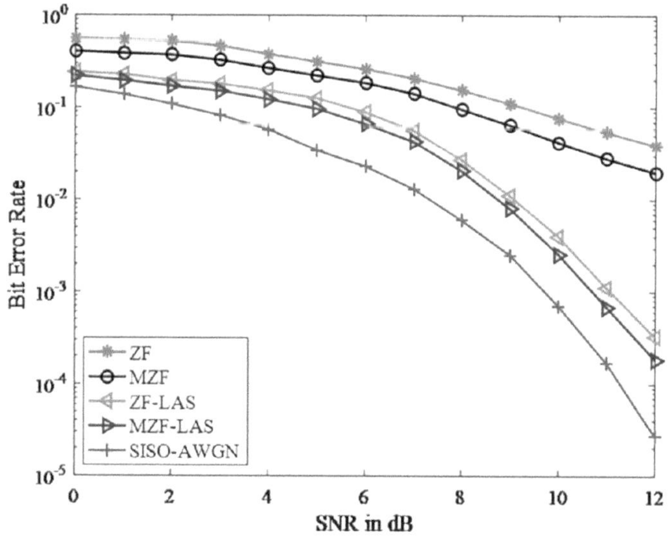

FIGURE 11.2 Comparison of bit error rate (BER) performance for MZF-LAS and classical detectors in the scenario of all users transmitting 16-QAM signals.

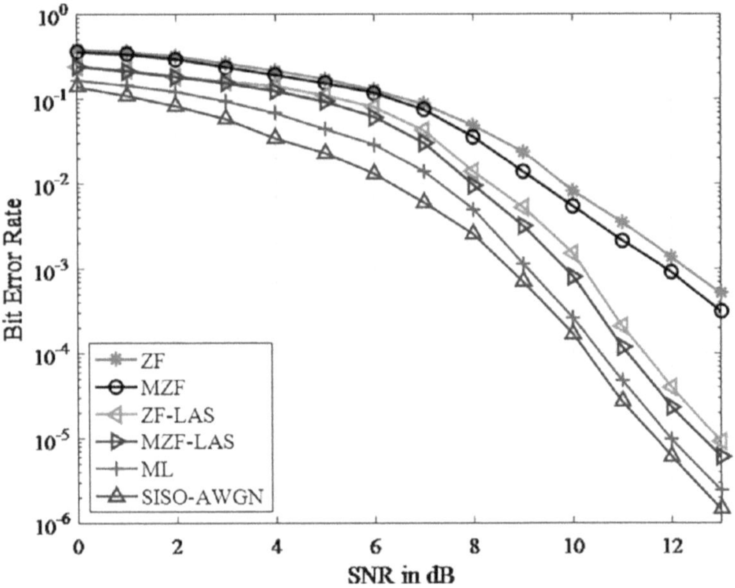

FIGURE 11.3 BER performance comparison of MZF-LAS with classical and optimal ML detectors while all users are transmitting BPSK signals.

made with the optimal ML detector, as depicted in Figure 11.3. In this evaluation, the average BER is computed under the scenario where users transmit BPSK signals to an e-MBB receiver. Once again, the results underscore the effectiveness of the proposed scheme, which closely approaches the performance of the optimal ML detector. For example, to achieve a 0.1 error rate, the optimal ML detector necessitates an SNR of 3 dB, while the proposed MZF-LAS, ZF-LAS, MZF, and ZF schemes achieve comparable error rates with SNR values of 4 dB, 6 dB, 7 dB, and 7.5 dB, respectively, as illustrated in Figure 11 3.

The performance analysis of neighbourhood search-based detection schemes, such as LAS and the newly proposed MZF-LAS, is further examined across different modulation configurations, as illustrated in Figure 11 4. Specifically, the study considers 4-QAM, 16-QAM, and 64-QAM modulation schemes, each with its corresponding performance results, displayed in Figure 11.4. As the modulation scheme becomes more complex, the presence of multiple-user interference (MUI) in the system also increases. Consequently, the bit error rate (BER) gradually rises with higher constellation schemes.

In contrast to the ZF-LAS detector, the MZF-LAS detector effectively mitigates MUI and approaches near-optimal performance. This trend is evident in Figure 11.4. For instance, when operating at an SNR of 10 dB and accommodating 4-QAM, 16-QAM, and 64-QAM modulation schemes, the SISO system yields BER values of 0.00081, 0.057, and 0.151, respectively. In contrast, the proposed MZF-LAS detector achieves BER values of 0.0201, 0.221, and 0.305, respectively, closely approximating the error rate performance of the SISO system, as demonstrated in Figure 11.4. Furthermore, Figure 11.4 illustrates how the variations in SNR between the proposed MZF-LAS

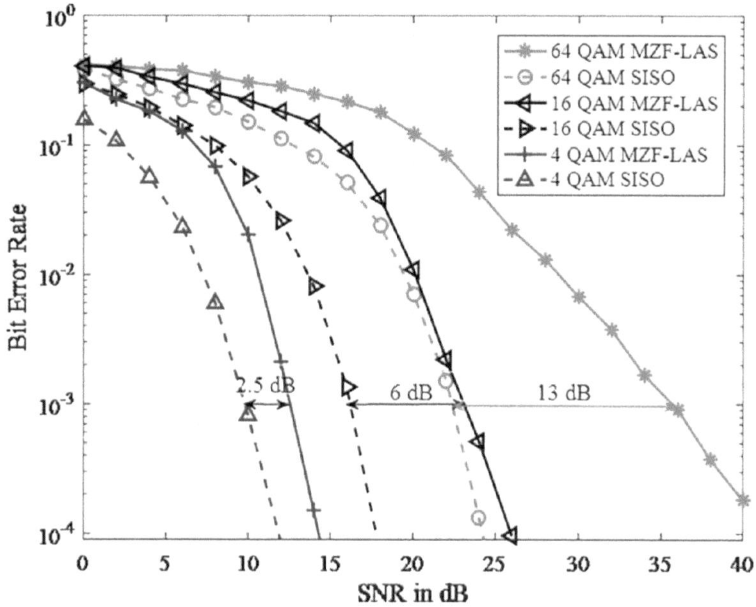

FIGURE 11.4 BER performance comparison of MZF-LAS with SISO-AWGN detector while all users are transmitting different modulated signals.

and SISO detectors are influenced by different QAM techniques. For the 4-QAM, 16-QAM, and 64-QAM modulation schemes, Figure 11.4 reveals that the SNR variations at the 10^{-3} BER threshold amount to 2.5 dB, 6 dB, and 13 dB, respectively.

11.5.2 Complexity Analysis

Figure 11.5 illustrates the computational complexity concerning the number of multiplications for an increasing number of users. This figure compares the complexity of the proposed MZF-LAS with that of traditional detectors, including ZF, MZF, and ZF-LAS. This comparison illustrates that the proposed MZF-LAS technique is somewhat more complex than traditional detectors.

However, it is crucial to emphasize that this increase in complexity is minimal when compared to the complexity of the ML detector, as depicted in Figure 11.5. This additional complexity is a necessary trade-off to achieve nearly optimal performance with the suggested MZF-LAS approach.

11.6 CONCLUSION

In conclusion, this research has provided valuable insights into the realm of large-scale massive MU-MIMO systems, a pivotal component of modern wireless communication technology. The study has underscored the significance of efficient multiuser detection techniques in unlocking the full potential of these advanced systems. The central focus of this investigation was the development and evaluation of the MZF-LAS algorithm, designed specifically to address the challenges presented by

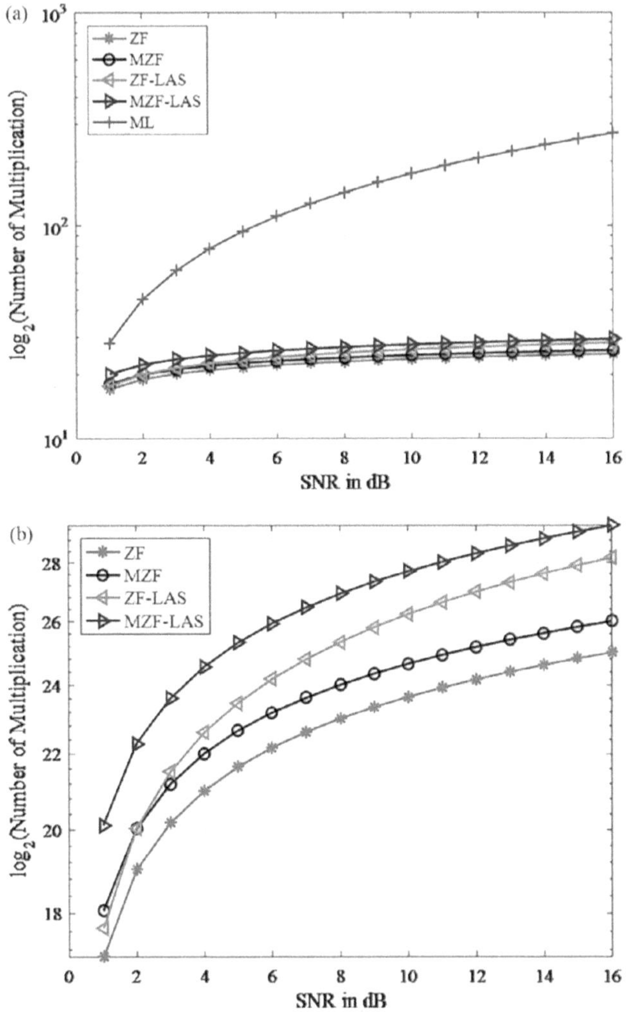

FIGURE 11.5 (a) Complexity comparison of various detector with ML, (b) Complexity comparison of various detector without ML.

large-scale MU-MIMO systems. Through systematic comparisons with traditional signal detection methods, including ZF, LAS, and ZF-LAS, the study aimed to assess the efficacy of MZF-LAS in terms of detection accuracy and computational complexity. The results of extensive simulations and performance analyses have provided compelling evidence of the superior performance of the MZF-LAS algorithm. It exhibited higher detection accuracy compared to the established techniques (ZF, LAS, and ZF-LAS), showcasing its potential to significantly improve the overall reliability of signal detection. Furthermore, MZF-LAS demonstrated a reduced computational load, which not only contributes to enhanced efficiency but also makes it an attractive candidate for practical implementation.

Overall, this research sheds light on the promise of the MZF-LAS algorithm as a solution to address the evolving demands of massive MU-MIMO systems characterized by an increasing number of antennas and users. By enhancing detection accuracy and minimizing computational complexity, MZF-LAS opens the door to improved performance and efficiency in wireless communication systems. Consequently, this study makes a notable contribution to the advancement of critical technology within the telecommunications industry, marking an important step forward in harnessing the full potential of large-scale massive MU-MIMO systems.

REFERENCES

[1] M. A. Albreem, A. H. A. Habbash, A. M. Abu-Hudrouss and S. S. Ikki, "Overview of Precoding Techniques for Massive MIMO," in IEEE Access, vol. 9, pp. 60764–60801, 2021, doi: 10.1109/ACCESS.2021.3073325.

[2] N. R. Challa and K. Bagadi, "Lattice Reduction Assisted Likelihood Ascent Search Algorithm for Multiuser Detection in Massive MIMO System," in 2018 15th IEEE India Council International Conference (INDICON), Coimbatore, pp. 1–5, 2018, doi: 10.1109/INDICON45594.2018.8987139.

[3] E. Yaacoub and M.-S. Alouini, "A Key 6G Challenge and Opportunity—Connecting the Base of the Pyramid: A Survey on Rural Connectivity," in Proceedings of the IEEE, vol. 108, no. 4, pp. 533–582, 2020, doi: 10.1109/JPROC.2020.2976703.

[4] S. Garg, K. Kaur, G. S. Aujla, G. Kaddoum, P. Garigipati and M. Guizani, "Trusted Explainable AI for 6G-Enabled Edge Cloud Ecosystem," in IEEE Wireless Communications, vol. 30, no. 3, pp. 163–170, 2023, doi: 10.1109/MWC.016.220047.

[5] M. A. Habibi, B. Han, A. Fellan, W. Jiang, A. G. Sancheg, I. L. Pavon, A. Boubendir, H. D. Schotten., "Toward an Open, Intelligent, and End-to-End Architectural Framework for Network Slicing in 6G Communication Systems," in IEEE Open Journal of the Communications Society, vol. 4, pp. 1615–1658, 2023, doi: 10.1109/OJCOMS.2023.3294445.

[6] C.-X. Wang, X. You, X. Gao, X. Zhu, Z. Li, C. Zhang, H. Wang, Y. Huang, Y. Chen, H. Haas, J. S. Thompson, E. G. Larsson, M. Di Renzo, W. Tong, P. Zhu, X. Shen, H. V. Poor and L. Hanzo, "On the Road to 6G: Visions, Requirements, Key Technologies, and Testbeds," in IEEE Communications Surveys & Tutorials, vol. 25, no. 2, pp. 905–974, 2023, doi: 10.1109/COMST.2023.3249835.

[7] M. Vaezi, A. Azari, S. R. Khosravirad, M. Shirvanimoghaddam, M. M. Azari, D. Chasaki and P. Popovski, "Cellular, Wide-Area, and Non-Terrestrial IoT: A Survey on 5G Advances and the Road Toward 6G," in IEEE Communications Surveys & Tutorials, vol. 24, no. 2, pp. 1117–1174, 2022, doi: 10.1109/COMST.2022.3151028.

[8] B. S. Khan, S. Jangsher, A. Ahmed and A. Al-Dweik, "URLLC and eMBB in 5G Industrial IoT: A Survey," in IEEE Open Journal of the Communications Society, vol. 3, pp. 1134–1163, 2022, doi: 10.1109/OJCOMS.2022.3189013.

[9] M. A. Albreem, A. H. A. Habbash, A. M. Abu-Hudrouss and S. S. Ikki, "Overview of Precoding Techniques for Massive MIMO," in IEEE Access, vol. 9, pp. 60764–60801, 2021, doi: 10.1109/ACCESS.2021.3073325.

[10] M. Agiwal, A. Roy and N. Saxena, "Next Generation 5G Wireless Networks: A Comprehensive Survey," in IEEE Communications Surveys & Tutorials, vol. 18, no. 3, pp. 1617–1655, 2016, doi: 10.1109/COMST.2016.2532458.

[11] Y. He, Y. Chen, Y. Hu and B. Zeng, "WiFi Vision: Sensing, Recognition, and Detection with Commodity MIMO-OFDM WiFi," in IEEE Internet of Things Journal, vol. 7, no. 9, pp. 8296–8317, 2020, doi: 10.1109/JIOT.2020.2989426.

[12] N. Panwar, S. Sharma and A. Kumar, "A Survey on 5G: The Next Generation of Mobile Communication," in Physical Communication, vol. 18, pp. 64–84, 2016, doi: 10.1016/j.phycom.2015.10.006.
[13] U. V. Menon, N. R. Challa and K. Bagadi, "Lenstra Lenstra Lovász (LLL) Assisted Likelihood Ascent Search (LAS) Algorithm for Signal Detection in Massive MIMO," in 2019 International Conference on Vision Towards Emerging Trends in Communication and Networking (ViTECoN), Vellore, pp. 1–4, 2019, doi: 10.1109/ViTECoN.2019.8899594.
[14] N. Fatema, G. Hua, Y. Xiang, D. Peng and I. Natgunanathan, "Massive MIMO Linear Precoding: A Survey," in IEEE Systems Journal, vol. 12, no. 4, pp. 3920–3931, 2018, doi: 10.1109/JSYST.2017.2776401.
[15] R. Kumar and C. V. K. Praveen, "Design of MC-CDMA Receiver Using Radial Basis Function Network to Mitigate Multiple Access Interference and Nonlinear Distortion," in Neural Computing and Applications, vol. 31, no. s2, pp. 1263–1273, 2019, doi: 10.1007/s00521-017-3127-0.
[16] K. Praveen, V. Annepu and S. Das, "Recent Trends in Multiuser Detection Techniques for SDMA – OFDM Communication System," in Physical Communication, vol. 20, pp. 93–108, 2016.
[17] N. R. Challa and K. Bagadi, "Design of Massive Multiuser MIMO System to Mitigate Inter Antenna Interference and Multiuser Interference in 5G Wireless Networks," in Journal of Communications, vol. 15, no. 9, pp. 693–701, 2020, doi: 10.12720/jcm.15.9.693-701.
[18] N. R. Challa and K. Bagadi, "Likelihood Ascent Search Detection for Coded Massive MU-MIMO Systems to Mitigate IAI and MUI," in Radioelectronics and Communications Systems, vol. 63, no. 5, pp. 223–234, 2020, doi: 10.3103/S0735272720050015.
[19] H. Semira, F. Kara, S. Member and H. Kaya, "Multi-User Joint Maximum-Likelihood Detection in Uplink NOMA-IoT Networks: Removing the Error Floor," in IEEE Wireless Communications Letters, vol. 10, no. 11, pp. 2459–2463, 2021.
[20] Y. Liu, "A Massive MIMO Signal Detection Method Based on ZF Method," in 2021 5th International Conference on Imaging, Signal Processing and Communications (ICISPC), Kumamoto, pp. 71–76, 2021, doi: 10.1109/ICISPC53419.2021.00021.
[21] Y. Xiong, S. Sun, N. Wei, L. Liu and Z. Zhang, "Asymptotic Analysis for Cell-Free Massive MIMO with MMSE Combining and Low-Resolution ADCs," in IEEE Communications Letters, vol. 25, no. 10, pp. 3219–3223, 2021, doi: 10.1109/LCOMM.2021.3102021.
[22] I. Boukhedimi, A. Kammoun and M.-S. Alouini, "Multi-Cell MMSE Combining Over Correlated Rician Channels in Massive MIMO Systems," in IEEE Wireless Communications Letters, vol. 9, no. 1, pp. 12–16, 2020, doi: 10.1109/LWC.2019.2939305.
[23] N. R. Challa and K. Bagadi, "Design of Near-Optimal Local Likelihood Search-Based Detection Algorithm for Coded Large-Scale MU-MIMO System," in International Journal of Communication Systems, vol. 33, no. 12, p. e4436, 2020, doi: 10.1002/dac.4436.
[24] N. R. Challa and K. Bagadi, "Design of Large Scale MU-MIMO System with Joint Precoding and Detection Schemes for Beyond 5G Wireless Networks," in Wireless Personal Communications, vol. 121, pp. 1627–1646, 2021, doi: 10.1007/s11277-021-08688-6.
[25] C. Zhang, B. Wang, D. Li and X. Tan, "Low Complexity Multiuser Detection with Recursively Successive Zero-Forcing and SIC Based on Nullspace for Multiuser MIMO-OFDM System," in China Communications, vol. 12, no. 9, pp. 53–63, 2015, doi: 10.1109/CC.2015.7275259.
[26] Y. Liu, S. Zhang, X. Mu, Z. Ding, R. Schober, N. Al-Dhahir, E. Hossain and X. Shen, "Evolution of NOMA Toward Next Generation Multiple Access (NGMA) for 6G," in IEEE Journal on Selected Areas in Communications, vol. 40, no. 4, pp. 1037–1071, 2022, doi: 10.1109/JSAC.2022.3145234.
[27] C. Ding, J.-B. Wang, H. Zhang, M. Lin and J. Wang, "Joint MU-MIMO Precoding and Resource Allocation for Mobile-Edge Computing," in IEEE Transactions on Wireless Communications, vol. 20, no. 3, pp. 1639–1654, 2021, doi: 10.1109/TWC.2020.3035153.

[28] K. Bagadi, C. V. Ravikumar and M. Alibakhshikenari, "Precoded Large Scale Multi-User-MIMO System Using Likelihood Ascent Search for Signal Detection," in Radio Science, vol. 57, no. 12, p. e2022RS007573, 2022, doi: 10.1029/2022RS007573.
[29] A. K. Sah and A. K. Chaturvedi, "Sequential and Global Likelihood Ascent Search-Based Detection in Large MIMO Systems," in *IEEE Transactions on Communications*, vol. 66, no. 2, pp. 713–725, 2018.
[30] S. Chakraborty, "Low Complexity Hybrid Layered Tabu-Likelihood Ascent Search for Large MIMO Detection with Perfect and Estimated Channel State Information," in ETRI Journal, vol. 45, no. 3, pp. 418–432, 2022, doi: 10.4218/etrij.2022-0052.
[31] D. C. Araújo, T. Maksymyuk, A. L. F. De Almeida, T. Maciel, J. C. M. Mota and M. Jo, "Massive MIMO: Survey and Future Research Topics," in IET Communications, vol. 10, pp. 1938–1946, 2016, doi: 10.1049/iet-com.2015.1091.
[32] Y. Liu and S. Zhou, "Low Complexity Multi-User Detector for Grant-free Massive Access with Massive MIMO with Consideration of both Temporal and Spatial Correlation," in 2021 International Wireless Communications and Mobile Computing (IWCMC), Harbin City, pp. 1098–1104, 2021, doi: 10.1109/IWCMC51323.2021.9498897.
[33] H. J. Park and J. W. Lee, "LDPC Coded Multi-User Massive MIMO Systems with Low-Complexity Detection," in IEEE Access, vol. 10, pp. 25296–25308, 2022, doi: 10.1109/ACCESS.2022.3156285.
[34] A. K. Sah and A. K. Chaturvedi, "An Unconstrained Likelihood Ascent Based Detection Algorithm for Large MIMO Systems," in IEEE Transactions on Wireless Communications, vol. 16, no. 4, pp. 2262–2273, 2017.
[35] A. K. Sah and A. K. Chaturvedi, "Beyond Fixed Neighborhood Search in the Likelihood Ascent Algorithm for MIMO systems," in 2016 IEEE International Conference on Communications (ICC), Kuala Lumpur, pp. 1–6, 2016, doi: 10.1109/ICC.2016.7511055.

12 Flexible Antennas – A New Frontier in 5G Technology
A Review

Shamina S. S. and Christina Josephine Malathi A.

12.1 INTRODUCTION

5G technology is playing a key role within the area of mobile wireless communications in this age. A good radio link is characterized by good data transmission, power efficiency, and an increase in the total throughput. Because of increased data rate and bandwidth, 5G has come into existence. In addition, 5G provides high data throughput and ultralow latency of less than 1 ms. Higher frequencies, beamforming, power consumption, size restrictions, and signal interference are some of the design problems of 5G. 5G features include a wide range of applications with increased bandwidth, very low latency of about 1 ms, high data rates and throughput, high speed of up to 10 gigabits/s, and ability to connect a greater number of devices. Flexible electronics paved the way for a major outbreak in 5G technology. Currently, researchers are very much interested in flexible microstrip patch antennas due to its light weight, conformality, bending and twisting conditions, mechanical stability and portability when compared with conventional microstrip patch antenna. In terms of environmental conditions and frequency range, the flexibility antenna design is different.

12.2 FREQUENCY BANDS IN 5G

There are three frequency bands in the 5G spectrum. Figure 12.1 illustrates these bands as low band, sub-6 GHz band, and millimetre wave band. The low band spans the frequency ranges of 600 MHz to 700 MHz, sub-6 GHz spans 3,500 MHz to 5,000 MHz, and mmWave is over 24 GHz.

12.2.1 Low Band

Low-band 5G transmits at a frequency between 600 MHz and 700 MHz, offering broad coverage. Although the low band offers slower speeds of about 50 Mbps, it has the most coverage. Internet of Things (IoT), industry automation, and business-critical use cases are likely to have low band utilization. The bands 600 MHz, 700 MHz, 800 MHz, 900 MHz, 1.5 GHz, 2.1 GHz, 2.3 GHz, and 2.6 GHz are taken into account for these applications [1].

DOI: 10.1201/9781032625034-12

However, due to reduced bandwidths, the 5G network's speed and latency will only be somewhat better than what is offered by 4G networks. The performance of the 5G network will be influenced by how close your cell site is. Low-band spectrum does, however, make it simpler for the wireless signal to pass through walls and windows.

12.2.2 Sub-6 GHz Band

The phrase 'sub 6 GHz' is used to describe the portion of the radio spectrum with frequencies lower than 6 GHz. For the delivery of wireless services like 4G, LTE, 5G, and Wi-Fi, this frequency band is employed. For 5G, mid-band spectrum (1 GHz to 6 GHz) is ideal, since it can transmit an enormous amount of data over considerable distances [2]. The 3.3 GHz to 3.8 GHz band of spectrum is excellent, according to the GSMA.

While millimetre wave frequency bands are capable of providing better speed downlink and have a smaller geographical coverage area, the sub-6 GHz band is unable to do so, but it has a wider geographical coverage area. Most researchers opt for sub-6GHz for outdoor signal coverage and straightforward antenna placement because millimetre waves span shorter distances. Since 3.5 GHz provides wider geographical coverage area, this frequency range is suitable for the implementation purposes in the real world.

Commonly used sub-6GHz band frequencies include:

- n1: 2.1 GHz
- n2: 1.9 GHz
- n3: 1.8 GHz
- n5: 850 MHz
- n77: 3.7 GHz
- n78: 3.5 GHz
- n79: 3.8 GHz

FIGURE 12.1 Representation of 5G frequency range.

Source: [3].

12.2.3 High Band

Higher-frequency radio bands between 24 GHz and 40 GHz are referred to as mmWave. Signals cannot go as far in the millimetre wave (high-band) spectrum as they can in the mid- and low-band spectra. The signal can occasionally travel less than a mile and can be more affected by obstructions like glass, trees, and even buildings. However, the advantage of millimetre wave spectrum is that consumers can obtain connection rates of 1 Gbps to 3 Gbps or even higher, provided the transmission is unobstructed [4].

- n257: 26 GHz
- n260: 28 GHz
- n261: 39 GHz
- n262: 41 GHz
- n263: 42 GHz
- n264: 43 GHz
- n265: 45 GHz

12.3 MICROSTRIP ANTENNA

12.3.1 Flexible Antenna

The principal structure of the flexible microstrip antenna is similar to that of a standard microstrip antenna as far as its design is concerned. A metal patch printed on a flexible substrate is the radiating element. The substrate is also printed with the ground plane, and it is usually located below the radiating element. Substrates are the major differences between a flexible microstrip antenna and a conventional microstrip antenna. Flexible substrates are made of materials having a lower dielectric constant, and this allows the antenna to be more compact. The antenna is also becoming more efficient because of the lower dielectric constant, so that it is able to reach a higher level of power [5]. In comparison to traditional microstrip antenna, flexible microstrip antennas have a number of advantages. They offer the

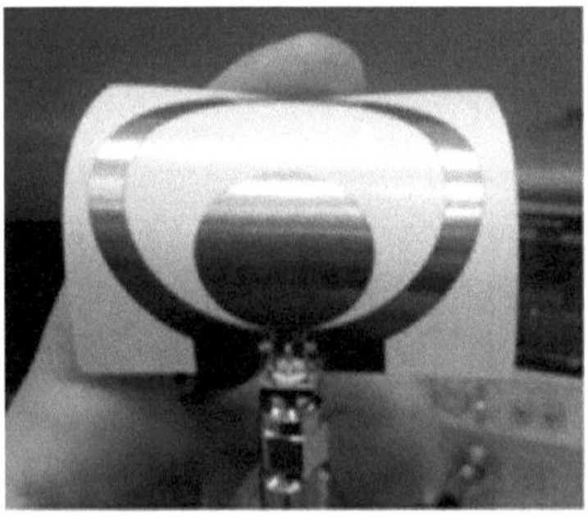

FIGURE 12.2 Flexible antenna.

Source: [5].

FIGURE 12.3 Flexible reconfigurable antenna.

Source: Reprinted by permission from *Flexible and Printed Electronics*, 'Low-profile flexible frequency-reconfigurable millimetre-wave antenna for 5G applications', Jilani, S. F.; Rahimian, A.; Alfadhl, Y.; Alomainy, A. Copyright 2018.

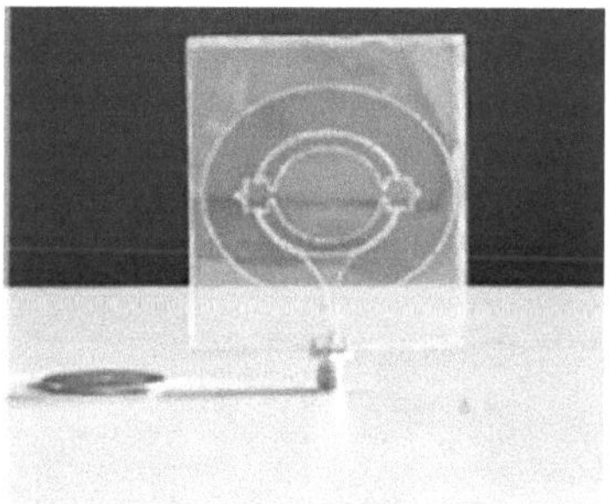

FIGURE 12.4 Flexible CPW-fed transparent antenna.

Source: Reprinted by permission from *Microwave and Optical Technology Letters*, 'Flexible CPW fed transparent antenna for WLAN and sub-6 GHz 5G applications', Arpan Desai, Trushit Upadhyaya, Jay Patel, Riki Patel. Copyright 2020.

flexibility and ease of bending or folding, enabling them to be applied in a variety of applications. Furthermore, compared to conventional microstrip antennas, it is more efficient in terms of energy efficiency and can transmit power.

12.3.2 Transparent Antenna

Transparent flexible antennas are an exciting new technology with a range of potential applications. Transparent flexible antennas are an antenna that is transparent to visible light and flexible at the same time. This type of antenna is made possible by the use of transparent conductive materials and are made up of sheets of conductive oxide and substrates which are flexible or non-flexible, such as indium tin oxide (ITO) or silver nano-oxides [7]. They have high electrical conductivity, which enables them to function as a beamforming element of the antennas, and are

FIGURE 12.5 Transparent antenna.

Source: [8].

also transparent with visible light, allowing an antenna to be seen through [8]. In order to provide wireless communications without disrupting the view, they may be placed in windows, displays and other transparent surfaces. In addition, to provide wireless connectivity without any degradation of appearance, they may be used in wearable devices, like smart glasses and contact lenses. An active field of research is the development of transparent flexible antennas. The researchers are working to increase the performance of this antenna, making it more efficient and extending its range.

12.4 FLEXIBLE MIMO IN 5G

MIMO antennas, which are designed for use in IEEE 802.11n wireless networks, have a minimum of two antennas in an individual physical structure. When compared to one antenna with the same radio transmission capacity, employing a large number of antennas increases data speed and range. The combination of the MIMO technology that contributes to enhanced reliability as well as helping in expanding communication coverage for the set of given metrices can achieve higher data rates. By combining the advanced MIMO and frequency configuration technologies, the conformal antenna is capable of providing frequency selection and multichannel transmissions [9]. Currently, there are a variety of advantages resulting in the widespread application of 5G MIMO antennas, such as increased data rate, beam directing, diversity, limited latency, and decreased multipath error. Mutual coupling of elements to provide a wide range of diversity characteristics is essential for the MIMO antenna design. It should have a value of less than –15dB in practical use. Several performance metrics, including the envelope correlation coefficient (ECC), channel capacity loss (CCL), and mean effective gain (MEG), have been used to assess the MIMO diversity performance. In contrast to the vertical placement of the antennas adjacent to each other in a vertical manner, the back coupling with a small separation distance allows less field interaction and yields to change in the phases of the intersected surface currents, which reduces the mutual coupling [10].

FIGURE 12.6 Decagon-shaped MIMO.

Source: [9].

FIGURE 12.7 Optically transparent MIMO.

Source: [10].

12.4.1 Performance Indicators of the MIMO Antenna

The indicators of performance for the MIMO antenna are as follows [11].

12.4.2 Envelope Correlation Coefficient

ECC stands for 'envelope correlation coefficient' in MIMO. It is a measurement of the correlation between envelopes containing signals that have been received in different MIMO antenna elements. A high ECC indicates that the received signals are highly correlated, which can reduce the diversity gain and throughput of the MIMO system. ECC is calculated using the following equation (1) [11]:

$$\rho_{e(i,j)} = \frac{\left|S_{ii}^{*}S_{ij} + S_{ji}^{*}S_{jj}\right|^{2}}{\left(1-\left|S_{ii}\right|^{2}-\left|S_{ji}\right|^{2}\right)\left(1-\left|S_{jj}\right|^{2}-\left|S_{ij}\right|^{2}\right)} \tag{12.1}$$

In MIMO systems, low ECC is desirable. It is technically hard to achieve zero ECC, but values below 0.5 are generally considered acceptable.

12.4.3 Total Active Reflection Coefficient

The total active reflection coefficient (TARC) is the amount of power reflected in a MIMO antenna array. A high TARC indicates that a significant amount of power is being reflected, which may reduce the efficiency of the MIMO system. The TARC is linked to the ECC, but it measures reflection of MIMO antennas in a broader way. The correlation between the reflected waves and individual antenna elements shall be considered by ECC, whereas TARC takes into account an overall amount of radiated power. TARC is calculated using the following equation (2) [11]:

$$TARC = \sqrt{\frac{((\left|s_{11} + s_{12}e^{j\theta}\right|^2) + (\left|s_{21} + s_{22}e^{j\theta}\right|^2)}{2}} \tag{12.2}$$

For MIMO systems, it is desirable to have a TARC of less than 2. It is difficult in practice to reach zero TARC, but values falling within the −10 dB range are generally accepted.

12.4.4 Diversity Gain

The diversity gain of MIMO antennas is an improvement in signal-to-noise ratio achieved by using multiple antenna elements at the transmitter and receiver. This is a measure of the system's ability to reduce the impact of fading.

$$DG = 10\sqrt{1-(ECC)^2} \tag{12.3}$$

Equation (3) is used for the calculation of diversity gain for MIMO [11]. The ideal value of diversity gain is approximately equal to 10. There are many factors which affect the diversity gain of MIMO antennas, including number of antenna arrays, their spacing and polarization, as well as surrounding environment and also mutual coupling between the antennas.

12.4.5 Mean Effective Gain

An antenna's capacity to improve the signal-to-noise ratio (SNR) of a wireless communication link is measured by the mean effective gain (MEG) of a MIMO antenna. It is computed by averaging the gain in each direction, that is, elevation and azimuth angles of the antenna. A higher SNR and more data transmitted over longer distances can be achieved by MIMO antenna with a high MEG. The ideal values of MIMO antenna are –3dB ≤ MEG ≤ –12dB [11].

$$MEG_i = 0.5\left[1 - \left|s_{ii}\right|^2 - \left|s_{ij}\right|^2\right] \tag{12.4}$$

$$MEG_j = 0.5\left[1 - \left|s_{ij}\right|^2 - \left|s_{jj}\right|^2\right] \tag{12.5}$$

S-parameters are utilized to calculate MEG, as shown in equations (4) and (5).

12.4.6 Channel Capacity Loss

The loss of channel capacity for MIMO antennas is due to nonideal antenna characteristics, which have resulted in reduced maximum achievable data rate for a MIMO system. It is a measure of the performance degradation of the MIMO system due to factors such as mutual coupling between the antennas, envelope correlation coefficient, and total active reflection coefficient.

$$C_{loss} = -log_2 \det\left(\psi^R\right) \tag{12.6}$$

where $\psi^R = \begin{bmatrix} \rho_{11} & \rho_{12} \\ \rho_{21} & \rho_{22} \end{bmatrix}$, $\rho_{11} = 1 - \left(\left|s_{ii}\right|^2 + \left|s_{ij}\right|^2\right)$, and $\rho_{ii} = -\left(s_{ii}^* s_{ij} + s_{ji}^* s_{ij}\right)$ for i, j = 1 or 2. The standard value of channel capacity loss is less than 0.4 bits/s/Hz [11].

12.4.7 Branch Power Ratio

The branch power ratio (BPR) of a MIMO antenna measures how much power is received at the strongest antenna element in relation to the weakest antenna element [12]. This is a measure of the uneven distribution of power between the antenna elements.

$$K = \frac{P_{min}}{P_{max}} \tag{12.7}$$

The value of K should be more than −3 dB. A very uneven distribution of power between the antenna elements is indicated by a high BPR, which may reduce the diversity gain and the speed of the MIMO system.

12.5 MATERIALS FOR FLEXIBLE ANTENNA

12.5.1 Selection of Materials for Flexible Antenna

For designing a flexible antenna, two types of materials are used for the design. They are *substrates* and *conductive material.* When selecting the materials for antenna, it should be selected in such a way that the antenna must perform without any deteriorations. The substrate material's dielectric constant is expected to range between $2.2 \leq \varepsilon_r \leq 12$ [13].

12.5.1.1 Dielectric Material

Substrates with low loss tangent dielectric should be chosen to reduce the dielectric losses. The characteristics of ideal microwave substrates are low relative permittivity, low dielectric loss, good flexibility, high thermal conductivity, low temperature coefficient of the resonant frequency, and low moisture absorption of the material. Thicker substrates increase radiation losses, and the surface waves are also more serious. It is observed that the smaller height of the dielectric substrate suppresses the higher mode and reduces radiation losses. The characteristics of the substrate are decisive in determining the flexibility of a thin microstrip antenna. The substrate shall be less flexible than the substrate having a low Young's modulus. The flexibility of the antenna is also influenced by the thickness of the substrate. The thin substrate will be more flexible, as opposed to the thicker substrate.

12.5.1.1.1 Polyimide

Polyimides are polymer-based substrates with low dielectric loss and mechanical robustness. Polyimides have high glass transition temperature. To incorporate conformal applications, MIMO antenna in flexible polyimide film is implemented [14]. The choice of polyimide Kapton substrate was made due to a good balance between its properties in terms of hardness, chemical, and electrical characteristics. In the whole range of frequencies, it has a relatively low loss tangent. Polyimide is a flexible and biocompatible material [15]. In conventional and bending conditions, the polyimide-based antenna is capable of satisfactory gain, bandwidth, and efficiency.

12.5.1.1.2 Polyethylene Terephthalate (PET)

Compared to other flexible materials, PET sheets are durable and have a relatively low tangent loss. In order to allow integration on both planar and non-planar surfaces, PET provides a low profile, low weight, compact and conformal two-element MIMO assembly.

In [16], PET with a height of 0.125 mm is used for mobile 5G applications, and it showed 60% of fractional bandwidth. They have outstanding conformal behaviour and mechanical stability and also used in high-temperature conditions, which are constrained by their low glass transition temperature.

12.5.1.1.3 Melinex

Melinex is the flexible substrate that emerges from PET which has high electrical conductivity and high dimensional stability. Because of its thin and flexible properties, Melinex is also a good candidate for antenna substrates. It also provides for the creation of conformable antenna that can be applied on a curved surface. Melinex has the optical transparency of 81% [17].

FIGURE 12.8 Flexible CP diversity antenna using Kapton polyimide.

Source: Reprinted by permission from Elsevier, *International Journal of Electronics and Communications*, 'Flexible CP diversity antenna for 5G cellular Vehicle-to-Everything applications', Sateesh Virothu, M. Satya Anuradha. Copyright 2022.

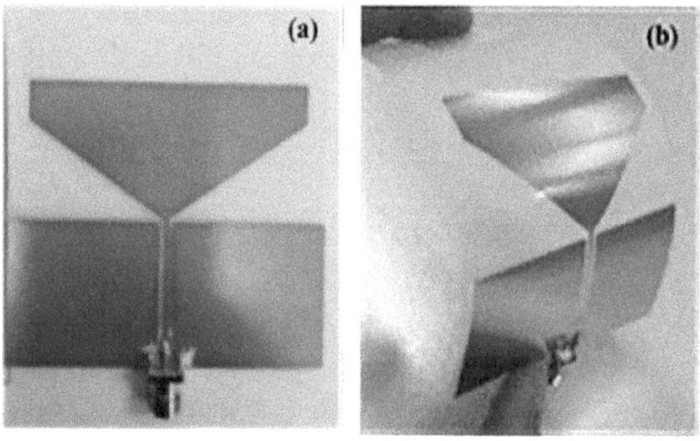

FIGURE 12.9 (a) Wideband transparent PET antenna. (b) Flexible wideband transparent antenna using PET.

Source: Reprinted by permission from *IEEE Access*, 'Dual Polarized, Multiband Four-Port Decagon Shaped Flexible MIMO Antenna for Next Generation Wireless Applications', Jayshri Kulkarni, Abdullah G. Alharbi, Chow-yen-Desmond Sim, Issa Elfergani, Jaume Anguera, Chemseddine Zebiri, Jonathan Rodriguez. Copyright 2022.

FIGURE 12.10 (a) Flexible MIMO wideband antenna using Melinex. (b) Flexible transparent Melinex antenna while bending.

Source: [17].

12.5.1.1.4 PDMS (Polydimethylsiloxane)

PDMS is a flexible polymer substrate based on high conformity and flexibility combined with low Young modulus, which results in an overall reduction of radiation efficiency. For its mechanical flexibility, water resistance, and inherent chemical stability, PDMS has been incorporated [18]. The antennas with flexible substrates are widely used for body-worn applications. As these applications work with low input power, the emission is comparatively low, and the effect on the human body is limited. Therefore, for the purpose of developing antennas in body-worn applications, it is necessary to use flexible PDMS [19].

12.5.1.1.5 LCP (Liquid Crystal Polymer)

LCP is a thin film substrate with low dielectric loss and thermal coefficient, which makes it attractive to high-frequency flexible antennas due to its ability to keep temperatures up to 30°C while also having less moisture absorption. Its dielectric constant is stable across a broad frequency range and can be fabricated easily.

The LCP structure is compact and simple, and the antenna has good gain and good radiation, even in different bending conditions. LCP is extremely thin, has better dimensional stability, and is capable of bending and stretching compared to traditional rigid dielectric substrates [20].

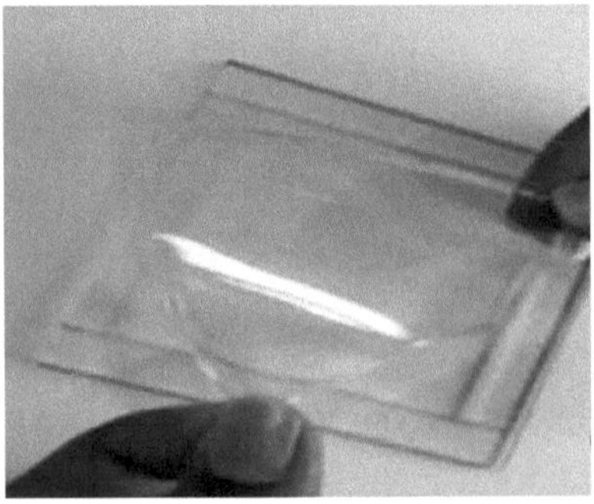

FIGURE 12.11 PDMS film.

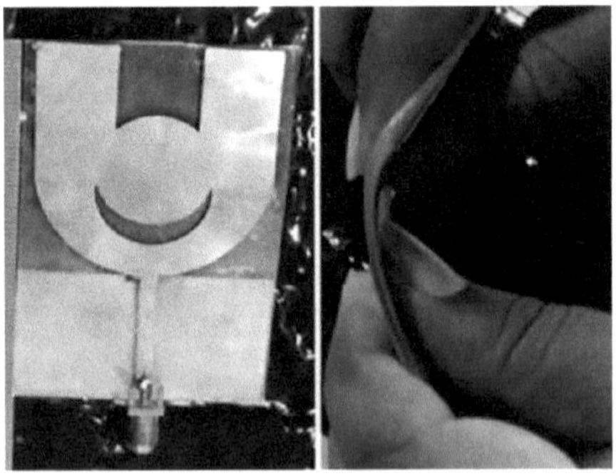

FIGURE 12.12 Flexible antenna using PDMS.

Source: Reprinted by permission from *Wireless Personal Communications*, 'Design and Development of Flexible PDMS Antenna for UWB-WBAN Applications', Doondi Kumar Janapala, M. Nesasudha, T. Mary Neebha, Raj Kumar. Copyright 2022.

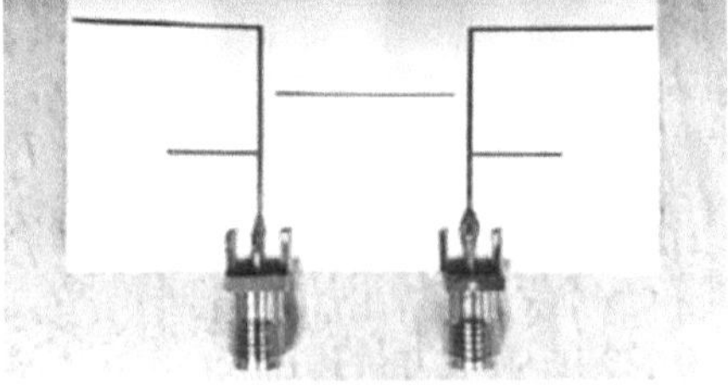

FIGURE 12.13 LCP-based flexible antenna.

Source: [20].

12.5.1.1.6 Paper

Paper substrate is low-cost, flexible, and environmentally protected [21]. Even though it is low-cost and flexible, it leads to impedance mismatch and inefficient efficiency of antenna.

12.5.1.1.7 Polyethylene Naphthalate (PEN)

Polyethylene naphthalate (PEN) is a thermoplastic polymer that has been shown to be a promising substrate for flexible antennas. It has excellent mechanical properties, including high rigidity and durability, making it a low-cost material. PEN also resists absorption of water and chemical substances, which makes it suitable for a broad range of applications [22].

The fact that it has a lower dielectric constant than the conventional substrates is one of the challenges in using PEN as an antenna substrate. The design of antenna that has a good impedance match will be hampered by this. However, it is possible to

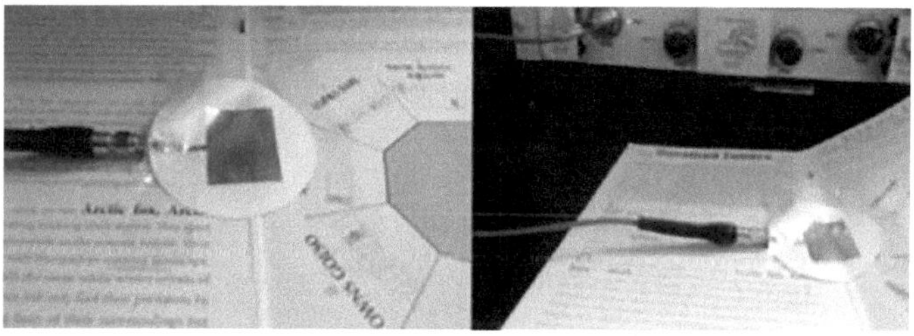

FIGURE 12.14 Paper-based eco-friendly antenna.

Source: Reprinted by permission from Elsevier, *Extreme Mechanics Letters*, 'Flexible, eco-friendly and highly sensitive paper antenna based electro mechanical sensor for wireless human motion detection and structural health monitoring', Srinivasulu Kanaparthi, Veerla Raja Sekhar, Sushmee Badhulika. Copyright 2016.

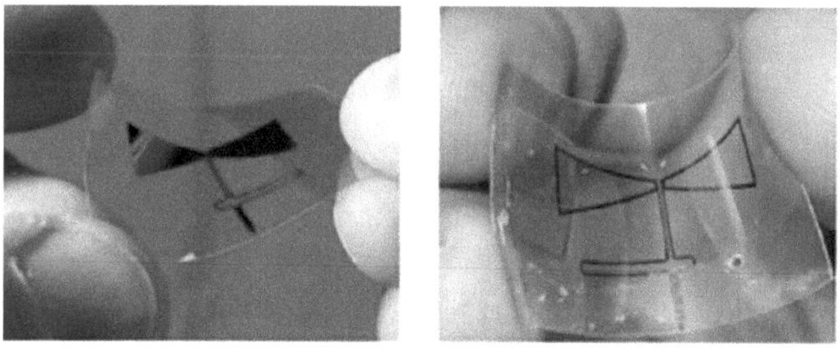

FIGURE 12.15 Flexible bow-tie PEN-based antenna.

Source: Reprinted by permission from *IEEE Transactions on Antennas and Propagation*, 'Design, Simulation, Fabrication and Testing of Flexible Bow-Tie Antennas', Ahmet Cemal Durgun, Constantine A. Balanis. Copyright 2011.

FIGURE 12.16 Flexible FR-4-based MIMO antenna.

overcome this problem through several techniques, including the use of more robust PEN substrates or a metamaterial.

12.5.1.1.8 FR-4 (Flame Retardant-4)

FR-4 is a more advanced technique, which can be easily achieved at lower costs. Compared to a large number of different flexible substrate materials, FR-4 also exhibits better electric and physical properties [23].

They are not as flexible as other substrate materials, so it can be harder to sort them out. In addition, in applications where strong flexibility or durability is needed, it will not be possible to use FR4 flexible substrates.

12.5.1.2 Conductive Material

In the performance of flexible antenna, conductive materials play an important role. Accordingly, it is necessary to select materials with high electrical conductivity, resistance, and good mechanical stability when designing the antenna. Materials with high mechanical stability withstand bending, twisting, and other mechanical deformations. Materials with good electrical properties achieve high gain, good bandwidth, and high radiation efficiency, and they also withstand mechanical properties, such as twisting, bending, and mechanical stresses.

12.5.1.2.1 Silver Nanoparticle

The most expensive conducting material, silver nanorod, has a high electrical conductivity. In order to prevent oxidation of silver nanoparticles, they are coated with an organic polymer layer and dispersed in an inert solvent, and the conductivity of the material depends on the number of printing iterations, the thickness of the layer, and the duration of the curing and sintering process. Silver nanoparticles are used to print the patch and feed, which improves the bandwidth, and silver nanoparticles achieve a consistent and continuous pattern [24].

12.5.1.2.2 Copper Nanoparticle

Copper (Cu) nanoparticles possess decent electrical conductivity, good mechanical stability, with low surface resistance. But these nanoparticles have less radiation efficiency [25].

12.5.1.2.3 *Silver Tin Oxide*

Silver tin oxides are conductive oxide sheets or fluid which have transparent property and have acceptable value of conductivity and low resistivity. AgHT-4 and AgHT-8 are the alternative for Indium, which is a rare earth metal and is in the verge of extinction. AgHT-4 has the optical transparency of 70% and is used to design conductive patch and ground plane, and it has low radiation efficiency, which causes more loss in power in non-radiating layers [26]. It is observed that upon exposure to hot soldering process, AgHT-8 loses its conductive property, and to prevent AgHT-8 from reducing conductivity, a conductive graphene silver adhesive is used [27].

12.5.1.2.4 *Graphene*

Graphene provides high electrical conductivity and sheet resistance [28]. But the fabrication of graphene is expensive. High reliability and quick data transmission are encouraged by graphene's flexibility and biodegradability.

12.5.1.2.5 *Silver Nanowires*

They are mechanically durable and robust, but silver nanowires have low durability and high roughness [29].

12.6 PARAMETERS TO BE CONSIDERED FOR FLEXIBLE ANTENNA

12.6.1 Return Loss

The reflection coefficient is one of the most important considerations in determining whether or not to make an adaptable antenna. It measures how much power is reflected off the antenna (also known as return loss, or gamma).

S_{11} = 0 dB if no power is radiated back and all power is reflected from the antenna [30]. The power emitted from port *i* to port *j* is measured as Sij. In order to maximize the input power and to ensure minimal power loss, the antenna's reflection coefficients must be better than or equal to −10 dB.

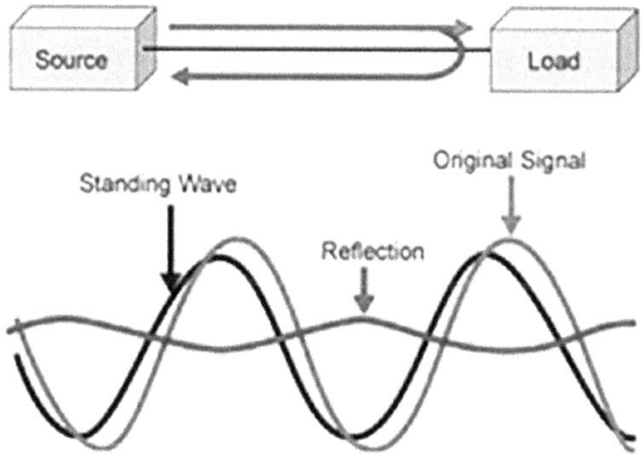

FIGURE 12.17 Representation of return loss.

$$Return\,loss = -20\log|\Gamma| \tag{12.8}$$

$$\Gamma = \frac{VSWR - 1}{VSWR + 1} \tag{12.9}$$

Where Γ denotes the reflection coefficient of the antenna [30].

12.6.2 VSWR

It stands for *voltage standing wave ratio.* It determines the manner in which successfully radio frequency energy is delivered from a power source to a load by means of transmission lines. The VSWR of the signal on the transmission line is a voltage ratio.

$$VSWR = \frac{|V_{max}|}{|V_{min}|} \tag{12.10}$$

where V_{max} denotes the signal's maximum voltage along the line, and V_{min} its minimum voltage. In the case of matching loads and transmission lines, Γ = 0, and VSWR = 1.0 [30] (or 1:1).

12.6.3 Radiation Efficiency

Antenna efficiency (or radiation efficiency) is the ratio of the power supplied to the antenna to the power it emits. In an anechoic chamber, where the antenna is powered up and the strength of the electromagnetic field it radiates into the surrounding environment is measured, the antenna's efficiency is often evaluated.

$$\text{Radiation efficiency} : \eta_r = \frac{P_{rad}}{P_{rad} + P_{loss}} \tag{12.11}$$

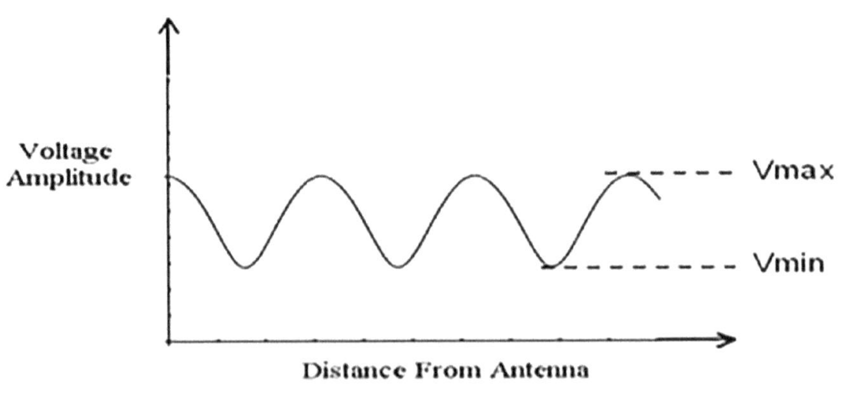

FIGURE 12.18 Representation of VSWR.

An ideal antenna has a 100% antenna efficiency, or it transmits all the power provided to it. However, in fact, a good antenna only transmits between 50% and 60% of the power provided to it. Another common unit of measurement for antenna efficiency is decibels (dB); an efficiency of 0.1 is 10% or (−10 dB), while an efficiency of 0.5 or 50% is −3 dB [31].

12.6.4 Gain

The performance of the flexible antenna shall be evaluated by gain. Generally, 3 dBi gain is needed in 5G applications. Using multiple antenna arrays and MIMO elements can maximize the gains of flexible antennas [32].

$$G_{dB} = 10log_{10}\left(\frac{4\pi\eta A}{\lambda^2}\right) \tag{12.12}$$

where antenna gain is denoted by G_{dB}, η is the efficiency, A is the physical aperture area, λ is the wavelength of the signal.

12.6.5 Radiation Pattern

As a function of the direction in which it is directed, an antenna's radiation pattern indicates how much energy flows into space. It is a three-dimensional pattern, but it is often depicted in two-dimensional plots, showing the radiation pattern in the azimuth or horizontal or H-planes and elevation or vertical or E-planes [33].

The E-plane is the plane that includes the antenna's radiated wave's electric field vector, and the H-plane is the plane that contains the magnetic field vector. At the antenna's far field, which is the area where the radiation pattern of the antenna is no longer influenced by the antenna's near field, the E-plane and H-plane radiation patterns of an antenna are commonly measured.

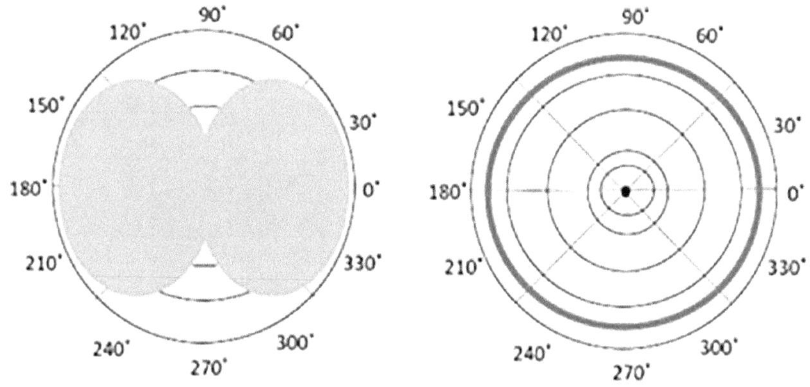

FIGURE 12.19 Representation of H-plane and E-plane.

Source: [33].

An antenna's coverage area, gain, and polarization can be calculated from its E- and H-plane radiation patterns. The area of space within which an antenna can send or receive signals at a specific power level is known as the coverage area of the antenna. An antenna's gain is a measurement of how much of the radiated energy it concentrates in a specific direction. The direction of the electric field vector of an antenna's radiated wave is known as polarization.

There are three main types of antenna radiation patterns:

- **Isotropic radiation pattern.** An isotropic antenna radiation pattern is a theoretical pattern that radiates energy equally in all directions. Although an isotropic radiation pattern is impossible for a real antenna to have, some antennas, such omnidirectional, do have radiation patterns that are nearly isotropic [34].
- **Omnidirectional radiation pattern.** An omnidirectional antenna wave pattern radiates energy in any horizontal direction, but it can vary depending on its vertical orientation [34]. Omnidirectional antennas are often used for applications such as broadcasting and cellular base stations.
- **Directional radiation pattern.** A directional antenna radiation pattern radiates energy in a specific direction or directions. Directional antennas are often used for applications such as satellite communication and point-to-point microwave links [34].

In designing and deploying wireless communications systems, the antenna radiation pattern is an important parameter to take into account. By choosing the right antenna for the application, it is possible to maximize the coverage and performance of the system.

The radiation characteristics of an antenna are measured using anechoic chamber. These chambers are particularly designed to absorb and reflect the radiation of electromagnetic waves. Pyramidal structures of carbon-impregnated foam, used as the absorbent material in the chamber, were measured in the anechoic chamber at the Antenna and Microwave Design Laboratory (AMD), SMVDU, Katra, J&K, India [35].

FIGURE 12.20 Anechoic chamber at Antenna and Microwave Design Laboratory.

Source: [35].

12.6.6 Bending Analysis

Bend analysis of the antenna is a key element for flexible antennas to function in all environments, even if their surfaces are not planar. The resonant frequency changes, in most cases resulting from a bending of the flexible antenna, changing its gain and radiation properties for specific applications. According to evaluations on the antenna's conformability and usability, it continues to function at bending angles lower than 160°, making it ideal for the majority of user case scenarios with strong adhesion and steady power transmission. Applications covered by the antenna's bandwidth include sub-6 GHz 5G mobile networks, WLAN, WiMAX, 2.45 GHz industrial, scientific, and medical applications, and 4G LTE telecommunications networks [36].

12.7 MINIATURIZATION OF FLEXIBLE ANTENNA

The task of miniaturizing flexible antennas is a difficult one, in view of the intrinsic distinction between size and performance. However, the use of a range of techniques that would not harm its performance could be achieved by limiting the size of flexible antennas. The particular use and desirable characteristics of the antenna shall be considered when selecting a miniaturization technique.

12.7.1 Metamaterial Surfaces

Metamaterials can be used to miniaturize antennas by exploiting their negative refractive index properties. Negative refractive index materials bend electromagnetic waves in the opposite direction to conventional materials, which allows them to be used to create smaller antennas that still have the same performance as larger

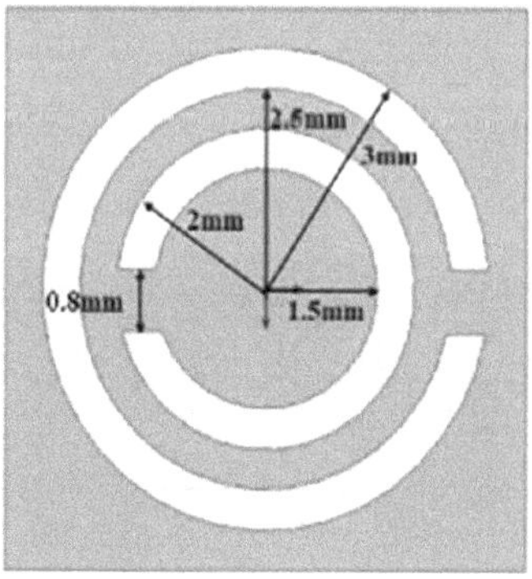

FIGURE 12.21 A circular SRR structure.

Source: Reprinted by permission from *IEEE Journal on Flexible Electronics*, 'A CPW-Fed Circular SRR-Inspired Flexible Antenna Using Polydimethylsiloxane (PDMS) Substrate for WLAN and WBAN Applications', Praveen Kumar Sharma, Navneet Gupta. Copyright 2022.

antennas. A reduction in the size of patch antennas may be achieved by using metamaterials split-ring resonators (SRR) [37].

Small resonant structures with a negatively refractive index at some frequencies are split-ring resonators. They can be used to increase the effective wavelength of the antenna, allowing it to be reduced, when they are placed on a patch antenna.

12.7.2 EBG Structures

EBG structures, which limit the propagation of electromagnetic waves at the required frequencies, can be applied for miniaturization of antennas. It is thus possible to reduce the size of antennas without sacrificing their performance [38]. By employing permittivity, electromagnetic bandgap structures can lower the antenna's physical surfaces while maintaining radiation efficiency. They may also reduce the mismatch that is caused by the structure. They also prohibit the propagation of electromagnetic waves, suppression of surface waves, low profile, high gain by acting as transmission/reflection surface.

EBG structures act as band-stop filter and suppress mutual coupling. EBG structures could be useful for miniaturizing antennas by providing a bandgap between the antenna's frequency of operation [39]. The design of an EBG structure that has a resonant frequency corresponding to the antenna's operating frequency could achieve miniaturization. This would cause the antenna to be unable to replicate at its intended frequency, resulting in a reduction of size.

12.7.3 PBG Structures

Photonic bandgap structures are three-dimensional structures that combine multi-layer metallic or tripod array with stacked EBG layers. The PBG substrate reduces the effect of surface wave loss, which leads to improved antenna electrical properties [40].

A photonic bandgap structure is a form of EBG structure designed specifically for controlling the propagation of electromagnetic waves in the optical spectrum.

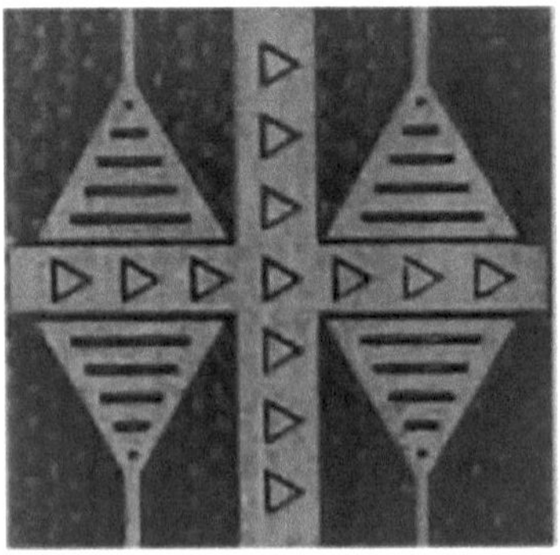

FIGURE 12.22 Prototype using EBG structures.

Source: [39].

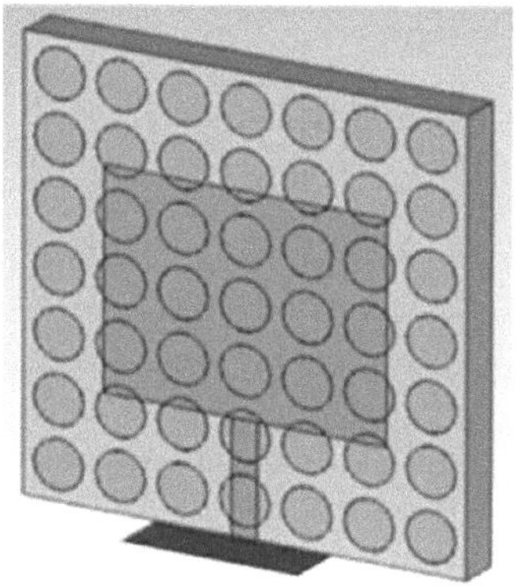

FIGURE 12.23 Microstrip patch using PBG.

Source: Reprinted by permission from *Photonic Network Communications*, 'Photonic band gap structure microstrip patch antenna for WiMAX and Wi-Fi application', Ritesh Sachan, Dhubkarya D. C. Copyright 2021.

They are made of tiny repeating unit cells arranged in a particular way. It is possible to achieve a bandgap in which the electromagnetic waves are not allowed to be transmitted, by arrangement of unit cells. These designs enhance gain and directivity while reducing the impact of radiation on the cylindrical curvature [41].

12.8 TECHNIQUES FOR FABRICATION

12.8.1 Inkjet Printing

A most promising fabrication technology for the fabrication of flexible antennas is inkjet printing technology. It is a flexible technology that enables the printing of many different conductive materials, such as copper, gold, and silver. Additionally, inkjet printing is a relatively inexpensive procedure, which makes it appealing for large production. Many research teams are attempting to enhance the performance of inkjet-printed antennas since inkjet printing is a rapidly developing technology. Inkjet technology is advancing as a result. Inkjet printing is a promising method for creating flexible antennas, with every aspect regarded. A number of flexible substrates can be printed on using this adaptable, affordable, and scalable technique. Printers such as Dimatrix materials printer (DMP-2831) [42], Dimatix DMP 2800 printer [43], Fujifilm Dimatix Materials Printer (DMP-2800) [44], and so on are used for inkjet printing of conductive materials over flexible substrates. Inkjet printing is anticipated to replace other fabrication techniques for flexible antennas as technology progresses.

12.8.2 3D Printing

A wide range of things, including antennas, are being created using 3D printing, a fast-growing technology. Flexible antennas may be produced using 3D printing in a range of sizes and shapes. This enables the antenna to be customized to the particular needs of the application [45, 46].

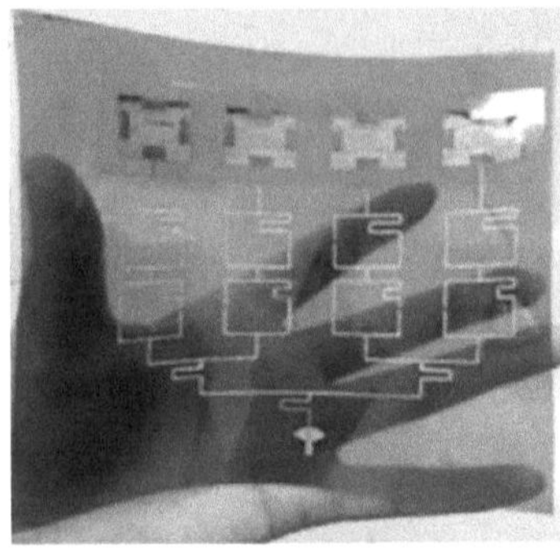

FIGURE 12.24 Inkjet-printed antenna.

Source: Reprinted by permission from *IEEE Transactions on Components, Packaging and Manufacturing Technology*, 'Inkjet Printing of Wideband Stacked Microstrip Patch Array Antenna on Ultrathin Flexible Substrates', Wen Tao Li, Yong Qiang Hei, Peter Mack Grubb, Xiao-Wei Shi, Ray T. Chen. Copyright 2018.

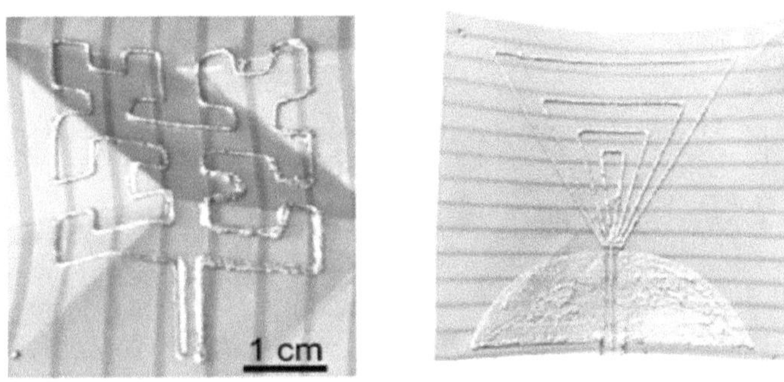

FIGURE 12.25 3D-printed antennas.

Source: Reprinted by permission from *Flexible and Printed Electronics*, 'Conformal 3D printing of non-planar antennas on wrinkled and folded Kapton films using point cloud data', Ezgi Kucukdeger, Yuxin Tong, Manjot Singh, Junru Zhang, Leon K Harding, Alejandro Salado, Steven W. Ellingson, Blake N. Johnson. Copyright 2021.

12.8.3 Screen-Printing

The printing method of screen-printing is flexible and can be used to produce a wide range of patterns and images. A variety of substrates, including flexible materials, can be printed on using this method because it is both scalable and affordable. Since screen-printing can be used to print conductive materials like silver, gold, and copper, it is a viable option for creating flexible antennas [47]. Through the use of a mesh screen, the conductive material is deposited onto a flexible substrate, resulting in a pattern of the material. Some of the advantages of screen-printing include low-cost process, mass production, and will be normally robust [48].

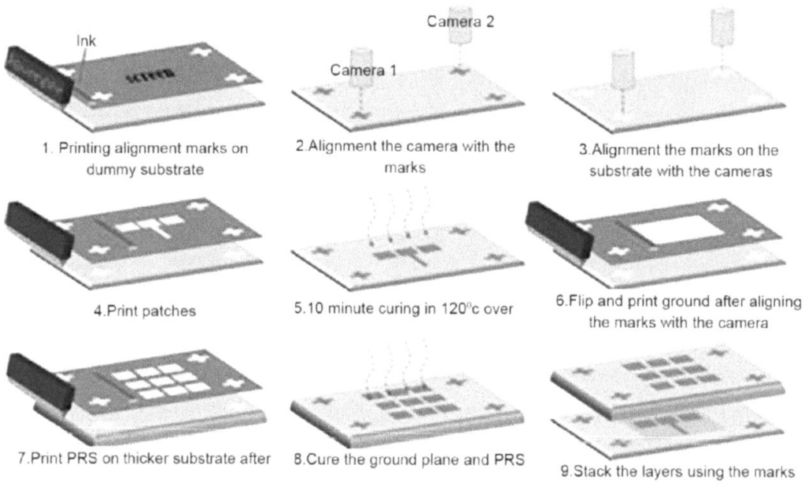

FIGURE 12.26 Screen-printing process.

12.8.4 Etching over Flexible Substrates

In order to remove material from the surface of a substrate, etching is a technique. It is a typical procedure in the production of printed circuit boards (PCBs), and flexible antennas can also be produced using it.

- **Chemical etching.** The material on the substrate is removed using a mixture of acids and bases, which make up a chemical solution in the process of chemical etching [49].
- **Plasma etching.** In plasma etching, the material is removed from the substrate by the use of plasma, which is produced when a high-frequency voltage is applied to a gas [50].
- **Ion beam etching.** This technique removes material from the substrate by using an ion beam which is accelerated by an electric field [51].

When etching flexible substrates, the following considerations should be made: the kind of the substrate material, the thickness of the substrate, and the antenna pattern that will be used for the design. Although etching flexible substrates might be difficult, it is an essential step in the creation of flexible antennas. One can successfully etch flexible substrates to produce high-quality antennas by paying close attention to the previously stated criteria. For instance, plasma etching is a suitable option for little features, whereas chemical etching is a good option for broad areas.

12.8.5 Embroidery

A relatively new technique which is gaining popularity due to its advantages over other fabrication techniques is the embroidery technique for flexible antennas. These

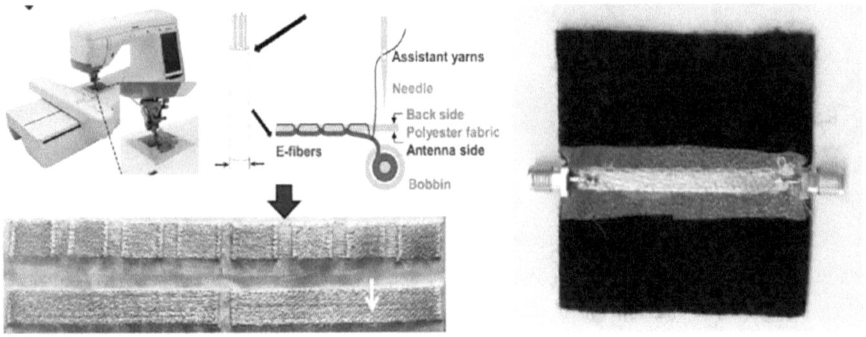

FIGURE 12.27 Embroidery process and antenna fabricated using embroidery technique.

Source: *IEEE Antennas and Wireless Propagation Letters*, 'E-Textile Origami Dipole Antennas with Graded Embroidery for Adaptive RF Performance', Saad Alharbi, Shreyas Chaudhari, Abdullahi Inshaar, Hamil Shah, Chengzhe Zou, Ryan L. Harne, and Asimina Kiourti. Copyright 2018.

advantages include the possibility to achieve high precision, the creation of complicated antenna geometries, a capability to manufacture rigid antennas that are resistant to deformation, and the ability to use a wide range of conductive materials, including silver, copper, and carbon nanotubes.

The method of embroidery involves the use of a computer-assisted embroidery machine to embroider conductive threads onto the textile substrate. The conductive threads are commonly made of silver or copper, and they have been selected due to their high conductivity and mechanical strength. In order to facilitate the folds and maintain RF functionality, the density of embroidery was decreased at the crease areas [52]. To maximize electrical performance and minimize fabrication complexity, the choice of stitch spacing is important [53].

Embroidery is a flexible method that can be used to produce different types of flexible antennas. It has been used in the production of antennas for wearables, medical implants, and environmental monitoring systems. The technique is in its very early development, but it is a potential game changer for the manufacture of bendable antennas.

12.9 APPLICATIONS FOR FLEXIBLE ANTENNA

Numerous applications for the flexible antennas exist at present in various fields. The following includes some of the applications for the flexible antennas.

The antenna should have a low gain, radiate in a wider angle, have an omnidirectional pattern, and be circularly polarized for cellular applications. The cellular applications will be in the range of 3.3 GHz–3.8 GHz [54]. For automotive applications, antenna should be circularly polarized. The DSRC and C-V2X works in the range of 5.8 GHz–5.9 GHz [55]. Flexible antenna systems are used to connect Internet of Things devices so they can be remotely monitored and controlled [56].

Flexible antennas are being used for improving the performance of satellite communication systems in aviation applications. Flexible antennas are being used for

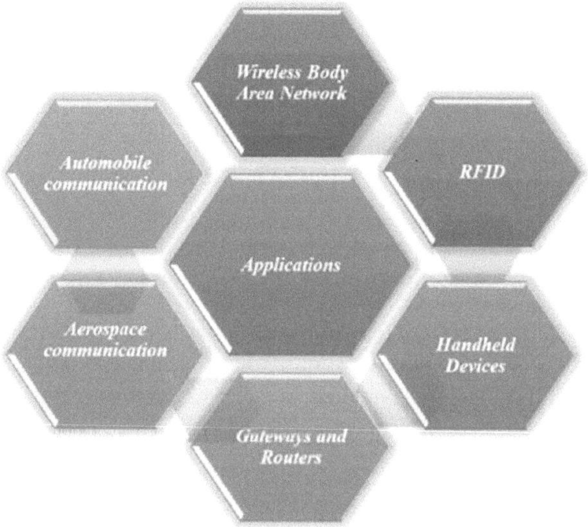

FIGURE 12.28 Applications of flexible antenna.

improving the performance of radar and communication systems in military applications. In defence and aeronautics applications, for example, radars or satellite communications, thin antennas are used to communicate and receive radio waves [57]. The antennas tend to be composed of strong and lasting materials, such as metal. With regard to transmission and reception of radio waves, flexibility antennas are used for vehicle applications, such as keyless entry systems or blind spot monitoring system. In particular, thin flexible materials such as plastic or metal are used for antennas. In wearable devices, such as smartwatches or fitness trackers, flexible antennas can be used. For sending and receiving data from the device to a user's mobile phone or another device, flexible antennas are applied. Flexible implantable antennas are used to monitor the patient's vital signs, for example, heart rate and blood pressure [58, 59].

12.10 CONCLUSION

Flexible antennas are an innovative, new technology that has a broad range of potential applications. They offer a number of advantages, notably minimal profile, low weight, and conformability, which are in comparison to traditional flexible antennas. It is foreseen that, as technology improves and makes these flexible antennas more accessible, smart patch antennas will have an increasingly important role in many sectors. Flexible antennas are going to be very bright in the future. They are on the verge of revolutionizing the way we interact with the world around us with continued research and development. In conclusion, flexible patch antennas represent a promising new technology that could have a wide range of applications. They are suitable for use in wearables, implantable, and other difficult environments because of a number of advantages compared to traditional rigid patch antennas.

REFERENCES

[1] www.celona.io/5g-lan/5g-bands

[2] www.gsma.com/spectrum/5g-spectrum-guide

[3] www.wilsonamplifiers.com/blog/5g-frequency-bands-explained

[4] Ramraj Dangi, Praveen Lalwani, Gaurav Choudhary, Ilsun You, Giovanni Pau, 'Study and Investigation on 5G Technology: A Systematic Review', Sensors, Vol. 26, pp. 1–32, 2022.

[5] Rafal Lech, Wojciech Marynowski, Adam Kusiek, Jerzy Mazur, 'Coplanar Waveguide-Fed Broadband Microwave Devices With (or Without) a Thin Dielectric Substrate for Use in Flexible Electronic Systems', International Journal of Antennas and Propagation, Vol. 2014, p. 379379, 2014.

[6] Syeda Fizzah Jilani, Ardavan Rahimian, Yasir Alfadhl, Akram Alomainy, 'Low-Profile Flexible Frequency-Reconfigurable Millimetre-Wave Antenna for 5G Applications', Flexible and Printed Electronics, Vol. 3, No. 3, pp. 1–8, 2018.

[7] Arpan Desai, Trushit Upadhyaya, Jay Patel, Riki Patel, 'Flexible CPW Fed Transparent Antenna for WLAN and Sub-6 GHz 5G Applications', Microwave and Optical Technology Letters, Vol. 62, pp. 2090–2103, 2020.

[8] Jorge Ivan Trujillo-Flores, Richard Torrealba-Meléndez, Jesús Manuel Muñoz-Pacheco, Marco Antonio Vásquez-Agustín, Edna Iliana Tamariz-Flores, Edgar Colín-Beltrán, ario López-López, 'CPW-Fed Transparent Antenna for Vehicle Communications', Applied Sciences, Vol. 10, pp. 1–10, 2020.

[9] Jayshri Kulkarni, Abdullah G. Alharbi, Chow-Yen-Desmond Sim, Issa Elfergani, Jaume Anguera, Chemseddine Zebiri, Jonathan Rodriguez, 'Dual Polarized, Multiband Four-Port Decagon Shaped Flexible MIMO Antenna for Next Generation Wireless Applications', IEEE Access, Vol. 10, pp. 128133–128150, 2022.

[10] Arpan Desai, Cong Danh Bui, Jay Patel, Trushit Upadhyaya, Gangil Byun, Truong Khang Nguyen, 'Compact Wideband Four Element Optically Transparent MIMO Antenna for mm-Wave 5G Applications', IEEE Access, Vol. 8, pp. 194206–194217, 2020.

[11] A. W. Mohammad Saadha, Kavitha Ashwathb, Poonkuzhali Ramaswamya, Tanweer Alic, Jaume Anguera, 'A Uniquely Shaped MIMO Antenna on FR4 Material to Enhance Isolation and Bandwidth for Wireless Applications', International Journal of Communications, Network and System Sciences, Vol. 123, pp. 1–13, 2020.

[12] Jamal Nasir, Mohd. Haizal Jamaluddin, Aftab Ahmad Khan, Muhammad Ramlee Kamarudin, Chee Yen Leow, Owais Owais, 'Throughput Measurement of a Dual-Band MIMO Rectangular Dielectric Resonator Antenna for LTE Applications', Sensors, Vol. 148, pp. 1–15, 2017.

[13] Constantine A. Balanis, Modern Antenna Handbook, Wiley, 2008.

[14] Sateesh Virothu, M. Satya Anuradha, 'Flexible CP Diversity Antenna for 5G Cellular Vehicle-to-Everything Applications', International of Electronics and Communications, Vol. 152, pp. 1–17, 2022.

[15] Farooq Faisal, Yasar Amin, Youngdae Cho, Hyoungsuk Yoo, 'Compact and Flexible Novel Wideband Flower-Shaped CPW-Fed Antennas for High Data Wireless Applications', IEEE Transactions on Antennas and Propagation, Vol. 67, No. 6, pp. 4184–4188, 2019.

[16] Mohammed Tighezza, S. K. A. Rahim, M. T. Islam, 'Flexible Wideband Antenna for 5G Applications', Microwave and Optical Technology Letters, Vol. 60, pp. 38–44, 2017.

[17] Arpan Desai, Merih Palandoken, Jayshri Kulkarni, Gangil Byun, Truong Khang Nguyen, 'Wideband Flexible/Transparent Connected-Ground MIMO Antennas for Sub-6 GHz 5G and WLAN applications', IEEE Access, Vol. 9, pp. 147003–147015, 2021.

[18] Syed Muzahir Abbas, Sharvil C. Desai, Karu P. Esselle, John Yiannis Volakis, Raheel M. Hashmi, 'Design and Characterization of a Flexible Wideband Antenna Using

Polydimethylsiloxane Composite Substrate', International Journal of Antennas and Propagation, Vol. 2018, p. 4095765, 2018.
[19] Doondi Kumar Janapala, M. Nesasudha, T. Mary Neebha, Raj Kumar, 'Design and Development of Flexible PDMS Antenna for UWB-WBAN Applications', Wireless Personal Communications, Vol. 122, pp. 3467–3483, 2022.
[20] Chengzhu Du, Xun Wang, Gaoya Jin, Shunshi Zhong, 'A Compact Tri-Band Flexible MIMO Antenna Based on Liquid Crystal Polymer for Wearable Applications', Progress in Electromagnetics Research, Vol. 102, pp. 217–232, 2021.
[21] Srinivasulu Kanaparthi, Veerla Raja Sekhar, Sushmee Badhulika, 'Flexible, Eco-Friendly and Highly Sensitive Paper Antenna Based Electro Mechanical Sensor for Wireless Human Motion Detection and Structural Health Monitoring', Extreme Mechanics Letters, Vol. 9, pp. 324–330, 2016.
[22] Ahmet Cemal Durgun, Constantine A. Balanis, 'Design, Simulation, Fabrication and Testing of Flexible Bow-Tie Antennas', IEEE Transactions on Antennas and Propagation, Vol. 59, No. 12, pp. 4425–4434, 2011.
[23] Yasin Kabiri, Alejandro L. Borja, James R. Kelly, Pei Xiao, 'A Technique for MIMO Antenna Design with Flexible Element Number and Pattern Diversity', Vol. 7, pp. 86157–86167, 2019.
[24] Haider Raad, 'An UWB Antenna Array for Flexible IoT Wireless Systems', Progress in Electromagnetics Research, Vol. 162, pp. 109–121, 2018.
[25] Daisuke Tomotoshi, Hideya Kawasaki, 'Surface and Interface Designs in Copper-Based Conductive Inks for Printed/Flexible Electronics', Nanomaterials, Vol. 10, pp. 1–53, 2020.
[26] Trushit Upadhyaya, Arpan Desai, Riki Patel, Upesh Patel, Kanwar Preet Kaur, Killol Pandya, 'Compact Transparent Conductive Oxide Based Dual Band Antenna for Wireless Applications', Progress in Electromagnetics Research Symposium, pp. 41–45, 2017.
[27] Arpan Desai, Trushit Upadhyaya, Riki Patel, 'Compact Wideband Transparent Antenna for 5G Communication Systems', Microwave Optical Technology Letters, Vol. 61, pp. 781–786, 2018.
[28] A. Scida, S. Haque, E. Treossi, A. Robinson, S. Smerzi, S. Ravesi, S. Borini, V. Palermo, 'Application of Graphene-Based Flexible Antennas in Consumer Electronic Devices', Materials Today, Vol. 21, No. 3, pp. 223–230, 2018.
[29] Tanminder Rai, Paolo Dantes, Behraad Bahreyni, Woo Soo Kim, 'A Stretchable RF Antenna with Silver Nanowires', IEEE Electron Device Letters, Vol. 34, No. 4, pp. 544–546, 2013.
[30] www.electronics-notes.com/articles/antennas-propagation
[31] www.antenna-theory.com/basics/efficiency.php
[32] www.geeksforgeeks.org
[33] https://edurev.in/t/245053/Radiation-Pattern
[34] Constantine A. Balanis, Antenna Theory, Analysis and Design, 3rd Edition, John Wiley & Sons, Inc, 2005.
[35] Kumud Ranjan Jha, Z. A. Pandit Jibran, Chitra Singh, Satish Kumar Sharma, '4-Port MIMO Antenna Using Common Radiator on a Flexible Substrate for Sub-1GHz, Sub-6GHz 5G NR, and Wi-Fi 6 Applications', IEEE Open Journal of Antennas and Propagation, Vol. 2, pp. 689–701, 2021.
[36] Muhammad Usman Ali Khan, Raad, Faisel Tubbal, Panagiotis Ioannis Theoharis, Sining Liu, Javad Foroughi, 'Bending Analysis of Polymer-Based Flexible Antennas for Wearable, General IoT Applications: A Review', Polymers, Vol. 13, pp. 1–34, 2021.
[37] Praveen Kumar Sharma, Navneet Gupta, 'A CPW-Fed Circular SRR-Inspired Flexible Antenna Using Polydimethylsiloxane (PDMS) Substrate for WLAN and WBAN Applications', IEEE Journal on Flexible Electronics, Vol. 1, No. 1, pp. 39–46, 2022.

[38] Anees Abbas, Niamat Hussain, Jaemin Lee, Seong Gyoon Park, Nam Kim, 'Triple Rectangular Notch UWB Antenna Using EBG and SRR', IEEE Access, Vol. 9, pp. 2508–2515, 2021.

[39] Ayman A. Althuwayb, Mohammad Alibakhshikenari, Bal S. Virdee, Nasr Rashid, Khaled Kaaniche, Ahmed Ben Atitallah, Ammar Armghan, Osama I. Elhamrawy, Chan Hwang See, Francisco Falcone, 'Metasurface-Inspired Flexible Wearable MIMO Antenna Array for Wireless Body Area Network Applications and Biomedical Telemetry Devices', IEEE Access, Vol. 11, pp. 1039–1056, 2023.

[40] Mohamed Nasr Eddine Temmar, Abdesselam Hocini, Djamel Khedrouche, Mehdi Zaman, 'Analysis and Design of a Terahertz Microstrip Antenna Based on a Synthesized Photonic Bandgap Substrate Using BPSO', Journal of Computational Electronics, Vol. 18, pp. 231–240, 2019.

[41] Ritesh Sachan, D. C. Dhubkarya, 'Photonic Band Gap Structure Microstrip Patch Antenna for WiMAX and Wi-Fi Application', Photonic Network Communications, Vol. 41, pp. 280–286, 2021.

[42] Isidoro Ibanez Labiano, Akram Alomainy, 'Flexible Inkjet-Printed Graphene Antenna on Kapton', Flexible and Printed Electronics, Vol. 6, pp. 1–8, 2021.

[43] Wen Tao Li, Yong Qiang Hei, Peter Mack Grubb, Xiao-Wei Shi, Ray T. Chen, 'Inkjet Printing of Wideband Stacked Microstrip Patch Array Antenna on Ultrathin Flexible Substrates', IEEE Transactions on Components, Packaging and Manufacturing Technology, Vol. 8, No. 9, pp. 1695–1702, 2018.

[44] Harish Subbaraman, Daniel T. Pham, Xiaochuan Xu, Maggie Yihong Chen, Amir Hosseini, Xuejun Lu, Ray T. Chen, 'Inkjet-Printed Two-Dimensional Phased-Array Antenna on a Flexible Substrate', IEEE Antennas and Wireless Propagation Letters, Vol. 12, pp. 170–173, 2013.

[45] Z. Hamouda, J. L. Wojkiewicz, A. A. Pud, L. Kone, B. Belaabed, S. Bergheul, T. Lasri, 'Dual-Band Elliptical Planar Conductive Polymer Antenna Printed on a Flexible Substrate', IEEE Transactions on Antennas and Propagation, Vol. 63, No. 12, pp. 5864–5867, 2015.

[46] Ezgi Kucukdeger, Yuxin Tong, Manjot Singh, Junru Zhang, Leon K. Harding, Alejandro Salado, Steven W. Ellingson, Blake N. Johnson, 'Conformal 3D Printing of Non-Planar Antennas on Wrinkled and Folded Kapton Films Using Point Cloud Data', Flexible and Printed Electronics, Vol. 6, pp. 1–15, 2021.

[47] Mohammad Vaseem, Zubair Akhter, WeiweiLi1, Emre Yarali, Thomas D. Anthopoulos, Atif Shamim, 'High-Conductivity Screen-Printable Silver Nanowire Ink for Optically Transparent Flexible Radio Frequency Electronics', Flexible Printed Electronics, Vol. 7, pp. 1–9, 2022.

[48] Azat Meredov, Kirill Klionovski, Atif Shamim, 'Screen-Printed, Flexible, Parasitic Beam-Switching Millimeter-Wave Antenna Array for Wearable Applications', IEEE Open Journal of Antennas and Propagation, Vol. 1, pp. 2–10, 2020.

[49] Nhan Ai Tran, Huy Nam Tran, Mau Chien Dang, Eric Fribourg-Blanc, 'Copper Thin Film for RFID UHF Antenna on Flexible Substrate', IOP Publishing Advances in Natural Sciences: Nanoscience and Nanotechnology, Vol. 1, pp. 1–6, 2010.

[50] A. Weber-Bargioni, A. Schwartzberg, M. Schmidt, B. Harteneck, D. Fogletree, P. J. Schuck, S. Cabrini, 'Functional Plasmonic Antenna Scanning Probes Fabricated by Induced-Deposition Mask Lithography', IOP Publishing, Nanotechnology, Vol. 21, pp. 1–6, 2010.

[51] B. Abasah, C. Santschi, T. V. Raziman, O. J. F. Martin, 'Fabrication of Plasmonic Structures with Well-Controlled Nanometric Features: A Comparison Between Lift-Off and Ion Beam Etching', IOP Publishing, Nanotechnology, Vol. 32, pp. 1–7, 2021.

[52] Saad Alharbi, Shreyas Chaudhari, Abdullahi Inshaar, Hamil Shah, Chengzhe Zou, Ryan L. Harne, Asimina Kiourti, 'E-Textile Origami Dipole Antennas with Graded

Embroidery for Adaptive RF Performance', IEEE Antennas and Wireless Propagation Letters, Vol. 17, No. 12, pp. 2218–2222, 2018.

[53] Sree Pramod Pinapati, Shengjian Jammy Chen, Joshua Brittain, Adrian Caldow, Christophe Fumeaux, 'Embroidered Ground Planes for Wearable Antennas', IEEE Transactions on Components, Packaging and Manufacturing Technology, Vol. 12, No. 6, pp. 1029–1039, 2022.

[54] Sateesh Virothu, M. Satya Anuradha, 'Flexible Circularly Polarized Antenna for LTEV2X/5G Vehicular Safety Applications', 2021 International Conference on Computing, Communication, and Intelligent Systems, pp. 814–821, 2021.

[55] Adamantia Chletsou, Yuxiao He, John F. Locke, John Papapolymerou, 'Multi-Band, Flexible, Lightweight Antenna on LCP for Automotive Applications', APS, pp. 1507–1508, 2020.

[56] Sharadindu Gopal Kirtania, Bachir Adham Younes, Abdul Rakib Hossain, Tutku Karacolak, Praveen Kumar Sekhar, 'CPW-Fed Flexible Ultra-Wideband Antenna for IoT Applications', Micromachines, Vol. 12, pp. 1–13, 2021.

[57] I. Naga Venkata Phani Kumar, K. Sk. Mujeeb Rahaman, K. Bhargav, 'Design of Flexible Antenna for S-Band Satellite Communication Applications', Journal of Emerging Technologies and Innovative Research, Vol. 10, No. 3, pp. 309–313, 2023.

[58] Kailash V. Karad, Vaibhav S. Hendre, 'A Flower Bud-Shaped Flexible UWB Antenna for Healthcare Applications', EURASIP Journal on Wireless Communications and Networking, Vol. 2023, p. 27, 2023.

[59] Mariella Särestöniemi, Marko Sonkki, Sami Myllymäki, Carlos Pomalaza-Raez, 'Wearable Flexible Antenna for UWB On-Body and Implant Communications', Telecom, Vol. 2, pp. 285–301, 2021.

13 Flexible MIMO Antennas for UWB Applications

G. Rajesh and R. Poonkuzhali

13.1 INTRODUCTION

The most contemporary international standard for wireless body area networks (WBAN) is IEEE 802.15.6. Many consumer electronics and real-time health monitoring applications are supported by WBAN. Its goal is to create a standard for low-power, short-range, and highly reliable wireless communication within the range of the human body [1]. The International Diabetes Federation (IDF) recently conducted a worldwide study which produced the results in advance of World Diabetes Day on Tuesday, 14 November. Seven out of ten people with diabetes (72%) only discovered they had the disease until experiencing difficulties. Furthermore, nearly all (94%) of the respondents reported having dealt with one or more diabetes-related issues at some point in their lives. People with diabetes from Africa, Asia, Europe, and South America participated in the study to learn more about the effect and knowledge of problems associated to diabetes [2]. Wearable monitoring technologies are an essential aspect of cost-effective preventive healthcare systems and significantly enhance the value of life through early identification of abnormal illnesses of human life [3]. Body area networks (BAN) have recently appeared due to increased demand for track and monitoring the individual's human medical conditions. BANs are an innovative generation of wireless sensor networks (WSN) and used for health monitoring systems. BAN uses nodes equipped with biomedical sensors, temperature sensors, motion detectors, and wireless communication devices. The nodes are located on human clothes, body, or implanted within the body. While a person is in regular activities, sensors or devices monitor physiological signals and record body health parameters, such as blood pressure, sugar level, glucose level, heartbeat rate, body temperature, and so on. The collected parameters from the human body are transmitted wirelessly and processed. These processed data are transferred to medical servers. A patient-specific individual monitoring system has replaced the conventional medical system, with recent advancements usually referred to as wireless body area networks (WBAN), which can be either wearable or implanted. WBAN is capable of several tasks, including collecting data from the human body using sensor nodes, data processing, and sensing. The primary target of WBANs is to protect people's health by transferring physiological information to medical servers via sensors on the body and allowing physicians to analyze the patient's health. The low-power portable and wearable devices are more popular nowadays to meet the requirements of consumer electronics and WBAN implementation. It is important to make wireless devices compact in size and flexible, which may be integrated into the human body and wearable devices. The microstrip patch antenna structure is well suited for

 DOI: 10.1201/9781032625034-13

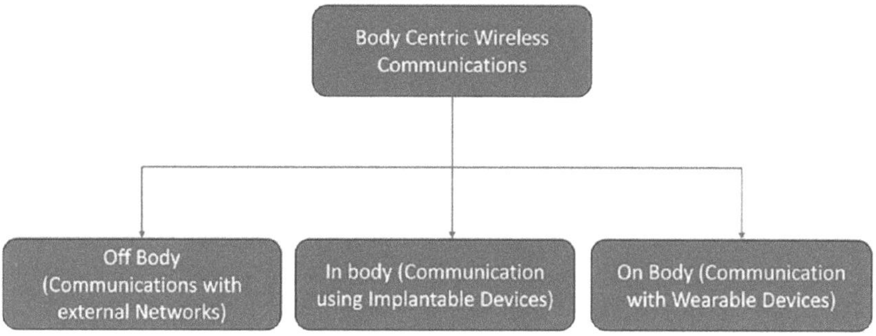

FIGURE 13.1 Different types of body-centric communication links.

wearable applications as it can be low-profile, conformal and easily integrated with those devices. As a part, the needs of flexible antennas have grown in recent years, particularly for biomedical applications [4].

Body-centric communication exists to define the different types of body area networks. To satisfy the real-world problems and wireless communication requirements, body-centric communication is classified into off-body, in-body, and on-body. Figure 13.1 depicts the different types of body-centric communication.

Based upon the location of the transmitter and the receiver, the antenna is placed to track and monitor healthcare applications; the different types of communications are defined. The description of the body-centric communication is depicted in Figure 13.2.

On-body communication describes the connection between devices worn on the body. The transmitter – Tx – and the receiver – Rx – antenna elements are installed on the body; communication occurs between the two antennas, with most of the channel located on the body's surface.

Off-body communication refers to the radio frequency (RF) connection between the devices worn on the body and the external base units placed in the environments.

In-body communication states communication between the implanted devices and on-body sensors. A notable amount of the channel is within the body, and implanted transceivers are employed. This form of connection has various medical uses, such as the identification of any diseases and abnormal conditions of the human body.

13.2 APPLICATIONS OF WBAN

The IEEE 802.15.6 is the most recent standard to the WBAN that allows for communication both inside and around the human body [5], and it can be utilized for a range of medical (health monitoring) and non-medical (consumer electronics) uses, like healthcare monitoring, lifestyle and entertainment, automatic operations, authentication, and security, respectively [3].

The standard that categorizes WBAN applications and some of the more significant applications are displayed in Figure 13.3.

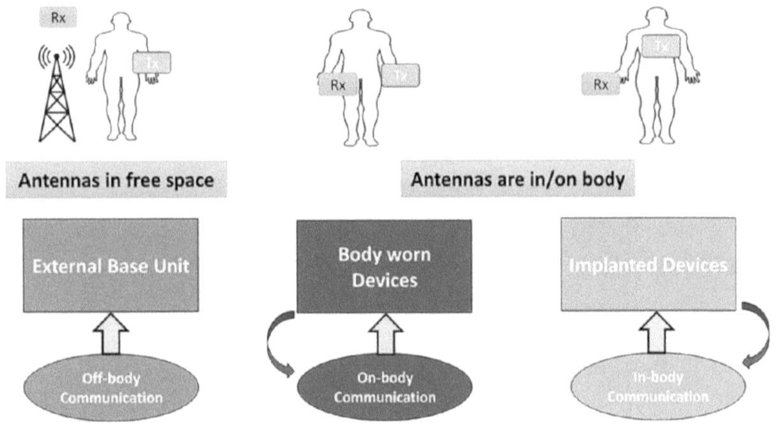

FIGURE 13.2 Description of body-centric communications.

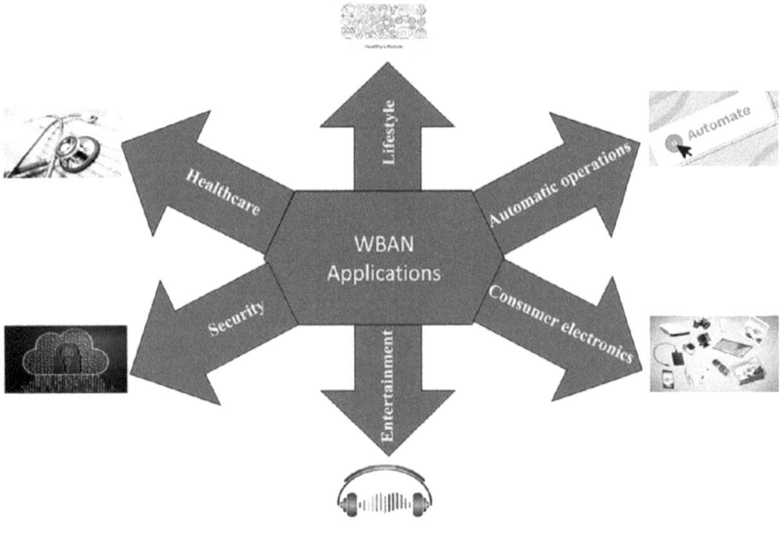

FIGURE 13.3 IEEE 802.15.6 – WBAN applications.

13.2.1 Medical Applications

Implantable/wearable devices are responsible for monitoring human health conditions. When an individual engages in daily activities, the sensors or devices monitor biological signals and record body health statistics, such as blood pressure, sugar levels, glucose level, heartbeat rate, body temperature, and so on [6]. All the sensors in the human body constitute a node. A collection of nodes like motion detectors, biological sensors in the human body connected from sensors, and a control device form the WBAN network. These nodes are implanted within the body or distributed across human body, woven with fabrics or body, which can be implemented in real-time telemedicine systems.

13.2.2 Non-Medical Applications

The non-medical applications of WBANs include consumer electronics, entertainment, automatic operations, security, and so on [3]. Entertainment, such as gaming, real-time streaming, also can be employed. Automatic operations system is responsible in making an efficient response in case any crisis occurs, such as fire or any disaster.

It is used in security-like authentication entry, defence security in particular, accessing the activities of defence personnel. This can be accomplished via WBAN sensors like cameras, biometric sensors, and GPS. Table 13.1 shows the WBAN applications utilized as medical (health monitoring) and non-medical (consumer electronics) categories. The common characteristics of WBAN applications are enhancing the user value of life [7].

The WBAN standard of IEEE 802.15.6 outlines medium access control (MAC) and supports three physical (PHYs) layers, namely, the narrow band (NB), human body communication (HBC), and ultrawideband communication (UWB) [1]. The WBAN frequency bands are depicted in Figure 13.4.

The HBC PHY layer employs a data connection through the human body, and it uses electric field communication (EFC) technology. It runs in two frequency bands, namely, centred frequency of 16 and 27 MHz with a bandwidth of 4 MHz [8].

The next approved band in WBAN is the NB PHY layer, which communicates between the devices on the body and implanted devices. It has followed the major three responsibilities, activation or inactivation of radio transceiver, clear channel assessment (CCA) inside the current channel, and collection of data transmission/reception. This layer holds medical implant communication services (MICS), licenced band, and is used for implanted communication in the frequency band of

TABLE 13.1
Categories of WBAN Applications

Category	Applications	Specific applications
Medical WBAN (wearable and implanted)	Healthcare	Glucose monitoring, ECG, blood sugar level, blood pressure monitoring, and so on
	Lifestyle	Activity tracking, sports, heart rate, tracking, and localization
Non-medical WBAN	Entertainment	Gaming and entertainment
	Consumer electronics	Smartphones, headphones, music player, laptops
	Automatic operations	Disaster response, fire sensors, and rescue operations
	Security	Authentication entry, defence security

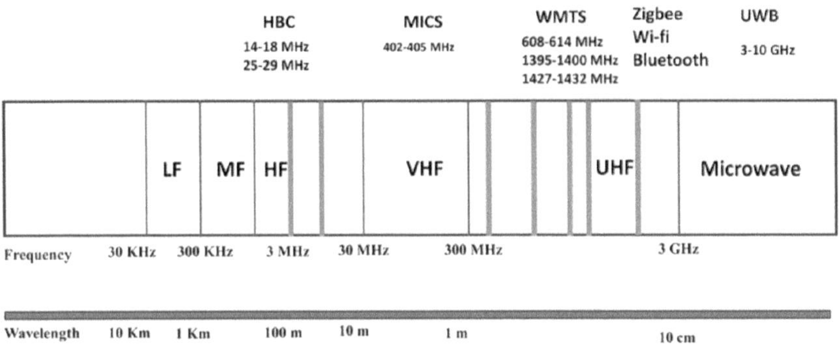

FIGURE 13.4 WBAN frequency bands.

402–405 MHz, with bandwidth of 300 KHz. The primary advantage of this layer is easy transmission into human tissue, but the disadvantages is its low bandwidth, which is inadequate for high-speed data transmission. Wireless medical telemetry system (WMTS) is a licenced band used for medical telemetry system. It employs frequency between 402 and 2483.5 MHz, with a bandwidth range of 300 KHz to 1 MHz. Data rate of this layer supports between 75.9 Kbps and 971.4 Kbps with receiver and is involved in different frequency bands: 608–614 MHz, 1,395–1,400 MHz, and 1,427–1,432 MHz. Both MICS and WMTS do not support high speed data rate applications.

In order to achieve high-speed data rate application, the industrial, scientific, and medical (ISM) are an unlicensed band, as (433.1–434.8 MHz, 868–868.6 MHz, 902.8–928 MHz, 2,400–2,500 MHz, and 5,725–5,850 MHz) is preferred, but interferences occurred with other radio receivers, like IEEE 802.11 and 802.15.4, which operate in 2,400 MHz and are used for communication between in-body and off-body/on-body devices. To increase high-speed data rates, ultrawide band (UWB) is preferred, due to its benefits like short range, low power consumption, and high-speed communication, and operates the frequency band between 3.1 and 10.6 GHz, with minimum bandwidth of 500 MHz. It supports two bands, namely, low and high bands, with a bandwidth of 499.2 MHz and operates in seven different channels. The low band of UWB consists of three channels (numbered as 1–3), and the high band consists of eight (numbered as 4–11), with a central frequency of 3,993.6 and 7,987.2 MHz, respectively. Out of these channels, the channel 2 in low band and channel 7 in high band are mandatory, and the remaining all are optional. In the PHYs layer, the frequency bands, like MICS, WTMS, ISM, and UWB bands, are used for medical applications, and the option of the PHYs layer depends on the applications.

In WBAN, for healthcare systems to function effectively, several requirements must be met [9], which are listed here.

1. **Reliability.** High reliability is required for the transmission of health-related data by WBAN sensors.
2. **Latency.** The response time to emergency situations should not be excessively slow. In this situation, transmission in real time is required.

3. **Security and QoS.** Sensitive and private information must be handled with care to ensure the confidentiality and safety of data. High level of quality of services (QoS) must be guaranteed in defence applications.
4. **Power consumption.** Generally, the WBAN devices/sensors are battery-powered, and it is required for a low-power device for radio transceiver.
5. **Accuracy.** The node data should be accurate and robust. The node may perform multiple tasks.
6. **Coverage.** To cover various surroundings and provide high data rates, WBAN can combine UWB technology with a narrow-band transmission. A UWB-based WBAN, for instance, might be necessary to handle larger data rates in some medical applications, such ECG monitoring.

Several issues must be addressed while developing efficient, precise, low-power wearable communication devices. Antenna design optimization is also to be considered for compact, lightweight interference with other equipment, for an effective body-centric communication network. Further, the characteristics of WBAN are most important to make a system efficient. WBAN communication architecture has been separated into three separate tiers [6, 10], as seen in Figure 13.5.

- Tier 1: intra-WBAN communication
- Tier 2: inter-WBAN communication
- Tier 3: beyond WBAN communication

In tier 1, the sensors are either connected to the body's surface or implanted within it. It collects and transmits biological information from the human body. The information supplied by tier 1 sensors is wirelessly relayed to and processed by the next layer, which comprises electronic devices, personal computers, and other intelligent

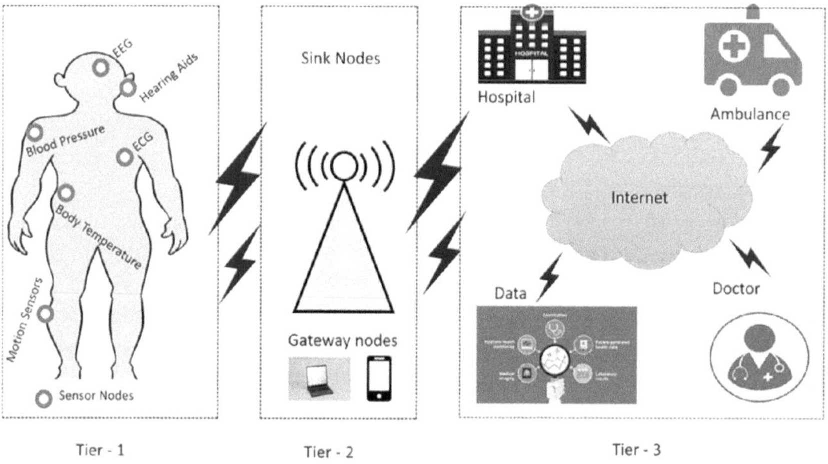

FIGURE 13.5 Communication architecture of WBAN.

electronic gadgets. The data centre at WBAN tier 3 is mostly made up of distant servers that support different applications. Its objective is to gather and assess received data to conduct an active reaction. When the sensor nodes collect unusual data, it will trigger an emergency broadcast and warning, which will speed up the emergency handling and rescue operations [10].

Tier 1. Body sensor devices are placed external to or within the human body to realize or continuously monitor vital biological signals, such as glucose level, body temperature, EEG, ECG, motion, etc., and then to transfer them via wireless medium to the next tier (tier 2). This category of communication is defined as intra-BAN communication.

Tier 2. Tier 2 takes the responsibility of collecting data from the sensor nodes of tier 1 to sink/gateway nodes, processing the data, and transferring it to the next tier (tier 3). The sink nodes bridge to another type of network. Wireless communication is available on the first and third tiers. This is commonly described as inter-BAN communication.

Tier 3. Tier 3 communication covers any network that occurs outside the BAN. It comprises internet-based connectivity. In a medical environment, tier 3 includes the medical history and database of the patient, which can be used for necessary response for the patient.

The following section covers the choice of flexible material, including conductive and dielectric substrates, while in Section 13.4, the different feeding mechanisms used for designing the flexible antenna and their characteristics are discussed. UWB in MIMO and their issues and challenges, including port parameters and decoupling methods, are discussed in Section 13.5. In Section 13.9, the various fabrication techniques are analyzed for the fabrication of flexible and wearable antenna. Finally, this chapter is concluded in Section 13.10.

13.3 CHOICE OF FLEXIBLE MATERIALS

Conductive materials and dielectric materials are the two key components that have a significant effect on the way flexible antennas operate [11].

1. The conductive substance is selected depending on the electrical conductivity property.
2. The dielectric material is chosen based on its dielectric qualities; tolerance to mechanical properties, such as bending, curving, and wrapping; durability in the outside environment; and sustainability to miniaturization.

The selection of conductive and dielectric materials is important to design an antenna to analyze its performance. A few applications that include flexibles/wearables are in sports, medicine, telecommunication, and health remote monitoring [12]. The classification of flexible materials used in designing flexible antennas is illustrated in Figure 13.6.

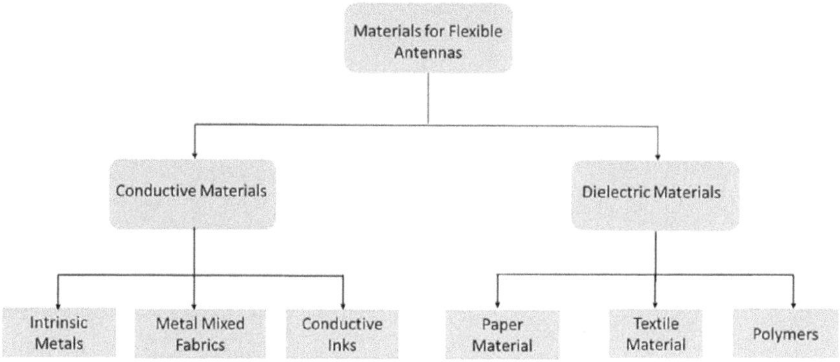

FIGURE 13.6 Flexible materials classification.

13.3.1 Conductive Materials

While designing the antenna, superior electrical conductivity is most needed for ensuring the performance of the antenna. Further, flexibility, such as bending, crumpling, and easy integration into the substrate materials of the conductive material, is also necessary to make the antenna efficient. The typical requirements of conductive materials for flexible designs are depicted in Figure 13.7. The materials frequently used as a conductive material for antennas include intrinsic metals, such as copper, silver, and metals-mixed fabrics, like nylon covered with Ni/Cu/Ag, zelt, nickel-copper-coated fibres, and conductive inks like silver nanoparticles, copper nanoparticle, and graphene-based ink. Some of the conductive materials, like silver and copper, are often used due to their low resistivity or high electrical conductivity for effective radiation [4]. The proposed antenna is fabricated with denim fabric substrate with the thickness of 2 mm and dielectric constant of 1.67, with the copper tape used as a radiating element operating at a frequency of 2.8 to 10.9 GHz UWB spectrum, suitable for WBAN application, as reported by S. Parameswari and C. Chitra [13]. Further, metal-mixed fabrics are used to design wearable antennas, like nylon covered with Ni/Cu/Ag conductive fabric placed on flexible PDMS substrate, which investigated the bending test on various places on human body–shaped phantom, and mechanical stability is carried out with this material. Hence, the material is mostly used to make an antenna that is robust, has frequency reconfigurability between 2.3 and 2.68 GHz, is flexible, and is wearable, as reported by [14]. Another conductive material, zelt, provides better protective fabric and flexibility for different properties, like bending, twisting, warping. The proposed antenna is made with a felt substrate, and the electro-textile zelt material is represented in [15] and offered cheap and low cost. The proposed antenna operated in X band frequency band (8–12 GHz) and is designed for application in medical imaging.

And nickel-copper-coated type fibre (nickel-copper-silver-coated nylon) is used to design an antenna of 0.08 mm thick with PDMS [16], and the proposed antenna design operates at a bandwidth of 2.85–8.6 GHz and suitable for conformal applications. Conductive inks are used as conducting material, such as silver nanoparticles,

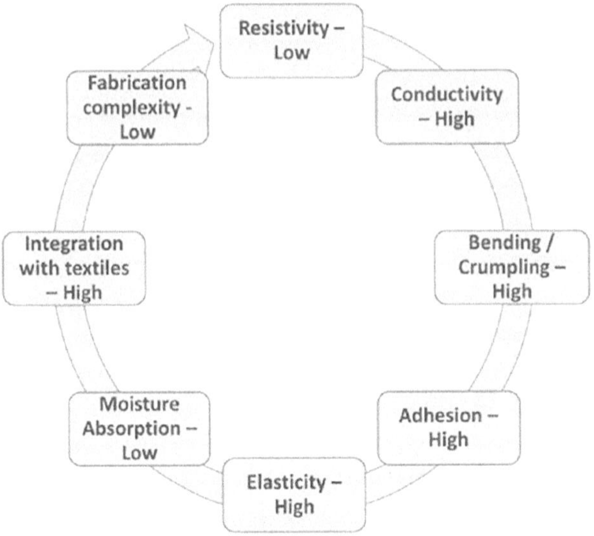

FIGURE 13.7 Requirements of conductive materials for flexible devices.

copper nanoparticles, and graphene-based inks. A silver nanoparticle ink is placed on a polymer substrate using inkjet printing to make the proposed antenna, as described in [17], and the sensing capabilities under tensile conditions are investigated along with mechanical characterization of strain sensors. The copper nanoparticle–based ink is investigated in [18]. The group of laser printing and laser sintering of copper nanoparticle ink is deposited onto the flexible substrate. It is used in the fabrication of a flexible substrate and suitable for small form factor and simplicity of integration-related applications. The proposed wideband elliptical dipole antenna is developed in [19] using graphene flakes–based ink as a conductive material. It has several advantages, like low cost, mechanical flexibility, chemical stability, and fabricated using screen-printing method for low-cost wireless communication applications.

Generally, the materials used for conductors can be categorized as intrinsic metals, metal-mixed fabrics, and conductive ink. Table 13.2 represents the conductivity value of several conductive materials.

13.2.2 Dielectric Materials

Dielectric materials mostly support the use of antenna radiation-emitting element and ground plane. The conductive material of an antenna is constructed on a substrate when the flexible substrate material is most likely to conform and has more stable performance than non-flexible substrates. The choice of the substrate typically depends on the application. To be able to incorporate electronic devices onto the human body, which has bending surfaces and constantly changing movements, the device must be flexible and physically flexible, even if it is stretchy. And the materials should have some characteristics, such as low resistivity and high conductivity, light weight, conformal, washable, stretchable, and robust [3], as shown in Figure 13.8.

TABLE 13.2
Conductivity and Properties of Conductive Materials

Reference	Conductive material	Conductivity, σ (S/m)
Intrinsic metals	Silver [4]	2.2×10^6
	Copper tape [13]	1×10^6
Metal-mixed fabrics	Nylon covered with Ni/Cu/Ag [14]	1.02×10^5
	Zelt [15]	1×10^6
	Nickel-copper-coated fibres [16]	5.4×10^6
Conductive inks	Silver nanoparticle [17]	2.2×10^6
	Copper nanoparticle [18]	5.813×10^7
	Graphene-based ink [19]	4.3×10^4

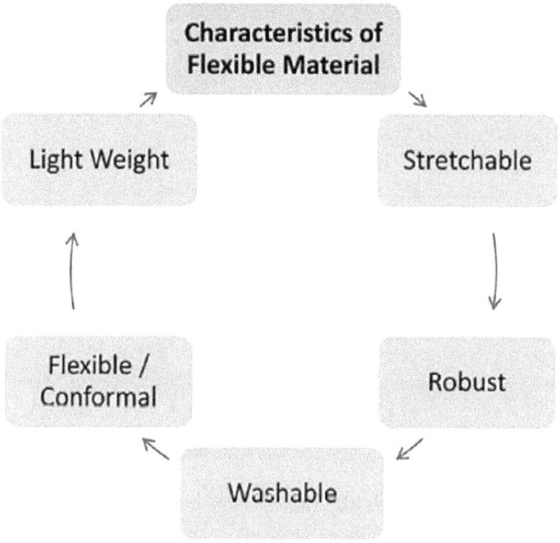

FIGURE 13.8 Characteristics of flexible materials.

The antenna uses a conductor as its radiating material or ground plane, with a dielectric substance as its substrate. The dielectric material is applied for the design of a flexible material, categorized as paper materials, textile materials, and polymer materials. Based on the applications, it is important to decide on a suitable substrate of the antenna. The antenna for WBAN applications must be made with a flexible substrate material. Various forms of flexible substrates have been reported in the report in [20]. Paper material is one of the most used organic materials for radio frequency applications, which is both renewable and eco-friendly. Paper material is quite inexpensive and has a wide range of uses. Paper material can be around ten times less expensive than polymers or plastic films; it is currently being investigated as an electronics substrate. Because the synthetic paper used in this material is not polymer-coated, the antenna retains its flexibility and could be twisted both forward and backward without breaking. The antenna substrate is made of Teslin-based synthetic paper material. Its thickness (h) is measured at 0.26 mm, its dielectric constant is 2, and its loss tangent is 0.0022 [21]. Another type of material used is Kodak picture paper of 0.254 mm thickness [22]. The dielectric characteristics are a loss tangent of 0.05 and relative permittivity (εr) of 2.85. Felt material with properties including dielectric permittivity of 1.22, loss tangent of 0.016, and thickness of 2 mm is also used as a substrate because the electro-textile and substrate are inexpensive, as reported in [15]. Another antenna is designed on a 0.7 mm thick denim flexible substrate with a dielectric constant value of 1.4 [23]. The antenna substrate is a typical, plain-woven polyester cloth with a thickness of 0.5 mm used for designing an antenna [24]. Polymer-based materials, such as Kapton polyimide film material, are preferred as substrate [25] because of their good mix of physical, chemical, and electrical characteristics, as well as their great flexibility and low loss tangent (tan□ = 0.002). In addition, the Kapton polyimide material has a low profile (50.8 μm) but is highly robust and has a tensile strength of 165 MPa at 73°F, a dielectric power of 3,500–7,000 volts/mil, and a temperature range of −65 to 150°C [26], and another flexible material, named PET glass substrate, has switching temperatures ranging between 80°C and 100°C [27]. Table 13.3 shows the dielectric materials with their properties used in the construction of flexible antenna system.

The type of flexible material is recognized in the requirements of flexible characteristics for various applications. Polymers used in flexible electronics and wearable applications include Kapton polyimide (PI), polyethylene terephthalate (PET), Teflon, polydimethylsiloxane (PDMS), and liquid crystal polymers (LCP). Table 13.4 depicts the most often utilized flexible materials with their various characteristics used in the construction of flexible antennas.

TABLE 13.3
Comparison of Different Substrate Materials Used in Flexible Antenna Design

Reference	Dielectric materials	Dielectric constant (ε_r)	Loss tangent (tan□)	Thickness (mm)
Paper material	Photo paper [28]	3.2	0.05	0.44
	Synthetic paper [21]	2	0.0022	0.26
	Kodak photo paper [22]	2.85	0.05	0.254
Textile material	Felt [15]	1.22	0.016	2
	Denim [23]	1.67	0.0035	2
	Plain woven polyester [24]	2.193	0.005	0.5
Polymers	Kapton polyimide (PI) [26]	3.5	0.007	0.05, 0125
	Liquid crystal polymers (LCP) [29]	2.9	0.002	0.1
	Polydimethylsiloxane (PDMS) [30]	2.7	0.134	1.5
	Polyethylene terephthalate (PET) [27]	3.16	0.0071	0.125

TABLE 13.4
Frequently Used Flexible Materials in Designing Flexible Antennas

Substrate/ Material	Density (g/cc)	Tensile Strength	Dielectric Constant	Dielectric Loss (tan□)	Thermal Stability	Deformability	Moist Sustainability	Electrical Stability	Fabrication Complexity
PI [26]	1.42–0.53	22–33	3.4–3 9	0.0013–0.0040	High	Low	Low	High	Simple
PET [27]	1.3–1.4	25–40	3.0	0.002	High	Low	High	High	Moderate
Teflon [31]	2.1–2.25	3.9–4.1	2.1–2.72	0.0002–0.0025	Low	Low	High	Low	Complex
PDMS [30]	0.97	0.25–1.3	2.3–2.8	0.0015–0.0035	Low	High	High	Low	Moderate
LCP [29]	1.4	29	2.9–3.14	0.0025	High	Low	High	High	Complex

13.4 FEEDING MECHANISMS

The feeding of microstrip antennas can be done in several methods. Microstrip line feed, co-axial probe feed, aperture coupling feed, and the proximity coupling feed are the most common designs. Even though the microstrip feed line is often much smaller than the patch line, it remains a conducting strip [32]. The microstrip patch antenna can be fed using different techniques, depicted in Figure 13.9. These methods can be categorized as either contacting or non-contacting. In the contacting approach, a connecting device, such as a microstrip line, is used to supply radio frequency power straight to the radiating patch, whereas in non-contacting method, the patch is not directly supplied with radio frequency power; as an alternative, electromagnetic coupling is used to transfer power from the supply line to the path. Aperture and proximity-coupled feed are the two most common non-contacting feeding methods.

Several antenna configurations are available with various feeding methods. Over alternative feeding systems, microstrip patch antennas with co-planar waveguide (CPW) are recommended because many beneficial characteristics exist, including reduced radiation losses, increased bandwidth, and better impedance matching; in fact, the ground plane and radiating elements are made on the same side of the substrate, which helps in cheap production costs, fabrication complexity, and low profile. Feeding the patch antenna with co-planar waveguide, which has a significant impact in achieving wider bandwidth, due to the special features of the CPW fed, better impedance may be controlled by optimizing the gap between two ground planes and not due to the thickness of the substrate. So this type of feeding is widely used to design such a kind of broad, wide, and ultrawideband antennas. Hence, the CPW fed is often used in the construction of flexible antennas.

A conventional co-planar waveguide on a dielectric substrate contains a centre strip conductor 'W' with a pair of ground planes on either side, divided by a small gap 'S', thickness of dielectric substrate 'h', as depicted in Figure 13.10, and the characteristic impedance (Z_0) is related to equation (1),

$$Z_0 = \frac{30\pi}{\sqrt{\varepsilon_{eff}}} \frac{K(k')}{K(k)} \tag{13.1}$$

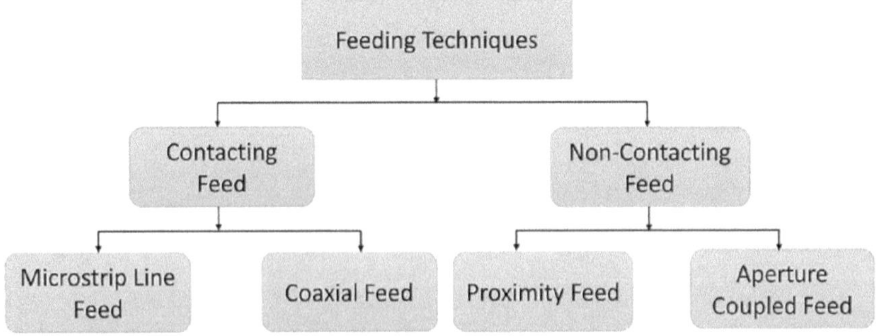

FIGURE 13.9 Types of feeding mechanisms.

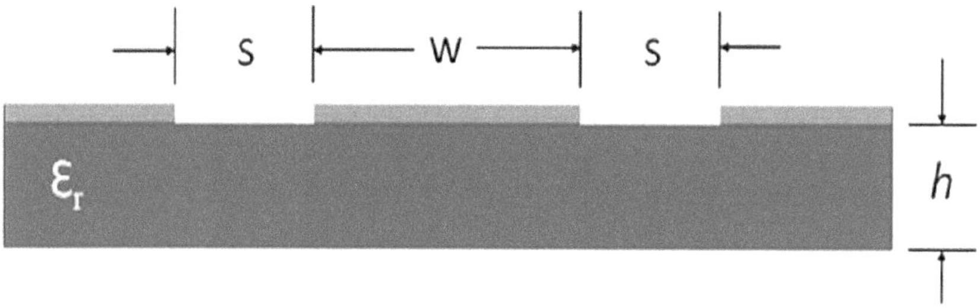

FIGURE 13.10 Co-planar waveguide feed.

where

$$\varepsilon_{eff} = 1 + \frac{\varepsilon_r - 1}{2} \frac{K(k')}{K(k)} \frac{K(k_1)}{K(k_1')} \tag{13.1a}$$

$$k = \frac{W}{W + 2S} \tag{13.1b}$$

$$k_1 = \frac{\sinh\left(\frac{\pi W}{4H}\right)}{\sinh\left(\frac{(W + 2S)\pi}{4H}\right)} \tag{13.1c}$$

where K is complete elliptical integral of the first kind and $k' = \sqrt{(1 - k^2)}$.

Practical antenna design for WBAN applications is more difficult because ultrawideband systems require extremely special antenna topologies and transmission characteristics than the narrowband systems. The antenna must function effectively throughout the bandwidth and be unaffected by the physical characteristics/parameters of the user in WBAN. As a result, it is critical to confirm that the antenna far-field radiation properties are essentially the same when bent for in-body/on-body use. When its dimensions are small, the antenna needs to be robust and reliable. Therefore, reduction in size has been vital for improving wearable antenna performance. Printed monopole patch antennas are frequently used in the construction of UWB communication systems due to their simple construction, significant bandwidth, and efficient radiation.

13.5 UWB MIMO FLEXIBLE ANTENNA

The ultrawideband (UWB) is a technology for short limit wireless communication systems, and it offers some benefits, such as higher data rates, low power consumption, and low-cost design [33, 34]. In the year 2002, the Federal Communication Commission (FCC) permitted the use of an unlicensed frequency spectrum spanning from 3.1 to 10.6 GHz of frequency, with a minimum bandwidth of 500 MHz [35]. However, it suffers from multipath fading and co-channel interferences. To resolve this issue, and to enhance the performance of the systems, the integration of the UWB with MIMO is introduced to provide reliability and make further improvements on wireless communication systems [36]. MIMO has multiple advantages, including reduced fading from multipath transmission, greater channel capacity, and improved link reliability through multiplexing and diversity gain [37]. The MIMO in UWB systems provides several advantages, but some challenges arise, including mutual coupling effect when placing the number of antenna elements close to each other while maintaining compactness. Several studies have been conducted for increasing the isolation between the elements in literature [38]. The basics of MIMO structures in wireless communication are discussed in Section 13.5.1, while the description of mutual coupling and the methods to reduction are presented in Sections 13.5.2 and 13.7, respectively.

13.5.1 MIMO Structures

A novel MIMO antenna has emerged as a viable choice for high-speed communication methods [39]. MIMO offers a fast wireless communication link that may support several applications without requiring an increase in bandwidth or transmission power. To send and receive data, two or many radiating elements are fed individually utilizing a co-planar or strip-line feed mechanism in such devices. However, port coupling is a notable issue in MIMO structure since it reduces the performance of MIMO antennas. In this section, MIMO systems and the types of MIMO antennas to be constructed are discussed, followed by performance evaluation and the various decoupling methods. The MIMO structures are generally described, as shown in Figure 13.11.

$$y = Hx + N \tag{13.2}$$

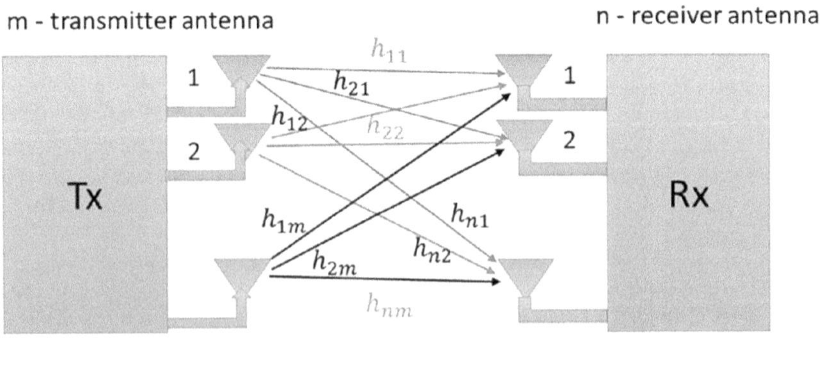

FIGURE 13.11 Basic MIMO structure.

where y is the received vector, x is the input vector, H is the channel transmission matrix consisting of N_T and M_R, which are the number of transmission and reception antennas, respectively, and N is the noise vector.

$$Channel\ Matrix, H = \begin{bmatrix} h_{11}, & h_{12} & \cdots & h_{n1} \\ h_{12}, & h_{22} & & h_{n2} \\ \vdots & & \ddots & \vdots \\ h_{1m}, & h_{2m} & \cdots & h_{nm} \end{bmatrix} \tag{13.3}$$

where h_{ij} is the channel gain between j^{th} transmitting antenna and i^{th} receiving antenna.

In a simple 2 × 2 MIMO configuration, the received signal vector is expressed as:

$$\begin{bmatrix} y_1 \\ \hline y_2 \end{bmatrix} = \begin{bmatrix} h_{11} & h_{12} \\ h_{21} & h_{22} \end{bmatrix} \begin{bmatrix} x_1 \\ \hline x_2 \end{bmatrix} + \begin{bmatrix} n_1 \\ \hline n_2 \end{bmatrix} \tag{13.4}$$

For MIMO antenna designs, particularly, the diversity techniques (spatial, polarization, and pattern) are employed. Therefore, in the extremely reduced space given by the small devices, the development of multiple antenna designs becomes more challenging. When the antennas are situated close together, it becomes difficult to take use of the diversity. Therefore, decorrelation of their patterns is necessary, or decrease of mutual coupling. The improvement of isolation between MIMO antenna access ports creates another difficulty. In Section 13.5.2, the description of mutual coupling that occurs in various types of MIMO antenna systems is discussed.

13.5.2 Mutual Coupling

The performance and capacity of MIMO systems are reduced by the correlation between their antenna parts in limited spaces and sizes. Mutual coupling implies the energy is attracted by the receiver of one antenna when the neighbouring antenna is radiating energy. Mutual coupling often impacts the input impedance, reflection coefficients, and radiation patterns of array elements. The mutual coupling of a MIMO system [37] is mathematically defined by the following equations (5) and (6):

$$MC_{mn} = \exp\left(-\frac{2x_{mn}}{\lambda}(\alpha + n\pi)\right), m \neq n \tag{13.5}$$

$$MC_{mn} = 1 - \frac{1}{N}\sum_{m}\sum_{m \neq n} MC_{mn} \tag{13.6}$$

where λ is the wavelength, N is the number of array components, α is the coupling level parameter, and $MCmn$ and x_{mn} define the mutual coupling and gap between the mth and n^{th} antenna elements, respectively. It is generally measured in dB. In addition to the array design, mutual coupling is affected by the excitations of other components. Typically, the isolation value should be $-20\log_{10}|S_{mn}|$ between m^{th} and n^{th} elements, or the dB valued S-parameter between them is used to quantify it.

In wireless communication systems, diversity antennas are used to enhance signal quality and reliability while multipath, interference, and fading are encountered. When the strength of the signal fluctuates because of environmental factors like movement, reflections, or obstructions, it is known as fading. When undesirable signals from outside sources interfere with a desired signal, interference expands. When a signal travels along many pathways to reach the receiver, it is said to be multipath and can result in delays and distortions. Diversity antennas can get around these problems and improve system performance by utilizing several antennas at the transmitter, receiver, or both. In MIMO systems, many forms of antenna diversity, including spatial, polarization, and pattern diversity, are utilized. *Spatial diversity* refers to the diversity techniques that employ two or more antennas to develop the reliability and quality of a wireless link. Specifically, the spatial multiplexing is based on MIMO technology, which combines multiple antenna elements at both transmitter and receiver to transmit different data streams simultaneously within the same frequency range. Antennas that have different polarization are used for transmitting and receiving multiple copies of a signal. When two or more co-located antennas with different radiation patterns exist, pattern diversity is achieved. Designing of MIMO antennas is important to satisfy the required parameters of MIMO antenna system.

Table 13.5 shows the various types of approaches employed, such as (1) based on radiator, (2) based on feeding mechanism, (3) based on diversity performance, and (4) based on polarization diversity.

In this list, the positioning of antenna elements, advantages, limitations, with antenna geometry, are implemented in the design of the MIMO antenna system. Figure 13.12 shows the types of MIMO antenna systems possible to design.

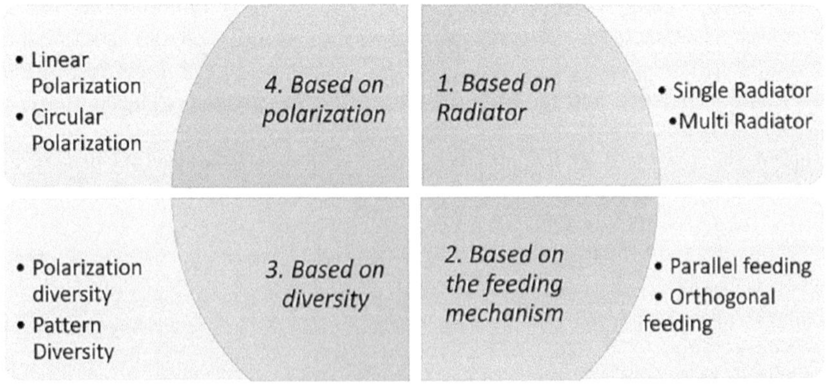

FIGURE 13.12 Types of MIMO antenna system design.

TABLE 13.5
Advantages and Limitations of Various Types of MIMO Antennas

Types of MIMO antenna	Description	Advantages	Limitations	Structure	
				(a) Top view	(b) Bottom view
Single radiator [40]	Multiport in one radiating element	Compact size	Low isolation	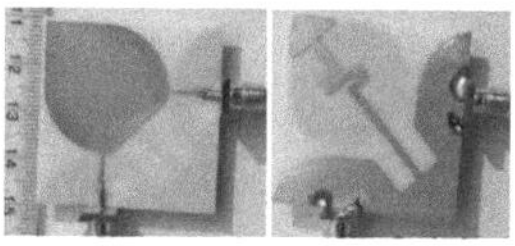 (a) and (b)	
Multiradiator [41]	Multi-antenna elements placed orthogonally	High isolation	More spacing between elements	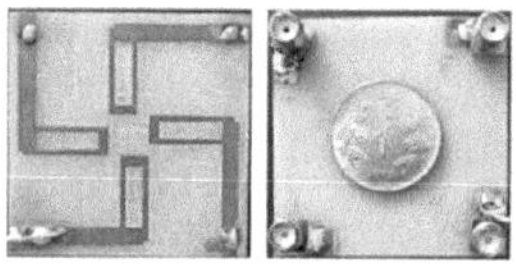 (a) and (b)	
Parallel feeding [42]	Parallel feed to antenna elements	Compact size	Polarization diversity not obtained (different polarization by elements)	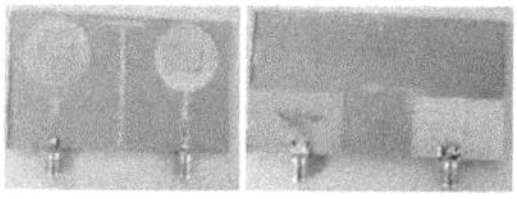 (a) and (b)	
Orthogonal feeding [43]	Perpendicular feed between elements	High isolation; polarization diversity	Reduction in spacing between radiators	 (a) and (b)	
Polarization diversity (LP and CP) [44]	Multiple versions of the signal transmitted via antennas	Diversity obtained	Obtaining polarization diversity in a single radiator difficult	 (a) and (b)	
Pattern diversity [45]	Co-located antennas producing different radiation patterns	Diversity obtained reducing ECC	Complex feeding and antenna structure	 (a) and (b)	
Linear polarization and circular polarization [46]	Generates linear polarization and circular polarization	Isolation can be improved	Obtaining the polarization with CP difficult	 (a) and (b)	

A multiport can be used from a single radiator to achieve compact dimensions. This method achieves poor isolation yet results in a compact design. Authors Situ Rani Patre and Surya P. Singh proposed a MIMO antenna with a dimension of

39 × 39 mm^2 and a low-cost FR-4 substrate for broadband applications operating between 2.4 and 12.7 GHz [40]. To attain adequate isolation between ports without adding a slot to the patch, a modified leaf-shaped radiating element shared equally at 45° by two tapering microstrip line and a universal curved ground plane make up the proposed antenna. In addition to that, with the use of an end-loaded meander line stub connected to the ground plane, low mutual coupling is attained. The top and bottom views of the fabricated structure are depicted in Figure 13.13(a) and (b).

Whereas the antenna uses multi-radiator, isolation can be achieved but occupies more space than the MIMO antenna with a single radiator. While designing a MIMO antenna system, a multiport can be used with multiple radiating elements to ensure effective isolation, but it consumes additional space. Authors E. Suganya et al. [41] proposed a MIMO antenna system using four radiators. An isolation between antenna ports is improved by utilizing $\lambda/4$ distance between antenna elements for WLAN applications with a dimension of 45.5 × 45.5 × 1.96 mm^3. The designed antenna is made up of four L-shaped monopole patch antennas, each of which has a corner truncated edge and a rectangular slot in the middle. Due to the separation of antenna elements, initially, the inter-element isolation of 12 dB is obtained. To increase the reflection coefficient and isolation parameter, a truncated corner is built at one edge for each radiating element. Hence, isolation of −58.37 dB is achieved in the proposed design. The fabricated MIMO antenna's top and bottom views are displayed in Figure 13.14(a) and (b), respectively. Parallel feeding techniques in MIMO antenna systems allow for small dimensions; however, polarization diversity is challenging to achieve. The parallel feeding provides compact, but not producing the polarization. Bazil Taha Ahmed et al. reported a triple-band notched superwideband monopole patch antenna [42]; the two-port and four-port parallel feeds are applied into the proposed MIMO antenna; the T-shaped metallic barrier is applied to improve isolation. Cross-line and circular ring are applied in four-port MIMO antennas for improvement of isolation; isolation is achieved at −12 dB in lower band and −15 dB, displayed in Figure 13.15(a) and (b).

Another method for achieving high isolation and polarization diversity in MIMO antennas is perpendicular feeding between multiple antenna elements. Mahmoud A. Abdelghany et al. [43] presented the four-port orthogonal-shaped MIMO antenna for UWB applications. An orthogonal-shaped four-port MIMO antenna with dimension of 62.5 mm × 60.5 mm of operating frequency range of 3.5–11 GHz with greater than −20 dB is obtained, as seen in Figure 13.16(a) and (b). The orthogonal layout produces a polarization diversity due to the current being restricted around an antenna and a little quantity of current dissipated to the other ports, which leads to increases in the isolation between the ports. Multiple signal versions are used to transmit and provide different polarization, which is difficult to achieve with a single radiator. The radiating antenna elements are replicated orthogonally to produce polarization diversity and improved isolation; all sides of the substrate are stripped of the L shape, as reported by V. R. Balaji et al. [44].

As seen in Figure 13.17(a) and (b), a hollow semi-hexagonal-shaped radiating element is used with modified substrate to obtain wideband spectrum. The four-port MIMO antenna is made with FR-4 substrate, and the operating frequency is 2.84–15.88 GHz. The modified substrate makes the antenna compact. The UWB MIMO is proposed for portable devices.

Co-located antennas in a MIMO antenna system generate different radiation patterns to achieve pattern diversity. Uzair Ahmad et al. proposed a MIMO antenna for sub-6 GHz band mobile phone applications, as depicted in Figure 13.18(a) and (b). Four elements are resonating at 2.5 GHz placed at the corners of the board, whereas another two port elements operating at 3.5 GHz are placed in the centre of the board, which leads to low mutual coupling [45].

Finally, the multi-radiating elements in the MIMO antenna system offer linear and circular polarization, resulting in strong diversity and isolation. However, achieving polarization with CP is challenging. Based on the polarization scheme, the proposed antenna achieved good isolation and generated the compact linear and circular polarization MIMO antenna for WLAN, WiMAX (3.4–5.44 GHz), and C band (5.9–6.3 GHz) application, with a total size of 60 mm × 60 mm, as reported by Qasim Hadi Kareem and Malik Jasim Farhan [46]. The proposed MIMO antenna performed in two configurations; the upper portion of the circular patch emits left-hand circular polarization (LHCP), while the lower part of the circular patch radiates right-hand circular polarization (RHCP), as displayed in Figure 13.19(a) and (b). The isolation value of the proposed MIMO antenna is attained due to the four antenna elements placed orthogonally, producing horizontal and vertical polarization.

Since MIMO antennas have multiple paths to send or receive data, they have drawn a lot of attention in current wireless communication systems, improving output performance and transmission range [47]. It is examined that for the MIMO antenna's components to operate independently and concurrently transmit and receive signals without compromising the antenna's performance, a large amount of isolation between the array elements is required. Mutual coupling reduction is a significant field of study that will have a direct influence on the extension of future generation wireless communication, including 5G, beyond 5G, 6G, and massive MIMO. In addition to the scattering parameters, the diversity parameters must also be considered to ensure the quality of the MIMO antenna system. Hence, some of the important diversity parameters and their acceptable range are listed in the Section 13.6.

13.6 MIMO PORT PARAMETERS

In addition, the reflection parameters, transmission characteristics, and several diversity parameters are considered to maintain the quality of a MIMO antenna. For practical applications, the MIMO antenna structures must meet the specified values for the diversity parameters [48]. In this section, some fundamental diversity factors for MIMO antennas are discussed, and Table 13.6 displays the acceptable ranges of MIMO parameters.

13.6.1 Isolation between Ports

The ratio between the power fed into one port of the designed antenna to the power received at another port [37], which is shown in equation (7):

$$\text{Isolation} = -10\,log_{10}\left(S_{ij}\right) \tag{13.7}$$

TABLE 13.6
Acceptable Ranges of MIMO Parameters

S. no.	Name of the MIMO parameter	Acceptable range
1	Isolation	>–15 dB
2	TARC	<–10 dB
3	ECC	<0.5
4	MEG	<–3 dB
5	DG	>9.95
6	CCL	<0.4 bps/Hz

where S_{ij} is a transmitted power from port *j* to port *i*. The isolation measures the coupling between antennas, which are closely located.

13.6.2 Total Active Reflection Coefficient (TARC)

The total power reflected by antenna elements to the total power impact on the patch has been referred to as the TARC of a MIMO system [47]. The general TARC expression for an N-port MIMO antenna can be found by equation (8).

That is, return loss of overall MIMO antenna elements.

For an N-element MIMO antenna: $$\Gamma_{MIMO} = \circ \frac{\sqrt{\Sigma_{i=1}^{N} \left|\boldsymbol{b}_i\right|^2}}{\sqrt{\Sigma_{i=1}^{N} \left|\boldsymbol{a}_i\right|^2}} \tag{13.8}$$

where *[b]* = *[S] [a]*; *S* is antenna scattering matrix, *b* is reflected signal, and *a* is incident signal. *S* is scattering matrix.

13.6.3 Envelope Correlation Coefficient (ECC)

The ECC parameter shows the correlation of adjacent MIMO antenna components. Radiation pattern or S-parameter can be used to calculate it [49]. The ECC value generated from the far-field radiation pattern, on the other hand, is preferred since it specifies the level to which different radiating modules in MIMO systems have separate radiation patterns. Furthermore, most planar antennas exhibit loss; hence, the S-parameters-based technique for calculating ECC should be prevented. Equation (9) gives the arithmetic equation for ECC utilizing MIMO radiation pattern.

In short, correlation between the radiation pattern of different antenna ports.

Using S-parameters:

$$\mathrm{ECC} = \frac{|S_{11}S_{12} + S_{21}S_{22}|^2}{(1-(|S_{11}|^2 + |S_{21}|^2))(1-(|S_{22}|^2 + |S_{12}|^2))} \tag{13.9}$$

where for two-port networks, the S_{11} and S_{22} are the reflection coefficient and S_{12} and S_{21} are called transmission coefficients.

13.6.4 Mean Effective Gain (MEG)

MEG is a significant diversity statistic for MIMO antenna systems, described as the ratio of the power received by an antenna with MIMO to the power received by an isotropic antenna. The ratio MEGj/MEGi should be less than 3 dB for optimum performance of a MIMO antenna with equal power levels. Equations (10a) and (10b), respectively, can be used to evaluate mean effective gain [47].

Gain value of antenna considering the environmental effects. MEG at single port must be around ±3 dB.

$$MEG_i = 0.5[1-|S_{ii}|^2 - |S_{ij}|^2] \tag{13.10a}$$

and

$$MEG_j = 0.5[1-|S_{ij}|^2 - |S_{jj}|^2] \tag{13.10b}$$

13.6.5 Diversity Gain (DG)

In wireless networks, DG represents the quality and durability of the MIMO antenna [47]. As a result, the DG of the MIMO system should be high (typically −10 dB) and within the permitted frequency range. The DG is determined using the ECC value and can be expressed mathematically, given by (11).

That is, the process of obtaining the powerful signal from a batch of 'N' signals.

$$\mathrm{DG} = 10\sqrt{1-(ECC)^2} \tag{13.11}$$

13.6.6 Channel Capacity Loss (CCL)

The largest rate at which information may be transferred via a transmission channel with nearly little loss is known as the CCL. The default CCL estimate for given MIMO antenna system is 0.4 bps/Hz [48].

Equations (12)–(16) give the representation of CCL using S-parameters.

$$CCL = -log_2 \det\left(\Psi^r\right) \tag{13.12}$$

where the psi Ψ is $\Psi^r = \begin{bmatrix} \rho_{11} & \rho_{12} \\ \rho_{21} & \rho_{22} \end{bmatrix}$ (13.13)

and the rho ρ is $\rho_{ii} = 1 - \left|\sum_{n=1}^{N} S_{in}^{*} S_{nj}\right|$ (13.14)

$$\rho_{ij} = -\left|\sum_{n=1}^{N} S_{in}^{*} S_{nj}\right|, \quad for\, i,\ j = 1, 2, 3, \text{ upto N} \tag{13.15}$$

The Shannon formula is managed to calculate channel capacity of the MIMO systems:

$$C = k\, log_2\left[\det\left(\left[I\right] + \eta \frac{SNR}{k}\left[H\right]\left[H^*\right]\right)\right] \tag{13.16}$$

where I is identity matrix, SNR is mean signal-to-noise ratio, and k is rank matrix (HH*). H is the channel matrix, and H^* is the matrix of Hermitian transpose.

Table 13.6 displays the acceptable ranges of different types of MIMO parameters.

13.7 EXISTING DECOUPLING METHODS

Various isolation improvement methodologies have been published in the literature to date. This section examines a variety of decoupling strategies for MIMO, including neutralization lines, parasitic elements, defective ground structure, metamaterials, electromagnetic bandgap structure, decoupling networks, and complementary split-ring resonators. The basic goal of the researchers is to minimize the mutual coupling as low as possible, with little influence on array performance and, if likely, without enlarging the physical antenna design. Decoupling techniques are crucial for optimizing the efficiency of MIMO antennas.

13.7.1 Neutralization Lines (NL)

An additional connection is provided in the neutralization procedure by joining two elements with a thin metal strip. By carefully regulating the length of the metal strip, the additional coupling balances out, and the coupling between the antennas, and the isolation between the two input ports, can be considerably improved [50]. NL can be employed to increase the isolation between antenna elements. As a result, the antenna elements current is neutralized. When the current and impedance are at their highest and lowest points, respectively, the current at the input elements is measured. The phase angle of the current is then updated by choosing the proper length of the NL. The linked current can be reduced by feeding the neighbouring antenna with this updated current. A pair of triband MIMO antenna resonating at 2.3 GHz, 3.5 GHz, and 5.7 GHz is proposed in [49]. The authors Ruipeng Liu et al. used two neutralization lines and analyzed them based on s-parameters and surface current distributions. In surface current distribution, one antenna, port 1, produces the coupling current named aI, and the same appeared on another antenna, port 2. Due to the NL, the coupling current bI is induced and cancels each other. So high isolation is achieved. So the U-shaped naturalization lines act as band-stop filter, which can decouple in mid- and high-frequency bands. The designed MIMO antenna structure and the variations of mutual coupling effects with and without NLs are depicted in Figure 13.20.

(a)

(b)

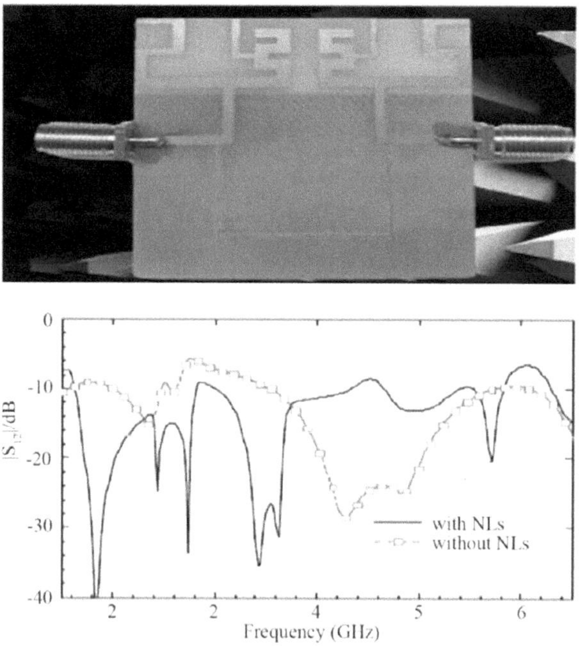

FIGURE 13.20 (a) Fabricated prototype and (b) S_{12} response of proposed antenna.

Source: [49].

13.7.2 Parasitic or Slot Elements

A *parasitic element* or passive radiator is a conductive material, in general a metallic bar, which is not linked to anything else electrically. The use of parasitic elements to enhance the efficiency, isolation, and correlation coefficient of the two adjacent MIMO antenna elements is another way of reducing mutual coupling between them. That is, reverse coupling occurs with the use of parasitic elements to cancel out the original coupling. For example, one antenna element produces a current called I_0; meanwhile, the aI_0 is coupled on another antenna element. When a parasitic component is added, two coupling routes are created: the original coupling path, *aI0*, and a second one, named *bI0*, which doubles the coupling path from parasitic components and continues till b^N if further parasitic elements are added up to *N*. The parasitic elements can produce an opposite coupling and reduce mutual coupling [51]. Authors Huy Hung Tran et al. [52] presented a MIMO antenna with parasitic elements to decrease mutual coupling in a way that added many square-shaped parasitic elements close to each of the patch antenna elements. These parasitic components have an impact on the distribution of electromagnetic fields, which lowers mutual coupling. Additionally, the wider bandwidth is also achieved. The two-element H-coupled and the multi-element E-coupled MIMO antennas have been tested using the proposed method. The structure of the designed H-coupled MIMO antenna and the isolation response are illustrated in Figure 13.21.

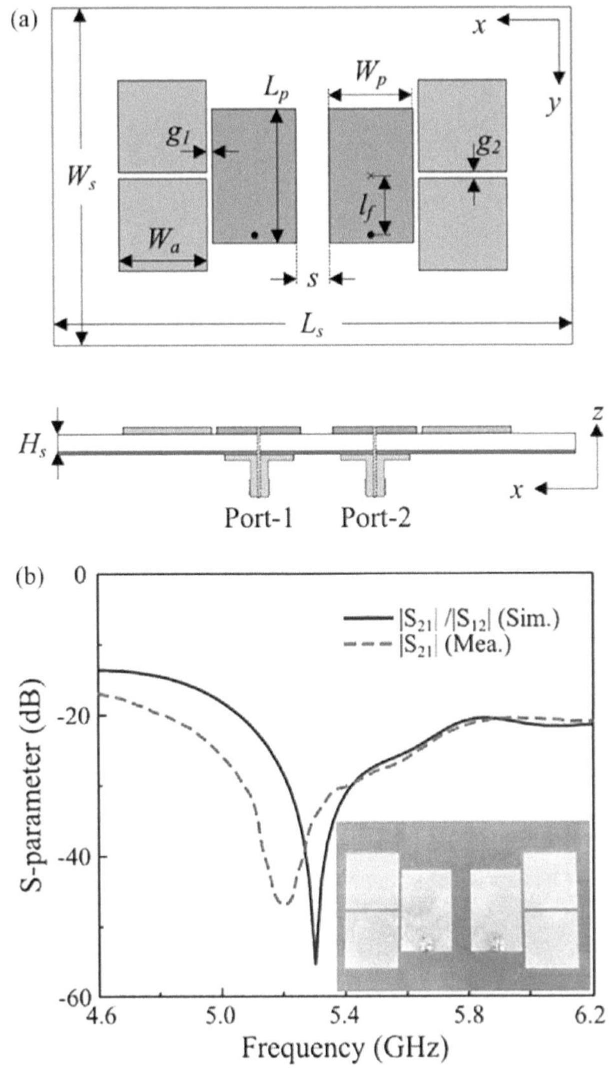

FIGURE 13.21 (a) Structure of MIMO antenna and (b) isolation response of proposed antenna.

Source: [52].

13.7.3 Defected Ground Structure (DGS)

High coupling between the adjacent antennas is created by the current that is generated on the ground plane. An alteration in the ground plane structure could reduce mutual interaction between antenna components, which creates a band-stop effect. For vehicle-to-everything (V2X) communications, a four-port MIMO antenna with circularly polarized (CP) working at 5.9 GHz is proposed [53]. For the drop of the mutual coupling between the antenna elements, a unique hybrid decoupling structure made of DGS and circular ring parasites is utilized, as shown in Figure 13.22. To decrease mutual interaction between perpendicular antennas, circular ring parasites are introduced among the antenna elements. The DGS in the structure and the circular ring parasites modify the distribution of electromagnetic fields, which lowers mutual coupling of greater than 34 dB.

Kun Wei et al. [54] proposed S-shaped periodic defective ground structure (PDGS) to decrease mutual coupling in microstrip patch antenna array. In this work, a novel S-shaped PDGS is introduced between two antenna elements. The proposed antenna was designed with Rogers TMM 6 substrate with εr = 6 and 3.81 mm thickness operating at 2.57 GHz. In the proposed work, two antenna elements are placed along the X-axis operating in a same band, and three DGSs are placed periodically along the Y-axis. Without PDGS, a significant amount of surface current is induced on another element 2 when the antenna element 1 is excited. After introducing the PDGS between elements, the current induced on another antenna element is significantly lowered. Hence, mutual coupling is lowered from −15 dB to −56 dB.

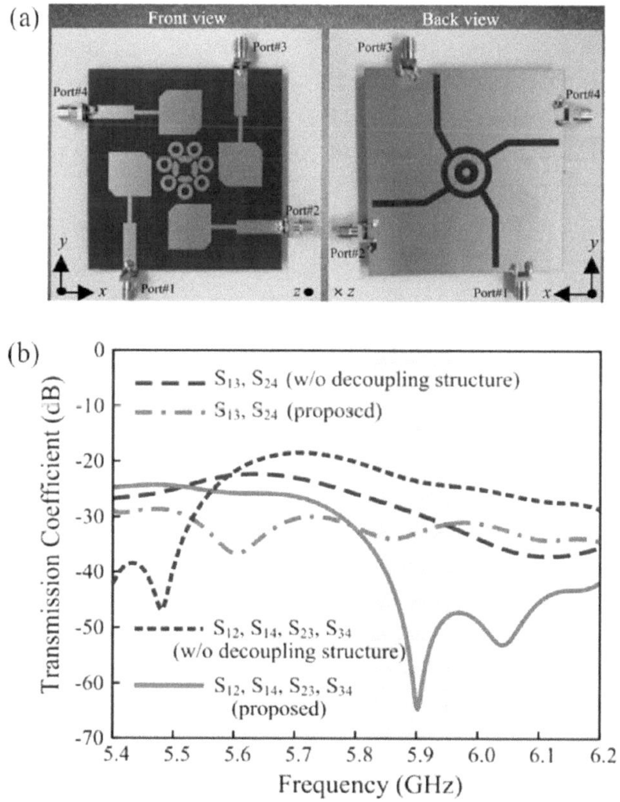

FIGURE 13.22 (a) Fabricated prototype and (b) S_{21} response of proposed antenna.

Source: [53].

13.7.4 Metamaterial

Metamaterials are artificial materials that have qualities that are not available in their original form. These metamaterials have either negative permittivity, permeability, or both. The metamaterial-based structure of split-ring resonator (SRR) is reported in [55]. Using a linear array of five identical SRR unit cells, a new metamaterial structure is constructed between the two antennas. The two radiating components' edge-to-edge spacing is fixed at 0.073λ (2.8 mm) at 7.8 GHz. The square SRR diameter is 2.4 mm, track width is 0.2 mm, and the gap between the two rings is 0.3 mm. The two rings are concentric and divided by 0.2 mm. The internal ring is 1.6 mm in length. A radiation box with dimensions of 2.8 × 2.8 × 5.6 mm^3 and electric and magnetic barriers as boundary conditions are placed. These barriers should validate the SRR's excitation conditions. An investigation is conducted on the variation and effect of the dimensions of the SRR opening for the two rings and to determine its effect on the resonant frequency. The initial value of 0.3 mm is used as a reference. Then, consider five different values around the nominal value while keeping all other factors constant. The five values are 0.3, 0.4, 0.5, 0.7, and 0.9 mm, used to investigate the reflection coefficient and transmission coefficient. The transmission coefficient S_{21} is observed at −42 dB with SRR, and without is −15 dB. The bandwidth is expanded from 2 to 18 GHz. The structure of an antenna and response of an antenna with/without SRR structure are displayed in Figure 13.23

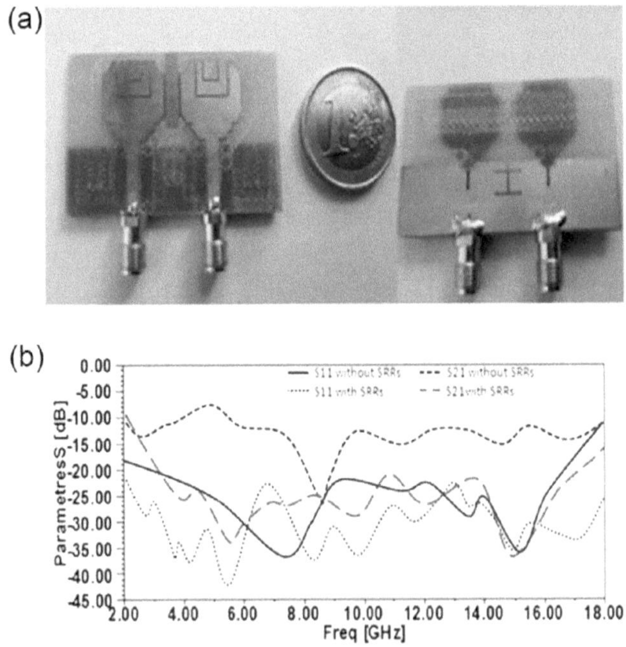

FIGURE 13.23 (a) Fabricated prototype and (b) S-parameters and transmission coefficient of proposed antenna.

Source: [55].

13.7.5 Electromagnetic Bandgap Structure (EBG)

Electromagnetic bandgaps (EBGs), influenced by the related concept of photonic crystals, are an interesting creativity in present antenna and microwave circuit design. EBGs contain defined frequency zones, where the surface wave propagation is banned, allowing them to be used as filters or to reduce coupling between circuit components. Various common EBG constructions, like cross EBG, circle EBG, star EBG, and square EBG, are used to design the EBG structure to minimize mutual coupling in printed antenna [56]. Niraj Kumar and Usha Kiran Kommuri [57] presented a two-port MIMO antenna with the isolation improvement by using EBG structure and a metal strip for WLAN applications. In this work, for closely spaced co-planar radiating patch antennas with centre-to-centre separation of 0.45 λ_0, the structure lowers mutual coupling in the H-plane. The suggested electromagnetic bandgap structure unit cell is created by combining the swastika structure with four square comb-shaped structures which are turned at 90° with each other. For EBG structure design, an FR-4 structure with a dielectric constant of 4.4 and a height of 1.6 mm is employed. The proposed EBG structure has an area of 6.8 × 6.8 mm^2, resulting in the compactness of the MIMO antenna system. EBG unit cell design parameters are optimized for resonance at WLAN (5.8 GHz) applications. A 46 mm × 20 mm MIMO antenna is built with two antennas with an edge-to-edge gap of 7.86 mm. The magnetic field of a MIMO antenna is connected through the non-radiating edges. A 3 × 1 unit cells are created from the ground surface between the non-radiating edges. The incorporation of EBG into the MIMO system has no effect on the frequency of resonance. The transmission coefficient between antennas that existed is lowered from 16.5 to 22.1 dB.

For further improvement of isolation, between two radiating patches, a 1 mm wide copper line strip is added, as seen in Figure 13.24.

13.7.6 Decoupling Network

The antenna components of the array mutually pair due to electromagnetic interaction. In the decoupling network method, two communication lines are linked to the input ports of the two closely coupled antennas. The length of the transmission lines is selected so that the trans-admittance among the ports shifts from complex to pure imaginary at antenna inputs. A shunt reactive element is then connected between the transmission lines ends to prevent the resulting imaginary trans-admittance. So, the mutual coupling is decreased by using a planar decoupling network which acts as a resonator. And mutual coupling between the surrounding antennas can be reduced by supplying the reverse coupling to disconnect the input ports between them as shown in Figure 13.25. The isolation between the nearby antennas has been achieved using distributed and coupled elements. Chenyin Yu et al. [58] proposed a novel spade shaped patch super wide band (SWB) MIMO antenna. An antenna prototype with four antenna elements with a dimension of 58 mm x 58 mm is fabricated using FR-4 substrate operating a frequency ranging from 2.9–40 GHz. Initially, the antenna is developed by inserting a staircase shaped structure at the bottom level of the antenna system. But the mutual coupling among ports is still high. Then the

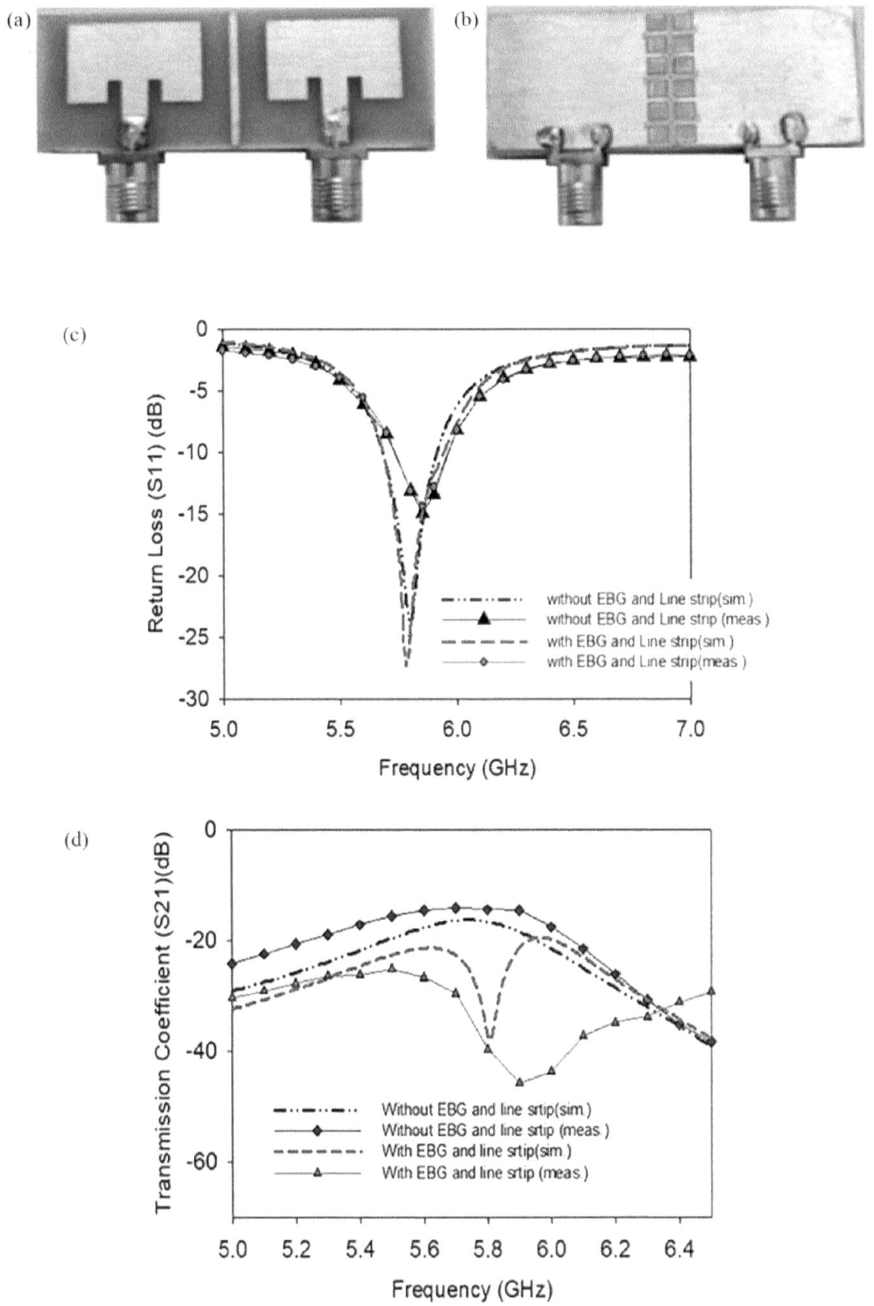

FIGURE 13.24 Fabricated antenna structure: (a) top view, (b) bottom view, (c) S-parameters of proposed antenna, and (d) transmission coefficient of proposed antenna [57].

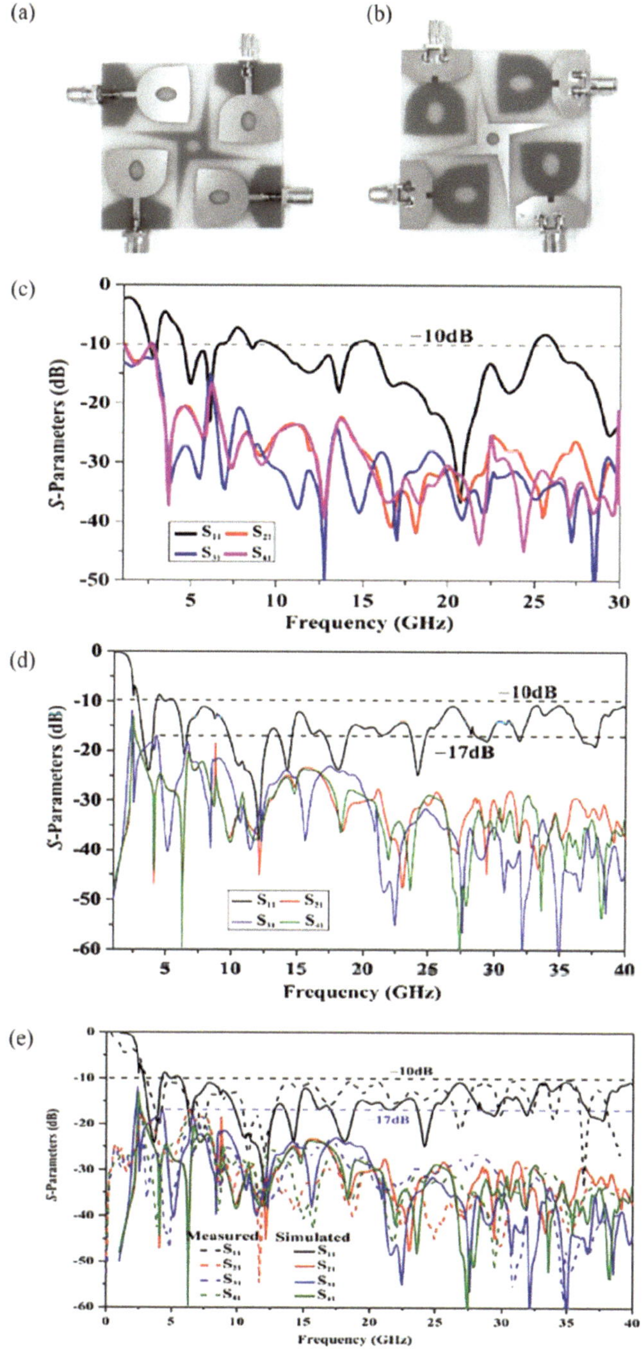

FIGURE 13.25 Fabricated structure (a) Top view (b) Bottom view (c) Simulated S-parameters with a staircase-shaped decoupling structure (d) Simulated S-parameters with decoupling structure (e) S-parameters for proposed four element SWB MIMO antennas [58].

modified structure using meander line is established at middle of the bottom layer of the proposed MIMO antenna system. Hence the isolation could be achieved. To improve port isolation at a level below –17 dB, a decoupling structure styled like a windmill is proposed in this work.

13.7.7 Complementary Split-Ring Resonators (CSRR)

To improve isolation, serve as a band-stop filter to supress surface waves, and reduce mutual coupling, the CSRR is employed. The CSRR is composed of two concentric ring arrangements with inverse spaces, as shown in Figure 13.26. The author proposed a MIMO antenna system operating at X-band (9.2 GHz). The array of three CSRR-based DGS is scratched in the ground plane among the two antenna elements [59]. In this work, the CSRR structure is proposed to operate in the same frequency of 9.2 GHz as the antenna array functioning at the same frequency band. The CSRR structure has dimensions of internal radius of 0.75 mm and c = g = d = 0.4 mm. Rogers RO 3003 has a thickness of 1.524 mm, a dielectric constant value of 3, and a loss tangent of 0.0013. The CSRR structure is stretched on the ground plane below the centre of a 3.4 mm wide microstrip line. The width of the feed line is chosen to correspond to 50 Ω. Through simulations, the proposed arrangement yields a mutual coupling of around −61 dB better than the standard array of the same size.

13.8 ANTENNA MEASUREMENT

13.8.1 Specific Absorption Rate (SAR)

While designing wearable and consumer electronics should adhere below the SAR safety limits, the specific absorption rate (SAR) describes the influence of electromagnetic (EM) radiation on the human body. The electromagnetic power is admitted or spent per unit mass of human tissue per unit time, given as W/kg or mW/g, known as SAR [60]. The International Commission for Non-Ionizing Radiation and Protection (ICNIRP) imposes regulatory restrictions to safeguard the human body from the effects of hazardous radiation. The maximum SAR for 10 g of tissue shall not exceed 2 W/kg, according to these regulations. In the meantime, the Federal Communications Commission (FCC) recommended that the SAR for 1g of tissue not go beyond 1.6 W/kg. The formula is used to calculate the SAR, as mentioned in the equation:

$$SAR = \frac{\sigma |E|^2}{\rho} \tag{13.17}$$

where σ represent the electrical conductivity of tissue (S/m) and ρ is the density of the tissue (kg/m^3), respectively. The density and electrical conductivity are known, so the SAR value can be computed by knowing the electric field intensity (E^2). The lower the SAR value, the less radiation the human body absorbs. Reduction of SAR is most important to implement the antenna in wearable devices and consumer electronics, as part of the reduction of SAR is investigated using

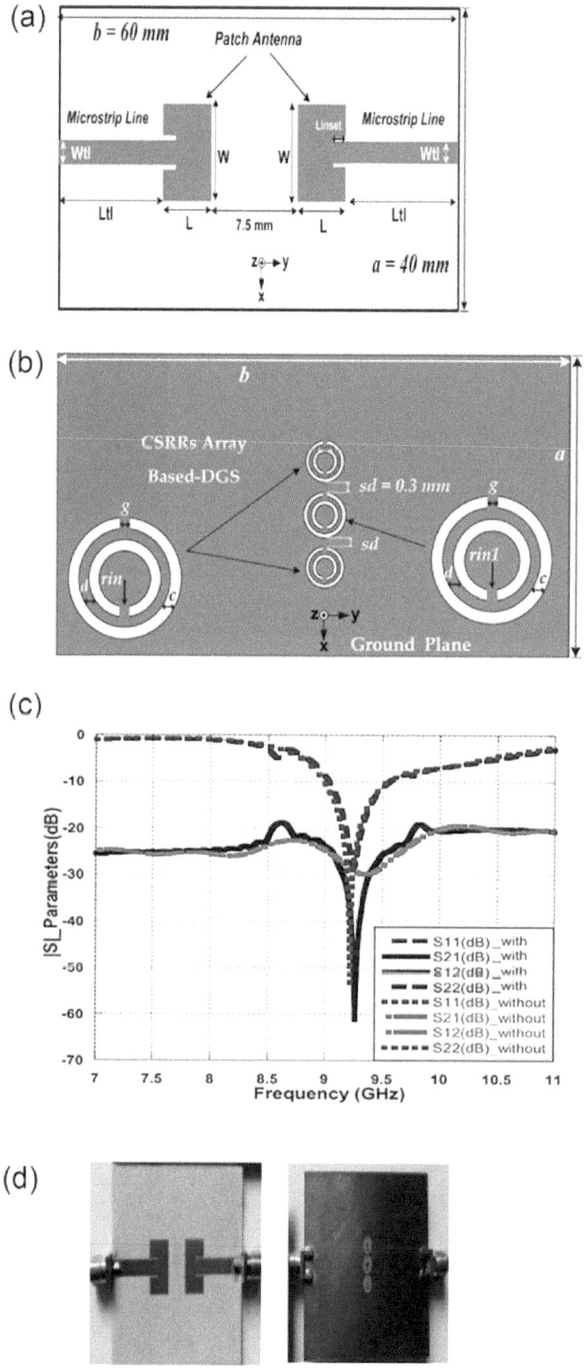

FIGURE 13.26 (a) Top view, (b) bottom view, (c) S-parameters of an antenna array, and (d) fabricated top view and fabricated bottom view [59].

metasurfaces. The spreading of SAR values of the antenna with the metasurface is proposed in [61].

13.8.2 Bending and Crumpling Effects

Flexible and wearable antennas are expected to bend, crumple, and twist while functioning on the human body. These effects alter the antenna performance for practical application. The flexible antenna bending performance is very important to analyze its aptness for conformal applications. During the bending operation in XZ plane, actual length is reduced; consequently, resonance is shifted to higher frequency, as well as while bending in YZ plane, the width of patch degrades the impedance matching; therefore, bandwidth is reduced [62]. The circular bending, folded bending, and human phantom modelling are carried out and investigated with different radii [63]. The difference in gain for various circular bending radii on both the X- and Y-axes is shown in Table 13.7. Hence, the flexible antenna must design without affecting the antenna performance when it is conformal.

13.8.3 Other Factors

13.8.3.1 Washability

Wearable textile-based antennas are regularly subjected to dust, dirt, and sweating, all of which can degrade performance. Textile antennas implanted in clothes may also be immersed in water and washed/laundered. After washing, the antenna's performance must be constant [11], and the preference of conductive and substrate materials has the greatest impact on performance stability.

13.8.3.2 Environmental Factors

The material qualities are affected by environmental factors, such as humidity and temperature. Because the wireless systems are sometimes expected to operate in difficult settings, performance strength measures are necessary. While an antenna is planned to function in a humid environment, the relative humidity should be investigated. Both the electromagnetic characteristics of the material, the relative permittivity, and loss tangent rise when relative humidity of the atmosphere in which the antennas are operated is studied [64]. Since flexible antennas are mounted on flexible substrates, the fabrication process is a requirement to the development of

TABLE 13.7
Gain Variations of Different Frequencies for Various Radii

Frequency	Flat	X-axis bending gain (dBi)				Y-axis bending gain (dBi)			
(GHz)	Radii (mm)	25	45	55	65	25	45	55	65
2.4	4.7	3.93	4.38	4.51	4.72	4.13	4.72	4.95	5.05
2.45	4.72	3.81	4.23	4.38	4.62	4.07	4.6	4.83	4.94
2.5	4.73	3.65	4.05	4.2	4.47	4.01	4.41	4.66	4.78

flexible antennas. The printed circuit board technology–made antennas can be hard to employ for on-body applications because they are challenging to incorporate within clothing, particularly for biomedical applications. Table 13.8 lists several UWB MIMO antennas that have been published along with their diversity characteristics, decoupling strategies, and applications.

13.9 FABRICATION OF FLEXIBLE ANTENNAS

Several commonly used fabrication methods exist for the fabrication of flexible and wearable antennas, which are based on the description of electrically conductive and a substrate material. This section discusses about the various fabricating methods which are used for fabricating wearable and flexible antennas, including screen-printing, inkjet printing, flexography, 3D printing, photolithography, sewing, and embroidering. Each approach is described in brief with examples.

13.9.1 Screen-Printing

One of the most effective printing methods is screen-printing; the screen-printing technology is applied to deposit antenna structures on flexible substrates. Umar Hasni et al. [76] presented a screen-printing approach in the fabrication of antenna design. Conductive inks are used to create conductive parts of an antenna on cloth in screen-printed antennas. Screen-printed antennas have a much greater print resolution. For antenna manufacturing, NovaCentrix (HPSFG32) conductive ink made up of 1.5 micron silver particles with a density of 25 mΩ/mm^2 is employed. For screen-printing the antennas on fabric (82% nylon and 18% spandex), with Ryonet screen-printing equipment with a 230 mesh screen is used. The fabric is chosen because of its high thread count and flexibility to stretch. Another wearable RFID tag antenna design that operates at 915 MHz (UHF) was presented. The fabric RFIDs are screen-printed with 60% cotton, 40% viscose fabric using a 156 mesh and Nova Centrix FLX15 conductive ink. The screen-printed antenna design and RFID tag structure are depicted in Figure 13.27(a) and (b).

The Monza R6 chip is rectangular in form, having exterior dimensions of 464.1 μm × 442 μm. The chip features two metal pads for connecting antennas. The pads are 166 m × 422 m in size, with a 112 m separation between them. Because the bonding pads are large enough, isotropic conductive adhesive (ICA) rather than anisotropic conductive adhesive (ACA) can be used to improve electrical contact between a chip and a graphene antenna.

13.9.2 Inkjet Printing

Another method of fabricating the flexible antenna is inkjet printing. Unlike the conventional etching method, which has been extensively employed in industry, inkjet printing is a direct-write process that transfers the design pattern to the substrate directly without the need of masks. In addition, inkjet printing creates no waste by jetting a specific droplet of ink from the jet to the required place, in contrast to

TABLE 13.8
Several UWB MIMO Antennas with Different MIMO Diversity Parameters

Reference	Dimensions (mm^3)	Ports	Operating range (GHz)	Substrate/ε_r	Decoupling methods	Isolation (dB)	ECC/DG	Gain (dBi)	SAR (W/kg)	Flexible	Applications
[65]	40 × 40 × 1.5	4	3.18–4.60, 8.89–12	Jeans/1.6	Left–right extended Slotted ground stub	20.5	<0.12/NR	5	1	Yes	Wearable RF circuits and UWB notch band applications
[66]	54 × 54 × 0.13	4	3.5–11	Rogers 3003/3	Orthogonal	≥17	<0.03/9.98	4.5	NR	Yes	UWB flexible networks
[67]	22 × 31 × 0.125	2	2.9–12	Kapton/3.4	Slot etched on T-shaped ground stub	<−15	0.3/NR	NR	NR	Yes	UWB applications
[68]	55 × 35 × 1	2	3.5/5.8/8/9/10.6	Jeans/2	E-shaped stub at ground surface	>19	< 0.06/<9.975	NR	1.27	NR	Downlink–uplink WiMAX, WLAN, C-band
[69]	65 × 65 × 0.1	4	2.9–10.86	LCP/2.9	Orthogonal, cross-shaped decoupling branches	<−22	<0.01/9.999	4	1.04 at 4 GHz	Yes	Wearable and IoT applications
[70]	50 × 50 mm^2	4	2.9–12	Cotton/1.6	Orthogonal, without decoupling structure	>17	<0.045/9.9	2.83 (3GHz), 3.54 (6GHz), 4.84 (10GHz)	0.229 (4 GHz), 0.253 (7 GHz), 0.463 (10 GHz)	NR	UWB, smart clothing applications

[71]	30 × 58 × 1	2	2.4–5.8	Textile/1.34	Π-shaped stub on ground plane	<23	0.005/NR	3	0.36 (5 GHz)	Yes	Wearable applications
[72]	25 × 35 × 1.6	2	2.9–10.4	FR-4/4.4	Stub and open slots/ ground plane connected	20	0.3/9.99	5	0.474 (6.5 GHz)	No	UWB, wearable and biomedical applications
[73]	185 × 111 × 4.21	2	2.34–2.46, 3.66–6 2–30	Polyester fabric/2.25	EBG structure	−15	0.5/NR	6.59,11.6	0.88 (2.4 GHz) and 0.070 (5.2 GHz)	Yes	WBAN applications 2.4 and 5.2 GHz
[74]	92 × 92 × 1.5	4	2–14	Polyethylene foam/2.5	Ground planes of antenna elements rotated 90°	>22	0.36/NR	7.2	Acceptable range	Yes	Footwear applications
[75]	21 × 16 × 1	2	3.35–4.12/6.07–11.67	Jeans/1.6	Slotted stub with DGS	20.3	0.2/NR	3.75	<1.0 (10.6GHz)	Yes	Wearable/ WLAN notch UWB applications

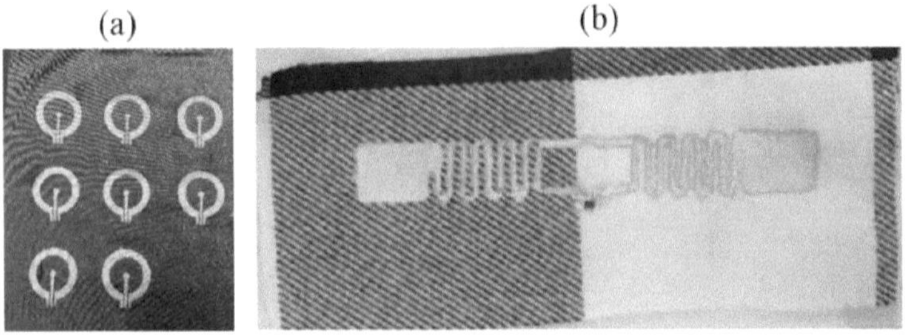

FIGURE 13.27 (a) Screen-printed on fabric structure; (b) screen-printed RFID design.

Source: [76].

etching, which is a subtractive procedure that removes undesirable metal from the substrate surface. Mahmuda Akter Monne et al. [77] presented a low-cost flexible phased antenna array using the inkjet printer graphene ink–based fabrication method. Every thin film and device are printed using a Fujifilm Dimatix (model DMP-2800 version) inkjet printer (Fujifilm, CA, USA). It came with a Dimatix cartridge (DMC-11610 version) made by Fujifilm. Each of the 16 piezoelectric jetting nozzles on the cartridge head has a length of 21.5 μm. From the cartridge head, the nozzles may discharge a droplet with a nominal volume of 10 pL. In this process, graphene ink is used, and six nozzles are utilized to print the graphene ink, which had an ideal drop spacing of 30 μm. Apart from a 0.3 sec eliminate per 10 printing passes, there was no pause in between successive printing passes. Using ACE 300 software, the design patterns have been converted into a printable version for the Fujifilm Dimatix printer. The Fujifilm Dimatix printer and image of printed antenna are shown in Figures 13.28(a), (b), and (c), respectively.

A Dimatix DMP 2800 model printer is employed to inkjet-print the designed antenna on a 0.125 mm flexible Kapton polyimide substrate [67]. A silver nanoparticle conductive ink is used. Three ink layers were jetted to provide the proper conductivity and continuous pattern. A 30 min post-sintering procedure at 150°C was used to maximize conductivity. A two-element UWB MIMO antenna with a dimension of $22 \times 31 \times 0.125$ mm^3 is presented by Wentao Li et al. In this work, the compact flexible antenna for UWB applications is created via inkjet printing on 0.125 mm thick flexible Kapton material substrate, as seen in Figure 13.28(d).

13.9.3 Flexography

Flexography printing is one of the most used printing processes. The basic concept of flexography is to transfer ink from an embossed cylinder to a substrate. Patterns are printed on individual plates, which are then joined to a plate cylinder. The plates are often made of rubber or photopolymers. An anilox roll is used to apply ink to the plate cylinder. An *anilox roll* is an etched cylinder filled with little cups. The cups are filled with ink, and the anilox roll transfers ink to the flexo cylinder. The cups are often etched with a laser, and the size varies depending on the design and ink. A doctor blade job is to remove excess ink from the anilox roll surface.

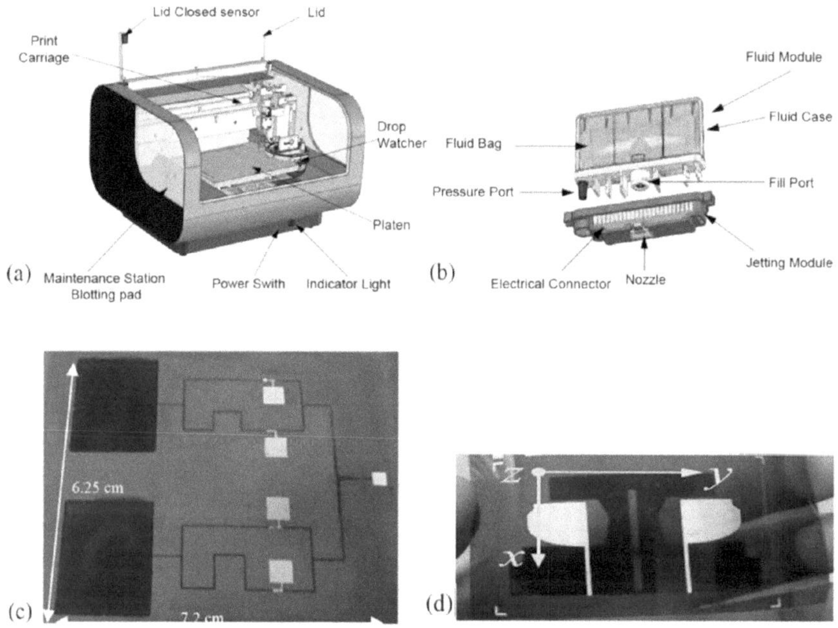

FIGURE 13.28 (a) DMP-2800 printer and (b) Fujifilm cartridge; (c) image of printed antenna; (d) proposed fabricated antenna structure.

Source: (a–b) [78], (c) [77], (d) [67].

The pressure between the substrate and the printing plate is set by the impression roll. The surface roughness of a substrate, the material of a printing plate and its mounting tape, the kind of an anilox roll, the printing speed, and the printing pressure are all factors that influence print quality. It is quite difficult to produce patterns with both big regions and delicate features with flexography printing. The settings must be modified one characteristic at a time. Smaller anilox roll cups are necessary to exact patterns or thin layers, whereas bigger cups are required for thicker layers [79]. The structure of flexography printer model is depicted in Figure 13.29(a). The flexible low-cost PET substrate printed antenna is designed for UHF applications by Diego Betancourt et al. [80]. In this chapter, the flexographic method is used to print the designed antenna. The semi-industrial roll-to-roll (R2R) machine is used for the printing process. The R2R printing machine and various parts are shown in Figure 13.29(b) and (c).

13.9.4 3D Printing

The most preferred printing technology is 3D printing technology, as seen in Figure 13.30. The most popular and affordable 3D printing methods, which have become more common recently, are especially well-suited for the creation of dielectric resonator antennas and horn antennas. The process of creating a solid item additively using a digital model is known as 3D printing. Compared to other conventional processes, this technology has the benefit of being able to construct structures

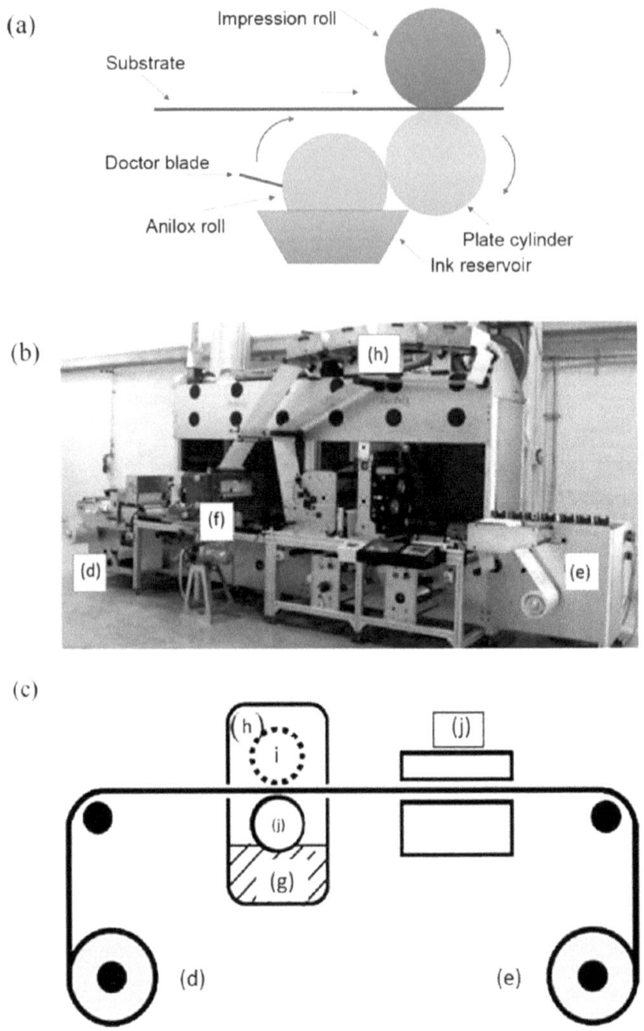

FIGURE 13.29 (a) Structure of flexography printer model; (b) and (c) R2R printing machine and internal structure; (d) un-winder roller; (e) re-winder roller; (f) flexographic printer unit; (g) ink chamber; (h) the plate; (i) print roller; (j) anilox roller; and (h) belt furnace.

Source: (a) [79], (b–c) [80].

with a wide range of forms, different angles of complexity, and levels of flexibility, all at a low price of manufacture. Layer by layer, a real thing is created by the additive process of 3D printing, which starts with a digital model and deposits materials. The flexible and wearable antenna system is printed on a 3D-printed substrate, as described in [81]; the authors Muhammad Rizwan et al. propose a fabrication and functioning analysis of flexible and stretchy wearable antenna on 3D-printed substrate. With NinjaFlex substance, the antenna patch is brush-painted with stretchy

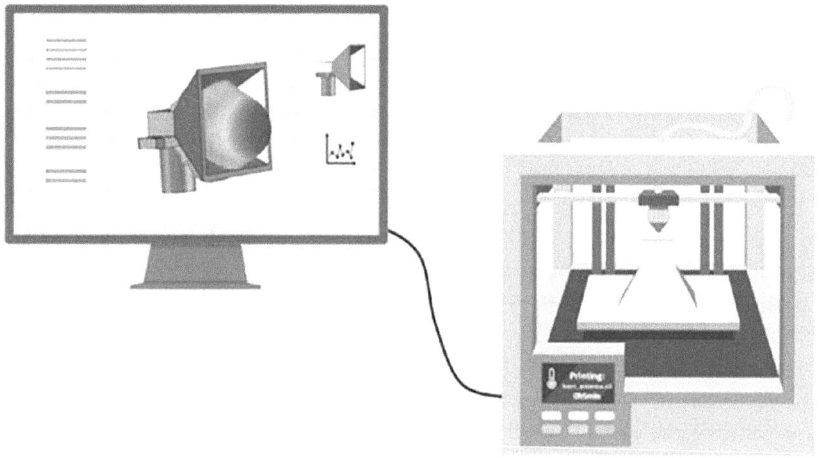

FIGURE 13.30 3D printer technology.

Source: [82].

silver conductive paste. The machine used for the fabrication of antenna is Prenta Duo 3D printer, which has two 1.75 mm printing nozzles. To print an antenna substance material, the printing speed is usually set to 35 mm/s, the bed temperature is set to 60°C, and the nozzle printing temperature is set to between 230 and 235°C. The substrate dimensions are configured to 65 × 54 × 1.2 mm^3, with a 100% infill and a 0.2 mm thickness for each printing layer. The entire six layers are printed, and the substrate model is transformed into a three-dimensional pattern using a slicer tool. Printing the antenna substrate requires 1,955 mm of filament and takes approximately 45 minutes. In order to calculate the infill percentage, the printed density and filament density are compared after the samples are printed. Brush painting is employed in the fabrication of the antenna. The NinjaFlex 3D-printed substrate is brush-painted with the elastic silver conductive paste before being thermally cured for 15 min at 110°C.

13.9.5 Photolithography

For the photolithographic process, the proper wavelength of ultraviolet (UV) light and photoresist sensitive to wavelength are required. There are two varieties of photoresist materials: positive and negative. Positive photoresist that has been exposed dissolves, and the negative photoresist hardens. Using UV, a type of light from an insect killer Philips TM fluorescent tube lamp and photoresist material in the form of the dry negative photoresist film to be placed as lamination to copper clad-type substrate are the two main tools employed in this investigation. This UV light has a wavelength of around 370 nm [83]. The authors Sanjay M. Palhade and S. P. Yawale presented the fabrication process step by step. The antenna geometry is constructed using computer simulation software as a first step. The mask is an image of the geometry printed on a transparent sheet in negative form. A double-sided 65 mm ×

65 mm copper-clad substrate (RO3003TM) is dried after being thoroughly cleaned with acetone. The etched pattern becomes discontinuous due to dust particles or other impurities on the copper-clad surface, which modifies the resonance frequency. The cleaned and dried-out copper-clad substrate is laminated with a negative photoresist film in the second phase. The photoresist laminated copper-clad substrate is securely covered with the negative mask made from the first stage. Ultraviolet (UV) light is applied to the copper-clad substrate that has been photoresist laminated and covered. The development of the UV-subjected photoresist laminated copper-clad substrate is the third stage. While photoresist that has not been exposed to UV light stays light blue and dissolves in the sodium carbonate solution, photoresist that has been subjected to UV light turns hard and dark blue in colour. Finally, a solution of ferric chloride (FeC13) is applied to chemically etch on the copper-clad substrate. FeC13 dissolves the copper components – all but those behind the hardened photoresist. After washing the etched substrate under running water to get rid of any etchant, it is then dried. Sodium hydroxide is used to remove the solidified photoresist. The stepwise view of fabricated antenna is seen in Figure 13.31.

13.9.6 Sewing and Embroidering

The majority of textile-based antennas use sewing or embroidery machines for fabrication methods, which are preferable to direct E-textile adhesion over cloth since no adhesive ingredients are used that might alter the material electrical characteristics.

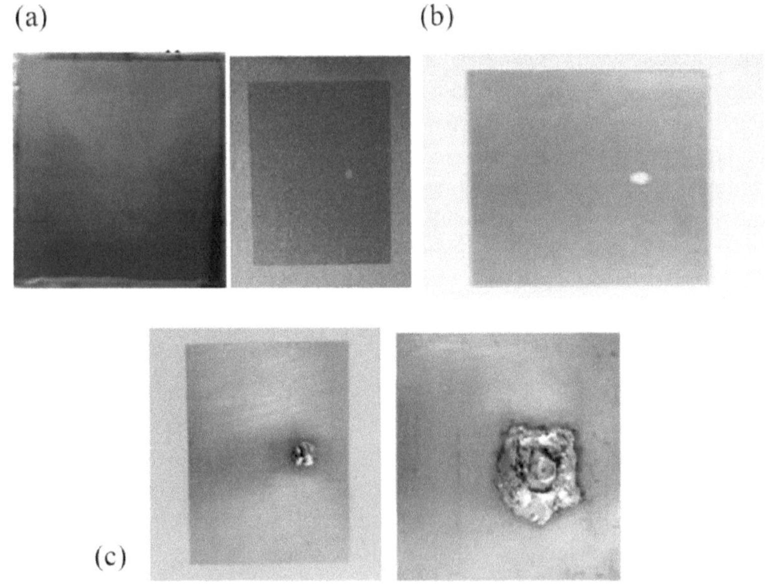

FIGURE 13.31 (a) Photoresist laminated UV-exposed laminated copper-clad RO3003; (b) final photoresist removed; and (c) final fabricated antenna, top and bottom view.

Source: [83].

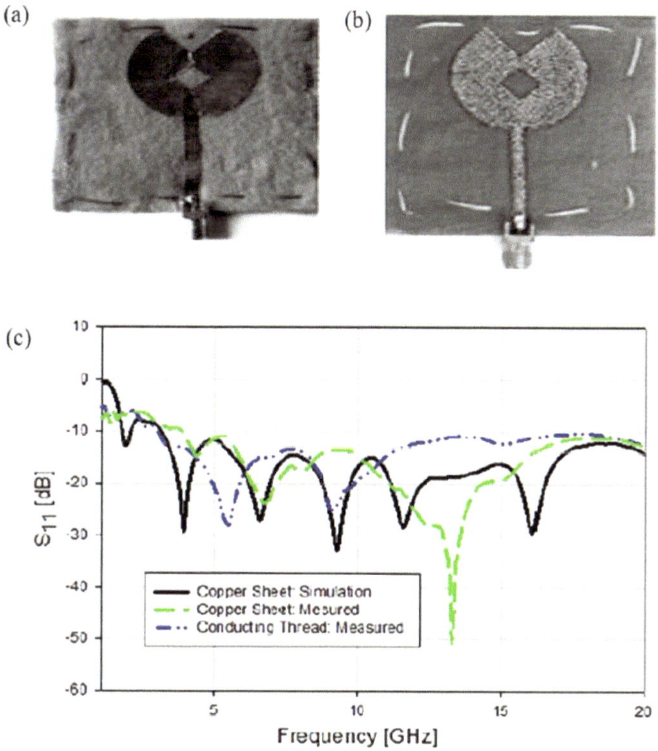

FIGURE 13.32 Fabricated prototype of antenna: (a) copper conductive sheet antenna and (b) conductive thread antenna; (c) S_{11} of textile UWB antenna.

Source: [85].

This is the most recommended manufacturing approach for wearable antennas on clothes [84]. The conducting thread and self-adhesive sheet conducting material are used on the radiative and ground elements [85]. As a substrate, three layers of flannel fabric are employed. The conductive thread is stitched on the solid surface of the flannel fabric substrate using a computer-controlled embroidery machine. Stitching the conducting thread on the surface of the fabric develops a conducting region on the top and bottom layers of the proposed antenna, as seen in Figure 13.32(a) and (b), and reflection coefficient of the designed textile UWB antenna.

13.10 CONCLUSION

The flexible multi-input, multi-output (MIMO) antennas for ultrawideband (UWB) applications are reviewed in this chapter. The need for flexible materials and the significance of the flexible materials are presented. The key challenges in creating MIMO antennas for portable wireless communication devices and a decrease in the mutual coupling between the antenna ports were discussed. Several flexible substrates, including

Kapton polyimide (PI), polydimethylsiloxane (PDMS), liquid crystal polymer (LCP), polyethylene terephthalate (PET), and textile substrate, have all been studied. Various aspects of flexible antenna, such as specific absorption ratio (SAR), durability, bending, and crumbling, are examined. It also encompasses flexible substrate production technologies, such as screen-printing, photolithography, inkjet printing, and so on. In this study, the mutual coupling in MIMO antennas and the several decoupling techniques for isolation improvements, such as neutralization lines, defective ground structures (DGS), decoupling networks, metamaterials, parasitic elements, electromagnetic bandgap, etc., have been presented. In addition, isolation, envelope correlation coefficient (ECC), total active reflection coefficient (TARC), mean effective gain (MEG), diversity gain (DG), and channel capacity loss have all been examined as MIMO characteristics.

REFERENCES

[1] S. Movassaghi, M. Abolhasan, J. Lipman, D. Smith and A. Jamalipour, *Wireless body area networks: A survey*, IEEE Communications Surveys and Tutorials 16 (2014), pp. 1658–1686.

[2] International Diabetes Federation (IDF) [Online], https://idf.org/news/more-than-two-in-three-people-with-diabetes-already-have-complications-at-diagnosis/

[3] V.K. Jhunjhunwala, T. Ali, P. Kumar, P. Kumar, P. Kumar, S. Shrivastava and A.A. Bhagwat, *Flexible UWB and MIMO antennas for wireless body area network: A review*, Sensors 22 (2022), p. 9549.

[4] S.G. Kirtania, A.W. Elger, M.R. Hasan, A. Wisniewska, K. Sekhar, T. Karacolak and P.K. Sekhar, *Flexible antennas: A review*, Micromachines 11 (2020), p. 847.

[5] A. Astrin, *IEEE standard for local and metropolitan area networks part 15.6: Wireless body area networks*, IEEE Std 802.15. 6 (2012).

[6] M. Yaghoubi, K. Ahmed and Y. Miao, *Wireless body area network (WBAN): A survey on architecture, technologies, energy consumption, and security challenges*. Journal of Sensor and Actuator Networks 11 (2022), p. 67.

[7] C.A. Tavera, J.H. Ortiz, O.I. Khalaf, D.F. Saavedra and T.H.H. Aldhyani, *Wearable wireless body area networks for medical applications*, Computational and Mathematical Methods in Medicine 2021 (2021).

[8] K. Hasan, K. Biswas, K. Ahmed, N.S. Nafi and M.S. Islam, *A comprehensive review of wireless body area network*. Journal of Network and Computer Applications 143 (2019), pp. 178–198.

[9] R. Cavallari, F. Martelli, R. Rosini, C. Buratti and R. Verdone, *A survey on wireless body area networks: Technologies and design challenges*, IEEE Communications Surveys and Tutorials 16 (2014), pp. 1635–1657.

[10] Y. Qu, G. Zheng, H. Ma, X. Wang, B. Ji and H. Wu, *A survey of routing protocols in WBAN for healthcare applications*, Sensors 19 (2019).

[11] H. Khaleel, *Innovation in Wearable and Flexible Antennas*, WIT Press 2014.

[12] M. Ikram, K. Sultan, M.F. Lateef and A.S. Alqadami, *A road towards 6G communication—A review of 5G antennas, arrays, and wearable devices*, Electronics 11 (2022), p. 169.

[13] S. Parameswari and C. Chitra, *Compact textile UWB antenna with hexagonal for biomedical communication*, Journal of Ambient Intelligence and Humanized Computing (2021), pp. 1–8. https://doi.org/10.1007/s12652-021-03228-3

[14] R.B.V.B. Simorangkir, Y. Yang, K.P. Esselle and B.A. Zeb, *A method to realize robust flexible electronically tunable antennas using polymer-embedded conductive fabric*, IEEE Transactions on Antennas and Propagation 66 (2018), pp. 50–58.

[15] A. Mersani, L. Osman and J.-M. Ribero, *Flexible UWB AMC antenna for early stage skin cancer identification*, Progress in Electromagnetics Research M 80 (2019), pp. 71–81.

[16] B. Mohamadzade, R.B.V.B. Simorangkir, R.M. Hashmi and A. Lalbakhsh, *A conformal ultrawideband antenna with monopole-like radiation patterns*, IEEE Transactions on Antennas and Propagation 68 (2020), pp. 6383–6388.

[17] M. Borghetti, M. Serpelloni, E. Sardini and S. Pandini, *Mechanical behavior of strain sensors based on PEDOT:PSS and silver nanoparticles inks deposited on polymer substrate by inkjet printing*, Sensors and Actuators A: Physical 243 (2016), pp. 71–80.

[18] O. Koritsoglou, I. Theodorakos, F. Zacharatos, M. Makrygianni, D. Kariyapperuma, R. Price, Copper micro-electrode fabrication using laser printing and laser sintering processes for on-chip antennas on flexible integrated circuits, Optical Materials Express 9 (2019), p. 3046.

[19] A. Lamminen, K. Arapov, G. De With, S. Haque, H.G.O. Sandberg, H. Friedrich, *Graphene-flakes printed wideband elliptical dipole antenna for low-cost wireless communications applications*, IEEE Antennas and Wireless Propagation Letters 16 (2017), pp. 1883–1886.

[20] M.A.S.M. Al-Haddad, N. Jamel and A.N. Nordin, *Flexible antenna: A review of design, materials, fabrication, and applications*, Journal of Physics: Conference Series 1878 (2021), p. 012068.

[21] H.F. Abutarboush, *Silver nanoparticle inkjet-printed multiband antenna on synthetic paper material for flexible devices*, Alexandria Engineering Journal 61 (2022), pp. 6349–6355.

[22] T.K. Saha, T.N. Knaus, A. Khosla and P.K. Sekhar, *A CPW-fed flexible UWB antenna for IoT applications*, Microsystem Technologies 28 (2022), pp. 5–11.

[23] S.N. Mahmood, A.J. Ishak, T. Saeidi, A.C. Soh, A. Jalal and M.A. Imran, *Full ground ultra-wideband wearable textile antenna for breast cancer and wireless body area network applications*, Micromachines 12 (2021).

[24] X. Lin, Y. Chen, Z. Gong, B.C. Seet, L. Huang and Y. Lu, *Ultrawideband textile antenna for wearable microwave medical imaging applications*, IEEE Transactions on Antennas and Propagation 68 (2020), pp. 4238–4249.

[25] H.R. Khaleel, H.M. Al-Rizzo and D.G. Rucker, *Compact polyimide-based antennas for flexible displays*, IEEE/OSA Journal of Display Technology 8 (2012), pp. 91–96.

[26] S. Ahmed, F.A. Tahir, A. Shamim and H.M. Cheema, *A compact Kapton-based inkjet-printed multiband antenna for flexible wireless devices*, IEEE Antennas and Wireless Propagation Letters 14 (2015), pp. 1802–1805.

[27] A. Hassan, S. Ali, G. Hassan, J. Bae and C.H. Lee, *Inkjet-printed antenna on thin PET substrate for dual band Wi-Fi communications*, Microsystem Technologies 23 (2017), pp. 3701–3709.

[28] H.F. Abutarboush and A. Shamim, *A reconfigurable inkjet-printed antenna on paper substrate for wireless applications*, IEEE Antennas and Wireless Propagation Letters 17 (2018), pp. 1648–1651.

[29] S.F. Jilani, M.O. Munoz, Q.H. Abbasi and A. Alomainy, *Millimeter-wave liquid crystal polymer based conformal antenna array for 5G applications*, IEEE Antennas and Wireless Propagation Letters 18 (2019), pp. 84–88.

[30] K.I. Jang, S.Y. Han, S. Xu, K.E. Mathewson, Y. Zhang, J.W. Jeong, *Rugged and breathable forms of stretchable electronics with adherent composite substrates for transcutaneous monitoring*, Nature Communications 5 (2014).

[31] A. Kavitha and J.N. Swaminathan, *Design of flexible textile antenna using FR4, jeans cotton and teflon substrates*, Microsystem Technologies 25 (2019), pp. 1311–1320.

[32] C.A. Balanis, *Antenna Theory Analysis and Design*, 2nd edition, Wiley (1997).

[33] G. Rajesh and R. Poonkuzhali, *A CPW fed flexible antenna for UWB applications*, in ViTECoN 2023–2nd IEEE International Conference on Vision Towards Emerging Trends in Communication and Networking Technologies, Proceedings, 2023.

[34] Z. Wang, L. Qin, Q. Chen, W. Yang and H. Qu, *Flexible UWB antenna fabricated on polyimide substrate by surface modification and in situ self-metallization technique*, Microelectronic Engineering 206 (2019), pp. 12–16.

[35] R. Zhang, Y.C. Liang, P.S.P. Hall, P. Gardner, J. Kelly and E. Ebrahimi, *Federal Communications commission revision of part 15 of the commission's rules regarding ultra-wideband transmission systems*, FCC, Washington, DC, First Report & Order 48 (2009), pp. 133–135.

[36] P. Kumar, S. Pathan, O.P. Kumar, S. Vincent, Y. Nanjappa, P. Kumar, *Design of a six-port compact UWB MIMO antenna with a distinctive DGS for improved isolation*, IEEE Access 10 (2022), pp. 112964–112974.

[37] S. Kumar, A.S. Dixit, R.R. Malekar, H.D. Raut and L.K. Shevada, *Fifth generation antennas: A comprehensive review of design and performance enhancement techniques*, IEEE Access 8 (2020), pp. 163568–163593.

[38] M. Alibakhshikenari, F. Babaeian, B.S. Virdee, S. Aissa, L. Azpilicueta, C.H. See, *A comprehensive survey on "Various decoupling mechanisms with focus on metamaterial and metasurface principles applicable to SAR and MIMO antenna systems,"* IEEE Access 8 (2020), pp. 192965–193004.

[39] P. Meerasri, P. Uthansakul and M. Uthansakul, *Self-interference cancellation-based mutual-coupling model for full-duplex single-channel MIMO systems*, International Journal of Antennas and Propagation 2014 (2014).

[40] S.R. Patre and S.P. Singh, *Shared radiator MIMO antenna for broadband applications*, IET Microwaves, Antennas and Propagation 12 (2018), pp. 1153–1159.

[41] E. Suganya, T. Prabhu, S. Palanisamy, P.K. Malik, N. Bilandi and A. Gehlot, *An isolation improvement for closely spaced MIMO antenna using/4 distance for WLAN applications*, International Journal of Antennas and Propagation 2023 (2023).

[42] B.T. Ahmed, D.C. Carreras and E.G. Marin, *Design and implementation of super wide band triple band-notched MIMO antennas*, Wireless Personal Communications 121 (2021), pp. 2757–2778.

[43] M.A. Abdelghany, M.F. Abo Sree, A. Desai and A.A. Ibrahim, *4-Port octagonal shaped MIMO antenna with low mutual coupling for UWB applications*, CMES—Computer Modeling in Engineering and Sciences 136 (2023), pp. 1999–2015.

[44] V.R. Balaji, T. Addepalli, A. Desai, A. Nella and T.K. Nguyen, *An inverted L-strip loaded ground with hollow semi-hexagonal four-element polarization diversity UWB-MIMO antenna*, Transactions on Emerging Telecommunications Technologies 33 (2022).

[45] U. Ahmad, S. Ullah, U. Rafique, D.Y. Choi, R. Ullah, B. Kamal, *MIMO antenna system with pattern diversity for sub-6 GHz mobile phone applications*, IEEE Access 9 (2021), pp. 149240–149249.

[46] Q.H. Kareem and M.J. Farhan, *Design a linear and circular polarization MIMO antennas based on compact size configurations with high isolation and stable gain characteristics for C-Band and WLAN/WiMAX applications*, International Journal of Intelligent Engineering and Systems 14 (2021), pp. 458–467.

[47] P. Sharma, R.N. Tiwari, P. Singh, P. Kumar and B.K. Kanaujia, *MIMO antennas: Design approaches, techniques and applications*, Sensors 22 (2022), p. 7813.

[48] A. Desai, J. Kulkarni, M.M. Kamruzzaman, S. Hubalovsky, H.T. Hsu and A.A. Ibrahim, *Interconnected CPW fed flexible 4-port MIMO antenna for UWB, X, and Ku band applications*, IEEE Access 10 (2022), pp. 57641–57654.

[49] R. Liu, X. An, H. Zheng, M. Wang, Z. Gao and E. Li, *Neutralization line decoupling tri-band multiple-input multiple-output antenna design*, IEEE Access 8 (2020), pp. 27018–27026.

[50] A. Diallo, C. Luxey, P. Le Thuc, R. Staraj and G. Kossiavas, *Study and reduction of the mutual coupling between two mobile phone PIFAs operating in the DCS 1800 and UMTS bands*, IEEE Transactions on Antennas and Propagation 54 (2006), pp. 3063–3074.

[51] Z. Li, Z. Du, M. Takahashi, K. Saito and K. Ito, *Reducing mutual coupling of MIMO antennas with parasitic elements for mobile terminals*, IEEE Transactions on Antennas and Propagation 60 (2012), pp. 473–481.

[52] H.H. Tran and N. Nguyen-Trong, *Performance enhancement of MIMO patch antenna using parasitic elements*, IEEE Access 9 (2021), pp. 30011–30016.

[53] M.A. Sufian, N. Hussain, A. Abbas, J. Lee, S.G. Park and N. Kim, *Mutual coupling reduction of a circularly polarized MIMO antenna using parasitic elements and DGS for V2X communications*, IEEE Access 10 (2022), pp. 56388–56400.

[54] K. Wei, J. Li, L. Wang, Z. Xing and R. Xu, *S-shaped periodic defected ground structures to reduce microstrip antenna array mutual coupling*, Electronics Letters 52 (2016), pp. 1288–1290.

[55] H. Sakli, C. Abdelhamid, C. Essid and N. Sakli, *Metamaterial-based antenna performance enhancement for MIMO system applications*, IEEE Access 9 (2021), pp. 38546–38556.

[56] S.D. Assimonis, T.V. Yioultsis and C.S. Antonopoulos, *Design and optimization of uniplanar EBG structures for low profile antenna applications and mutual coupling reduction*, IEEE Transactions on Antennas and Propagation 60 (2012), pp. 4944–4949.

[57] N. Kumar and U.K. Kommuri, *MIMO antenna H-plane isolation enhancement using UC-EBG structure and metal line strip for WLAN applications*, Radioengineering 27 (2019), pp. 399–406.

[58] C. Yu, S. Yang, Y. Chen, W. Wang, L. Zhang, B. Li., *A super-wideband and high isolation MIMO antenna system using a windmill-shaped decoupling structure*, IEEE Access 8 (2020), pp. 115767–115777.

[59] H.A. Atallah, A.B. Abdel-Rahman, K. Yoshitomi and R.K. Pokharel, *Mutual coupling reduction in MIMO patch antenna array using complementary split ring resonators defected ground structure*, Applied Computational Electromagnetics Society Journal 31(7) (2015).

[60] J.I. Bangash, A.H. Abdullah, M.H. Anisi and A.W. Khan, *A survey of routing protocols in wireless body sensor networks*, Sensors 14 (2014), pp. 1322–1357.

[61] M. Wang, Z. Yang, J. Wu, J. Bao, J. Liu and L. Cai, *Investigation of SAR reduction using flexible antenna with metamaterial structure in wireless body area network*, IEEE Transactions on Antennas and Propagation 66 (2018), pp. 3076–3086.

[62] A. Somasundaram, S.K. Thengalpalayam Rajamanickam and Z.C. Alex, *Development of wideband CPW-fed microstrip patch antenna on textile substrate*, Microwave and Optical Technology Letters 65 (2023), pp. 2352–2358.

[63] I. Adam, M.R. Kamarudin, A.H. Rambe, N. Haris, H.A. Rahim and W.Z.A. Wan Muhamad, *Investigation on wearable antenna under different bending conditions for wireless body area network (WBAN) applications*, International Journal of Antennas and Propagation 2021 (2021).

[64] C. Hertleer, A. Van Laere, H. Rogier and L. Van Langenhove, *Influence of relative humidity on textile antenna performance*, Textile Research Journal 80 (2010), pp. 177–183.

[65] A.K. Biswas, S. Biswas, S. Haldar and A. Nandi, *A highly decoupled flexible 4-element MIMO antenna with band notched characteristics for ultra wide-band wearable applications*, AEU—International Journal of Electronics and Communications 173 (2024).

[66] A.A. Ibrahim, M.I. Ahmed and M.F. Ahmed, *A systematic investigation of four ports MIMO antenna depending on flexible material for UWB networks*, Scientific Reports 12 (2022).

[67] W. Li, Y. Hei, P.M. Grubb, X. Shi and R.T. Chen, *Compact inkjet-printed flexible MIMO antenna for UWB applications*, IEEE Access 6 (2018), pp. 50290–50298.

[68] S. Rekha and G.S. Let, *Design and SAR analysis of wearable UWB MIMO antenna with enhanced isolation using a parasitic structure*, Iranian Journal of Science and Technology—Transactions of Electrical Engineering 46 (2022), pp. 291–301.

[69] J. Zhang, C. Du and R. Wang, *Design of a four-port flexible UWB-MIMO antenna with high isolation for wearable and IoT applications*, Micromachines 13 (2022).

[70] T. Govindan, S.K. Palaniswamy, M. Kanagasabai and S. Kumar, *Design and analysis of UWB MIMO antenna for smart fabric communications*, International Journal of Antennas and Propagation 2022 (2022).
[71] S. Kumar and B. Gupta, *Design of compact textile MIMO with high isolation for wearable applications*, Wireless Personal Communications 131 (2023), pp. 337–355.
[72] P. Gupta, M. Bharti and A. Kumar, *Two-element UWB antenna with multiple open slots in fountain-shaped ground for wearable and biomedical applications*, Mapan—Journal of Metrology Society of India 38 (2023), pp. 203–215.
[73] B. Kumkhet, P. Rakluea, N. Wongsin, P. Sangmahamad, W. Thaiwirot and C. Mahatthanajatuphat, *SAR reduction using dual band EBG method based on MIMO wearable antenna for WBAN applications*, AEU—International Journal of Electronics and Communications 160 (2023).
[74] S. Jayant, G. Srivastava and S. Kumar, *Quad-port UWB MIMO footwear antenna for wearable applications*, IEEE Transactions on Antennas and Propagation 70 (2022), pp. 7905–7913.
[75] A.K. Biswas, B. Basu and A. Nandi, *Compact wearable UWB MIMO antenna with reduced mutual coupling and notch characteristics of WLAN band*, Arabian Journal for Science and Engineering 47 (2022), pp. 14561–14569.
[76] U. Hasni, M.E. Piper, J. Lundquist and E. Topsakal, *Screen-printed fabric antennas for wearable applications*, IEEE Open Journal of Antennas and Propagation 2 (2021), pp. 591–598.
[77] M.A. Monne, P.M. Grubb, H. Stern, H. Subbaraman, R.T. Chen and M.Y. Chen, *Inkjet-printed graphene-based 1 x 2 phased array antenna*, Micromachines 11 (2020).
[78] FUJIFILM Dimatix Materials Printer DMP-2800 Series User Manual, FUJIFILM Dimatix Materials Printer, https://content.ilabsolutions.com/wp-content/uploads/2016/11/DMP-2800-Users-Guide-Version-2.0–6.pdf
[79] D.R. Gamota, P. Brazis, X. Kalyanasundaram and J. Zhang, *Printed Organic and Molecular Electronics,* Springer Science & Business Media (2004).
[80] D. Betancourt and J. Castan, *Printed antenna on flexible low-cost pet substrate for UHF applications*, Progress in Electromagnetics Research C 38 (2013), pp. 139–140.
[81] M. Rizwan, M.W.A. Khan, L. Sydanheimo, J. Virkki and L. Ukkonen, *Flexible and stretchable brush-painted wearable antenna on a three-dimensional (3-D) printed substrate*, IEEE Antennas and Wireless Propagation Letters 16 (2017), pp. 3108–3112.
[82] D. Helena, A. Ramos, T. Varum and J.N. Matos, *Antenna design using modern additive manufacturing technology: A review*, IEEE Access 8 (2020), pp. 177064–177083.
[83] S.M. Palhade and S.P. Yawale, *Design and photo-lithographic fabrication of microstrip patch antenna*, International Journal of Science and Research 4 (2013), pp. 2021–2024.
[84] I. Locher, M. Klemm, T. Kirstein and G. Tröster, *Design and characterization of purely textile patch antennas*, IEEE Transactions on Advanced Packaging 29 (2006), pp. 777–788.
[85] M.A.R. Osman, M.K.A. Rahim, N.A. Samsuri, H.A.M. Salim, M.F. Ali and M.A.R. Osman, *Embroidered fully textile wearable antenna for medical monitoring applications*, Progress in Electromagnetics Research 117 (2011), pp. 321–337.

14 Design of Inset-Fed Microstrip Patch Antenna Using Surrogate Model–Assisted Differential Evolution

Vishnu P. R., Gopi Ram, and E. Aparna

14.1 INTRODUCTION

In modern optimization algorithms, the concept of machine learning is also incorporated. Machine learning has become an important and interesting approach to design and optimization recently. They are either parametric or non-parametric in nature [1]. To construct non-linear mapping between input features and real values outputs, non-parametric regression provides a flexible approach. Some of the common examples involve kernel-based approaches, like Gaussian process regression (GPR) and support vector regression (SVR), and ensemble-based approaches, like random forest, neural networks–based methods, etc. [2]. Among these, GPR provides a scalable and novel approach. It is extensively used in various data analytics applications. This approach is only limited by its data handling capability, where the complexity becomes large when the dataset is high [3].

Patch antennas are among the antenna structures that find increasing use in the field of wireless technologies. They are very flexible in nature, namely, having easiness of construction, implementation, and they can be used in applications that involve frequency selectivity, sharp radiation pattern, and can also be used for lossy applications [4]. However, construction of this is often a cumbersome process due to complexities arising while designing various parameters associated with antenna synthesis [5].

Differential evolution is one among the evolutionary algorithms used for optimization of antenna parameters. Studies have proven that DE has very fast convergence and is simple and robust in nature. When global optimum is reached, this algorithm slows down. Also, when the number of parameters to be optimized is very large, DE alone would not be efficient enough to give the best values of parameters [6]. In the basic differential evolution algorithm, a trial vector is compared with the target vector, and this process is kept in recurrence as long as convergence is obtained. A cost function is defined for the given optimization problem, which is minimized or maximized according to the requirement. When convergence criterion is met, the obtained values are taken as the final and best result, thus stopping the algorithm [7].

DOI: 10.1201/9781032625034-14

In order to overcome the effects of such complexities and calculations, it is often advantageous to blend evolutionary algorithms with machine learning to get optimum results. Initially, whatever values were obtained as the result of evolutionary algorithm output are used to train the Gaussian regression model. Then, values of parameters to be optimized are taken as output of the model. By training this model for some range of input, optimum values of parameters can be obtained. On comparing this method over conventional evolutionary algorithms, it can be proved that both speedy and optimum solutions are obtained in this scenario [8].

14.2 DESIGN THEORY

The microstrip antenna with inset feed patch is made of a metallic rectangular patch mounted on an FR4 substrate, having dielectric constant value of 4.4 and loss tangent of 0.03 and a ground plane. Impedance matching is easily done by choosing optimum position of inset feed [9].

Figure 14.1 shows the structure of antenna to be designed. It should be taken care that the ratio of width (W) to height (h) is always greater than 1.

By taking the metal thickness 0.035 mm, the length of inset feed (Lin) for each substrate as L/2, and gap (g) between feed line and patch as 1 mm, the values of parameters and return loss of patch antenna can be calculated using the following formulas to resonate at 2.4 GHz.

The effective dielectric constant is calculated using:

$$\epsilon_{eff} = (\epsilon + 1)/2 + ((\epsilon - 1)/2) * ((1 + 12h/W)^{-1/2}) \tag{14.1}$$

The resonant frequency is calculated using

$$fr = v / ((L + 2\Delta L) * 2\sqrt{\epsilon_{eff}}) \tag{14.2}$$

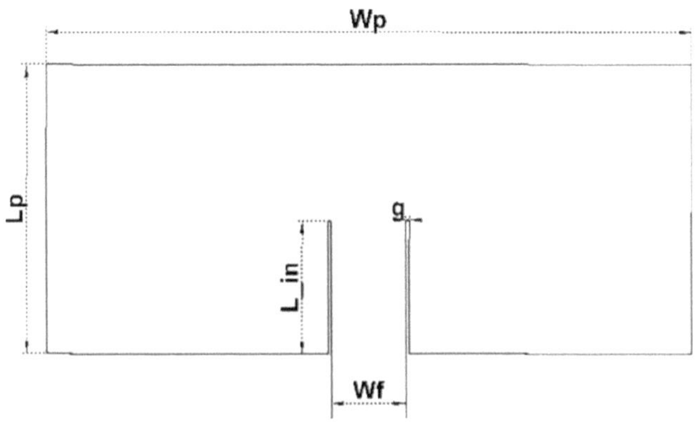

FIGURE 14.1 Microstrip antenna with inset feed patch structure.

where ε is the dielectric constant of material and *h*, *W* are height and width, respectively.

The change in length is calculated using:

$$\Delta L = (0.412h * (\epsilon_{eff} + 0.3)(W / h + 0.264)) / (\epsilon_{eff} - 0.258)(W / h + 0.8) \quad (14.3)$$

where v is the velocity of the wave.

The input impedance is calculated using:

$$Rin = Zo / (cos^2(\pi Lin / L)) \quad (14.4)$$

where *Zo* is the characteristic impedance.

The reflection coefficient is calculated using:

$$S11 = (\text{Rin} - 50)/(\text{Rin} + 50) \quad (14.5)$$

The return loss is calculated using:

$$\text{Lr} = 20\log(1/\text{S11}) \quad (14.6)$$

Values of *L*, *W*, and *Lin* are calculated using this method for a minimum value of return loss.

14.3 DIFFERENTIAL EVOLUTION ALGORITHM

Differential evolution is an evolutionary algorithm for optimization. DE dominates over other techniques in its uniqueness of simplicity, high reliability, good performance, and relatively low complexity. It can be applied for optimization of linear, non-linear, noisy, single, as well as multidimensional problems. It is based on real value operators. It constitutes mainly four different steps: initialization, mutation, crossover, and selection. The algorithm performance is judged based on crossover and mutation parameters as defined in the following section [10].

14.3.1 Initialization

This is the first step of DE, where initial population is generated randomly within the search limits. These values are iterated over and over using the given formula to generate newer population from the existing one.

$$x_{i,G}^{j} = x_{min}^{j} + random(0,1).(x_{max}^{j} - x_{min}^{j}) \quad (14.7)$$

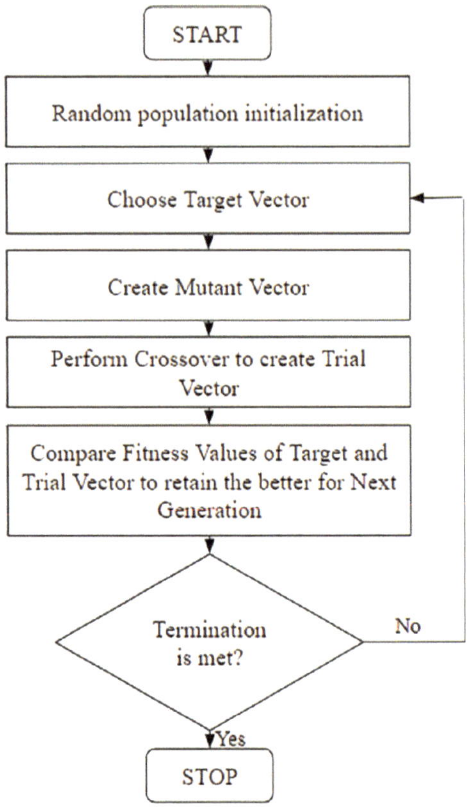

FIGURE 14.2 Differential evolution algorithm.

14.3.2 Mutation

Take a member from the population generated by the preceding process and select three vectors, such that *i*, *r1*, *r2*, and *r3* are distinct. From this, a mutant vector is generated using the following formula:

$$v_{i,G+1} = x_{r1,G} + F(x_{r2,G} - x_{r3,G}) \tag{14.8}$$

where F is a scaled parameter in the open interval (0,1).

14.3.3 Crossover

A trial vector is generated by performing a binomial crossover between the target and the mutant vectors generated previously. It is basically generated from the genes of the target and mutant vectors with the help of another user-defined parameter known as crossover probability parameter (CR). This crossover probability constant represents the probability that trial vector acquires and inherits the parameter values from the mutant vector.

$$u_{ji,G+1} = v_{ji,G+1} if (rand_j \leq CR) \tag{14.9}$$

$$u_{ji,G+1} = x_{i,G}; otherwise \tag{14.10}$$

Finally, the trial and target vector fitness values are compared to choose the fitness function which is better for next generation. This process is iterated as long as the convergence or termination criterion is met.

14.4 SURROGATE MODEL–ASSISTED DIFFERENTIAL EVOLUTION MODEL FOR ANTENNA SYNTHESIS

Optimization problem is to minimize a cost function. So normal evolutionary algorithms will not be sufficient enough when the number of parameters to be optimized is large. Initially, a dataset is generated randomly using Latin hypercube sampling. Then, DE algorithm is performed on this dataset for minimizing the cost function, and another dataset is generated from this for training the surrogate-assisted model. The surrogate model trained based on this dataset predicts the optimized parameter values based on the trained data for a given input [11].

The algorithm can be summarized as following:

Step 1: Initially, Latin hypercube sampling is used to randomly generate the population for starting the algorithm.

Step 2: If the already-defined convergence criterion is met, the SADEA ceases and the resultant values will be declared as the optimum value of parameters, or step 3 is taken.

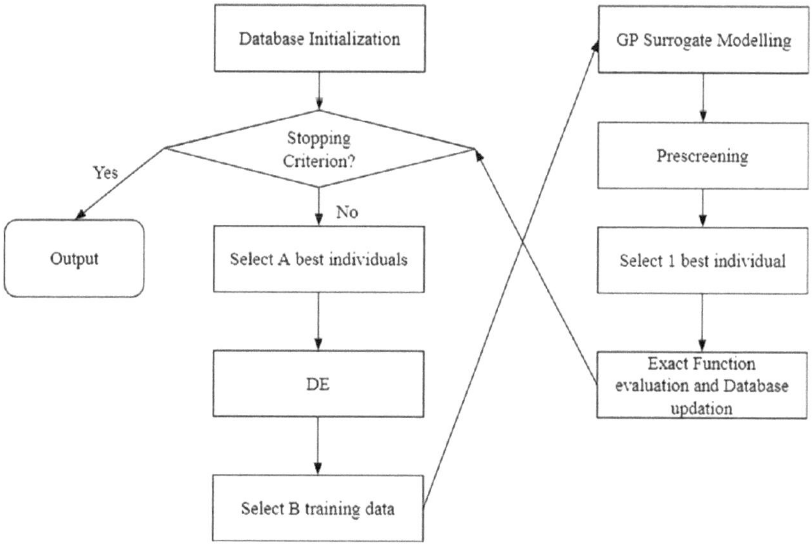

FIGURE 14.3 Surrogate model–assisted differential evolution model.

Step 3: From the preceding formed population, best solutions are taken to run the DE algorithm.
Step 4: DE algorithm is performed over it to minimize a defined cost function. Few child solutions are generated as a part of this.
Step 5: Take the best but not optimized solutions generated as a part of step 4 and use it to train the surrogate model.
Step 6: Using a Gaussian process model, screen the child solutions obtained in steps 4 and 5.
Step 7: Evaluate the pre-screened final results of this algorithm and check if convergence is met. In case the solutions are not found to be optimized, the obtained results are further fed as dataset to step 3.

14.5 METHODOLOGY

By using the formulas stated in Section 14.2 in this chapter, we might not be able to efficiently optimize the parameters, and hence, optimal design of antenna may not be possible to get the desired S_{11}. Optimum results can be obtained by this method after doing a large number of calculations. So MATLAB, Google Colab, and CST Microwave Studio are collaborated with each other to get the desired result. In this chapter, CST Microwave Studio was used to design the patch antenna and simulate the output S_{11} graph, while MATLAB was used for performing differential evolution (DE) algorithm on the randomly generated data. The initial dataset of length (*L*), width (*W*), and length of inset feed (*Lin*) was generated on MATLAB within the defined range of:

$$\text{Length} : L : \lambda / 3 - \lambda / 2 \tag{14.11}$$

$$\text{Width} : W : 2\lambda / 3 - \lambda \tag{14.12}$$

$$\text{Length of the inset feed} : \text{Lin} :\ 0\ -\ \text{L} / 2 \tag{14.13}$$

where λ is the wavelength of the wave.

Table 14.1 shows the parameters used in DE algorithm and the values taken for each of it.

TABLE 14.1
Differential Evolution Parameter Values

Parameters	Value
Crossover rate (CR)	0.6
Fitness parameter (F)	0.4
Initial population range	120
No. of iterations in one run	200

14.5.1 Cost Function

As mentioned in Section 14.3, a *cost function* or a *fitness function* is to be defined to run the DE algorithm. This function is minimized or maximized according to the termination criterion, and final parameter values generated will be the optimized values.

The cost function equipped for this chapter is:

$$CF = abs(S - S') + abs(f - f') \tag{14.14}$$

Where $S = S_{11}$ optimum value, that is, −30 dB, f = optimum frequency of resonance = 2.4 GHz, S' = minimum value of return loss, f' = frequency value at minimum return loss

DE algorithm is run to minimize this cost function, and the corresponding values of L, W, Lin, and S_{11} are taken as the optimized values of parameters. The cost function defined earlier is in terms of S and f. This will not be sufficient enough to do the process. Hence, the cost function variables are further represented using other equations. This process is done as long as the cost function contains the variables that are to be optimized. All other variables present in the equation needs to be assigned constant values, as shown in Table 14.2.

In the DE algorithm, a looping process is used where the loop runs for a defined number of iterations, as given in Table 14.1. In each iteration, a minimum or best value of cost function is calculated using the algorithm using the values of parameters from the defined range and the population. In the next loop, parameter values are changed and checked if the new cost function is lesser than the previously calculated one. If it is so, the new cost value is taken, and the parameter values will be taken as the more optimized values. This looping is continued as long as the minimum value is calculated, and the corresponding value of the parameters will be taken to be the better optimized values of the lot. These values will still not be the best optimized, since the number of iterations taken is 200. If the number of iterations is increased, the process gets slower. Hence, a new surrogate-assisted model using DE is equipped.

TABLE 14.2
Constant Values Given in DE

Constants	Value
Dielectric constant	2.2
Velocity in free space	3e8 m/s^2
Height	0.1588 cm
Frequency of operation	2.4 GHz

14.6 RESULTS

Initially, a random amount of parameter values for a random amount of S_{11} values were generated using MATLAB, each within the specified range. Then, the DE algorithm was run for 200 iterations to minimize the objective cost function. The values of L, W, Lin, and S_{11} for these 200 iterations carried out ten times are shown in Table 14.3.

TABLE 14.3
DE-Optimized Values for Ten Runs

No. of iterations	L (mm)	W (mm)	Lin (mm)	S_{11} (dB)
1	40.977	93.291	20.487	−32
2	40.932	103.154	20.478	−29
3	40.931	103.290	19.936	−31
4	40.945	100.236	20.438	−32
5	40.937	101.968	20.488	−29
6	40.906	109.763	20.349	−30
7	40.920	106.763	20.525	−30
8	41.954	98.151	18.549	−31
9	40.954	98.061	20.487	−30
10	40.955	97.893	20.491	−31

TABLE 14.4
Predicted Values of *L, W,* and *Lin* from SADEA

Input S_{11} (dB)	L (mm)	W (mm)	Lin (mm)	Feed width Wf (mm)	Gap G (mm)
-28	40.935	102.561	20.483	14	0.3
-33	40.968	95.154	25.5006	11	0.5

This data is fed as input to the Gaussian regression model to train it. After training, the required S_{11} was fed as input to the model to get the corresponding output as optimized values of L, W, and Lin. Two values of S_{11} were given to check the model at −26 dB and −33 dB. Corresponding values of L, W, and Lin as shown in Table 14.4 were taken, and the antenna is designed in CST Microwave Studio to simulate the desired output.

The simulated graphs of S_{11} = −26 dB and −33 dB are plotted and shown in Figures 14.4 and 14.6, respectively. It comes in accordance with our calculation that the peaks of S_{11} were obtained at 2.4 GHz for the corresponding values of S_{11} of −26 dB and −33 dB, respectively. Also, the three-dimensional radiation patterns of each of the preceding antennas are also plotted alongside, as shown in Figures 14.5 and Figure 14.7.

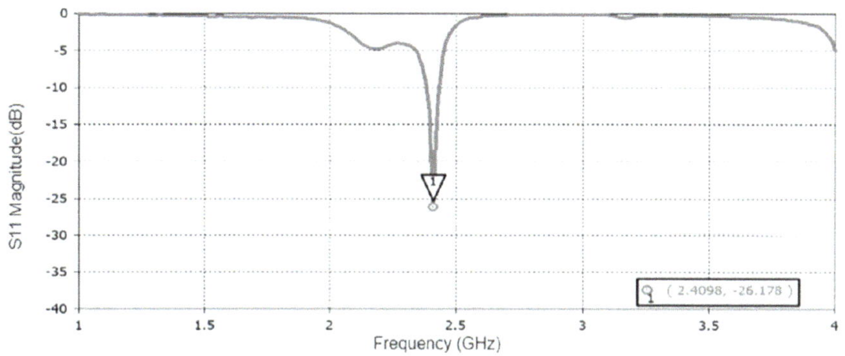

FIGURE 14.4 The simulated S_{11} response for L = 40.395, W = 102.561, and Lin = 20.483 obtained from SADEA.

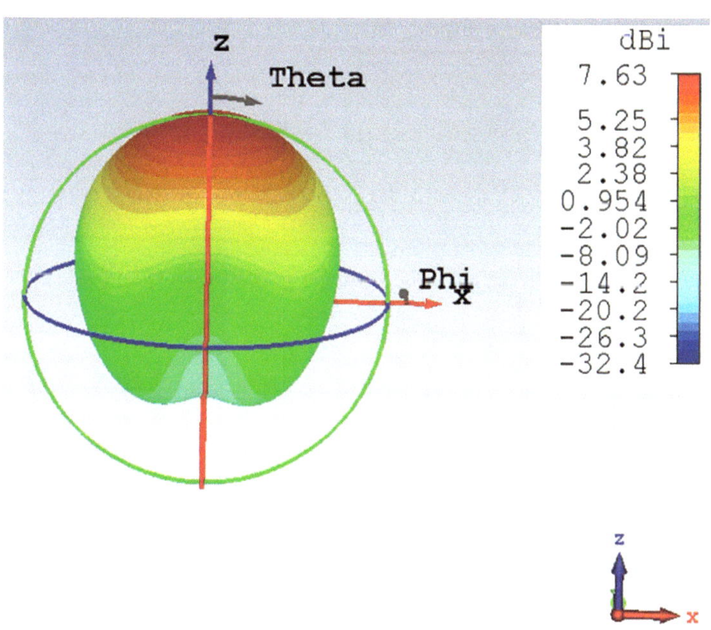

FIGURE 14.5 The antenna radiation patter for L = 40.968, W = 95.153, and Lin = 25.5006 obtained from SADEA.

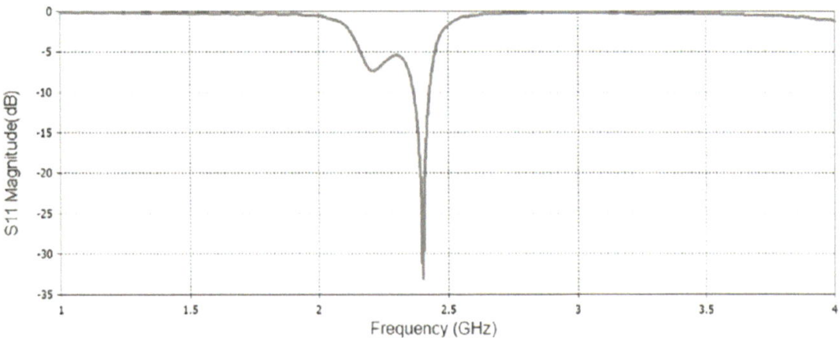

FIGURE 14.6 The simulated S_{11} response for $L = 40.968$, $W = 95.153$, and $Lin = 25.5006$ obtained from SADEA.

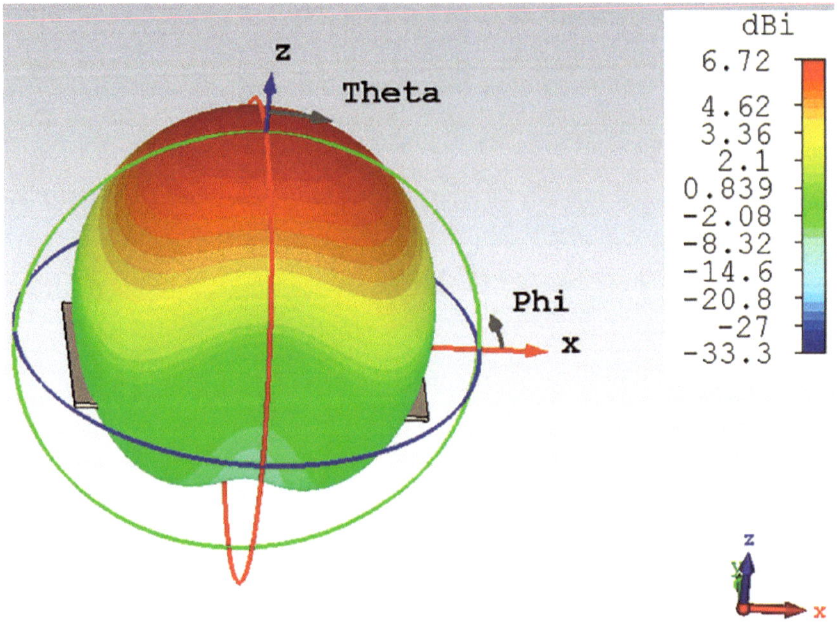

FIGURE 14.7 The antenna radiation pattern for for $L = 40.968$, $W = 95.153$, and $Lin = 25.5006$ obtained from SADEA.

14.7 CONCLUSION

This chapter initially gives a brief introduction to microstrip patch antennas, differential evolution algorithm, and surrogate-assisted model used for optimization purposes. Then, the methodology used for general microstrip patch antenna design is discussed with design equations. Following this, differential evolution algorithm and surrogate-assisted model using differential evolution is elaborated. Finally, a microstrip patch antenna with inset feed is designed to resonate at 2.4 GHz using SADEA, and the results obtained using MATLAB, Google Colab, and simulated S_{11} values generated from CST Microwave Studio are shown. Thus, it can be concluded that a design can be successfully optimized using SAEDA, which is a combination of evolutionary algorithm and machine learning, when the number of design parameters is high, and the results will be better when compared to design generated using evolutionary algorithm only.

REFERENCES

[1] Hsin-Min Lu, Jih-Shin Chen and Wei-Chun Liao, "Nonparametric Regression via Variance-Adjusted Gradient Boosting Gaussian Process Regression," IEEE Transactions on Knowledge and Data Engineering, Vol. 33, No. 6, 2021.

[2] Chih-Chung Chang and Chih-Jen Lin, "LIBSVM: A Library for Support Vector Machines," ACM Transactions on Intelligent Systems and Technology, Vol. 2, pp. 27:1–27:27, 2011.

[3] Robert R. Richardson, Michael A. Osborne and David A. Howey, "Gaussian Process Regression for Forecasting Battery State of Health," Journal of Power Sources, Vol. 357, pp. 209–219, 2017.

[4] Susmit Bhattacharya, Souti Chattopadhyay, Soham Talukder, Soumyadip Bag, Sourav Mishra and Malay Gangopadhyaya, "Optimization of Inset-Fed Microstrip Patch Antenna Using Genetic Algorithm," IEEE 2015 International Conference and Workshop on Computing and Communication (IEMCON), Vancouver, BC, 2015.

[5] K. Deb, A. Pratap, S. Agarwal and T. Meyarivan, "A Fast and Elitist Multiobjective Genetic Algorithm: NASA-II," IEEE Transactions on Evolutionary Computation, Vol. 6, No. 2, 2002.

[6] Hong-Kyu Kim, Jin-Kyo Chong, Kyong-Yop Park and David A. Lowther, "Differential Evolution Strategy for Constrained Optimization and Application to Practical Engineering Problems," IEEE Transactions on Magnetics, Vol. 43, No. 4, 2007.

[7] R. Storn, "System Design by Constraint Adaptation and Differential Evolution," IEEE Transactions on Evolutionary Computation, Vol. 3, No. 2, pp 22–34, 1999.

[8] Xuezhi Chen, Yubo Tian, Jing Gao and Tianliang Zhang, "An Antenna Design Method Based on Gaussian Process Surrogate Model and Differential Evolution Algorithm," IEEE MTT-S International Conference on Numerical Electromagnetic and Multiphysics Modeling and Optimization, 2020.

[9] Malay Gangopadhyaya, Bhaskar Gupta, Pinaki Mukherjee, Suvrajit Manna and Udit Sharma, "Design Optimization of Microstrip Fed Rectangular Microstrip Antenna Using Differential Evolution Algorithm," 2015 IEEE 2nd International Conference on Recent Trends in Information Systems (ReTIS), 2015.

[10] Tahzeeb Jamal, Talha Ahmed Khan, Shahzad Nasim and Raza Jafri, "Comparative Performance Analysis of Various Optimized Inset Fed Patch Antennas Using Differential Evolution (DE)," International Journal of Scientific Technology Research, Vol. 10, No. 4, 2021.

[11] Bo Liu, Hadi Aliakbarian, Zhongkun Ma, Guy A. E. Vandenbosch, Georges Gielen and Peter Excell, "An Efficient Method for Antenna Design Optimization Based on Evolutionary Computation and Machine-Learning Techniques," IEEE Transactions on Antennas and Propagation, Vol. 62, No. 1, 2014.

15 Millimetre Wave Antenna Design for 5G Applications

Dr Sangeetha Subbaraj, Dr Shweta B. Thomas, and Dr Saffrine Kingsly

15.1 INTRODUCTION TO MILLIMETRE WAVE SYSTEMS

The new emerging trend in 5G communication system has impact on numerous opportunities and challenges. The data rate, operating bandwidth, and low end-to-end delay are the important key requirements of users to connect simultaneously. The frequency range between 30 GHz and 300 GHz is occupied by the millimetre wave communication, which corresponds to the wavelength from 10 mm to 1 mm. Millimetre wave technology can support connectivity up to Gbps for the short-distance communication of electronic devices. Compared to the existing technology, data rate of millimetre communication is 100 times faster. This tremendous increase in the data rate leads to the development of applications like augmented reality, short-range radar, security scanners, radio astronomy, remote sensing, automotive radars, massive wireless access, etc. In order to support the large amount of traffic and seamless wireless connectivity, it is necessary to accommodate a large number of antennas at the transmitter and receiver sides, and to increase the resources of the spectrum by using the newly identified frequency bands. The various allocation of frequency bands by the International Telecommunication Union (ITU) is shown in Figure 15.1. The mmWave-allocated frequency ranges are 24.25–27.5 GHz, 37–40.5 GHz, 47.2–50.2 GHz, 50.4–52.6 GHz, 66–76 GHz, and 81–86 GHz. As the frequency range increases, the data rate is increased, but it suffers from high penetration loss when the signal path is blocked by an obstacle.

Due to these obstacles in the signal path, mmWave is affected by blockage and deafness problems. The misalignment of directional beam of transmitter and receiver leads to a situation of deafness. It also suffers from oxygen absorption, scattering environment. However, using beamforming techniques, a large number of antenna elements increase the gain, which compensates for these losses. The connection failure between the transmitter and receiver is avoided by increasing the number of links and quickly finding the alternative path for the communication. Thus, the interruptions due to blockages, misalignment of the antenna, increase in the distance between the transmitter and the receiver are reduced by combining various mechanisms. Multi-connectivity is another technique used in mmWave to establish multiple connections simultaneously at the same time. Two parameters associated

 DOI: 10.1201/9781032625034-15

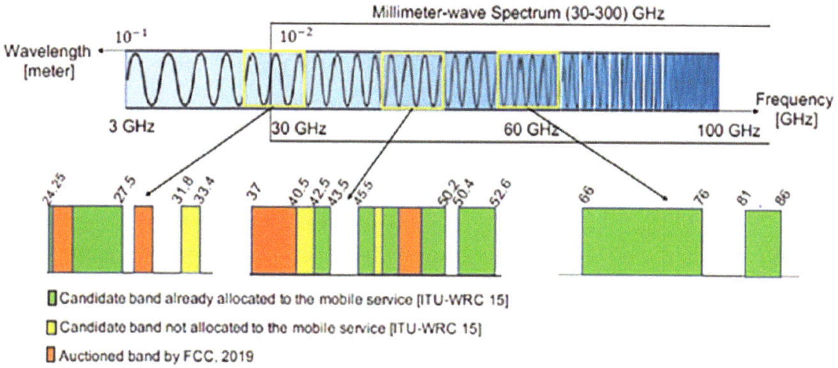

FIGURE 15.1 5G frequency characteristics.

with multiple connectivity are throughput and reliability. *Reliability* deals with sending the same data to the connected receiver, whereas *throughput* deals with sending different data to the associated receiver. Therefore, there exists a trade-off between the reliability and the throughput.

The frequency characteristics behaviour beyond 30 GHz is different from the sub-6 GHz frequency bands. The capacity of the channel is defined as

$$C = B \log_2 \left(1 + \frac{S}{NB} \right)$$

where S and N are the power of signal and noise spectral density.

The power of the signal is limited, and it restricts mmWave communication beyond the limit of safety. Since mmWave uses higher frequency for communication, it has the following advantages:

1. The higher frequency level is reused.
2. The larger bandwidth and high speed are possible in higher frequencies.
3. The higher frequencies can be used for secure communications.

The 3GPP and the International Telecommunication Union (ITU) have released two phases of 5G research standards. The first phase covers frequency range below 40 GHz, and the frequency range up to 100 GHz is handled in the second phase of the research. The most general 5G bands are 28 GHz, 37 GHz, 39 GHz, and 64–71 GHz bands. The WRC-19 identified new service frequency ranges, such as 24.5–27.5 GHz, 37–43.5 GHz, 45.5–47 GHz, 47.2–48.2 GHz, 66–71 GHz, etc. Due to the low water vapour absorption and low oxygen adsorption, a 28 GHz band is very efficient for data transmission at higher speeds, and it has low loss.

15.2 PROPAGATION LOSSES

The loss between the transmitter antenna and the receiver antenna is given as

$$\mathrm{L}(\mathrm{dB}) = 20\log_{10}\left(\frac{4\pi d}{\lambda}\right)$$

where d is the distance between the transmitter antenna and receiver antenna. λ is the operating wavelength. For mmWave communication, the propagation loss is higher even if it is a short-distance communication. According to the Friis formula, the received power is given as

$$\mathrm{P_R}(\mathrm{d}) = \mathrm{P_T G_T G_R}\frac{\lambda^2}{(4\pi)^2 d^n}$$

where P_R is received power, P_T is transmitted power, G_T is gain of the transmitter antenna, and G_R is gain of the receiver antenna. The distance of 10 m is kept as constant, and the path loss is calculated at various frequency bands, as depicted in Table 15.1. It is observed that as the frequency increases, the loss factor also increases. To compensate for the path loss, the gain of the antenna needs to be increased.

Along with path loss, other loss factors, such as gases, scattering effects, diffraction, and environment effects, such as water vapour, oxygen also affects the transmission path.

15.3 LINK BUDGET DESIGN

Link budget design is used determine the amount of gain and losses in the communication channel. For 28 GHz frequency band, the signal power received at the receiver is given as

$$\mathrm{P_r = P_t - P_l + G_t + G_r}$$

TABLE 15.1
Path Loss at Different Operating Bands

Frequency (GHz)	Path loss (dB)
2	58.5
5	66.4
20	78.5
30	82
60	88

where P_r is the received power, Pt is the transmitted power, P_l is path loss, and G_t and G_r are the gains of the transmitter and receiver antenna. The path loss at 1 m is 61 dB. For 10 dBm of the transmitter power, the received power $P_r = 10 - 61 = -51$ dBm.

If the noise figure of LNA is 6 dB and the bandwidth of the corresponding frequency is 1 GHz, then the noise power is calculated as

$$P_n = -174\,\text{dBm/Hz} + 10\log(1\,\text{GHz}) + 6\,\text{dB} = -78\,\text{dBm}$$

Then the signal-to-noise ratio is calculated as

$$\text{SNR}(\text{dB}) = -51 - (-78) = 27\text{dB}$$

The distance between the transmitter and the receiver is considered to be 1 m in the aforementioned scenario. Similarly, the link budget analysis can be extended for larger distances.

15.4 REQUIREMENTS OF 5G CHANNEL MODEL

The following points summarizes the requirement of the 5G channel model.

1. It should have a wide range of frequency bands, from 410 MHz up to 100 GHz.
2. It should have broad bandwidth.
3. It should be able to support a wide range of scenarios, such as indoor, rural, urban, subsurban, and high-speed environment.
4. It should have spatial consistency to support closely spaced transmitters and receivers.
5. The parameters related to the channel model should vary smoothly with the frequency.
6. It should support massive MIMO.
7. It should be able to support D2D/V2V communication scenario. In both the communication sides of the transmitter and the receiver, it should support mobility.

15.5 ANTENNA DESIGN CONSIDERATION AND REQUIREMENTS FOR 5G APPLICATIONS

In the communication system, the antenna is the most important element for communication. It transmits the signals in the form of electromagnetic waves, and also, it helps in receiving the electromagnetic waves. So it acts as a transducer to transmit/receive between the communication channels. Today, in all the electronic devices, antenna becomes as in-built component for signal reception/transmission. The size of the antenna depends on the operating frequency. As the size of the antenna increases, the frequency decreases. Since the size of the antenna reduces at mmWave communication. It makes the designer design multiple antennas in the small available space,

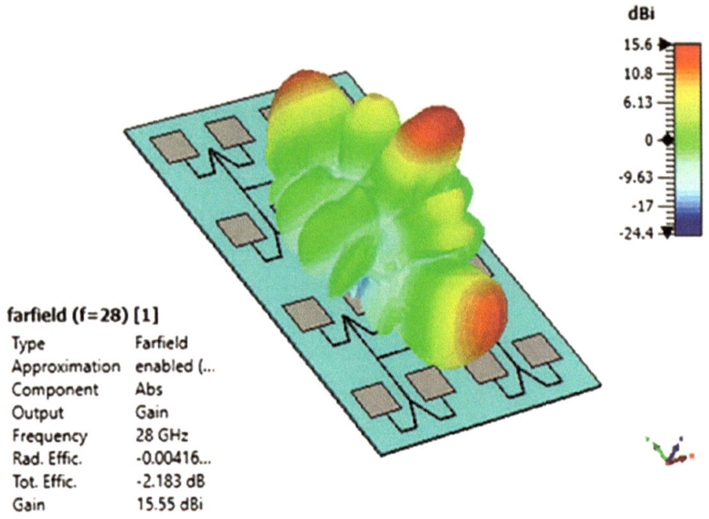

FIGURE 15.2 Radiation pattern of patch array.

which in turn reduces the size of the communication devices. One of the issues in mmWave communication is high path loss. This can be overcome by designing the high gain antennas for better SNR. Another issue is deafness, which can be avoided by implementing the wide beam scanning/switched beam array to search the required signals. The directionality of the mmWave communication is high, and this results in a narrow main beam pattern, as shown in Figure 15.2. When the main beam is narrow, the side lobes occur with respect to the main beam. Therefore, these effects also need to be considered while designing for mmWave communication.

15.5.1 Material Selection

The selection of suitable material has impact on the performance of the 5G antenna. For the frequency range below 6 GHz, it usually experiences low substrate loss. The properties required for choosing the suitable PCB at higher frequencies are dielectric constant, loss tangent, thermal conductivity, and thickness of the substrate. Higher dielectric values support smaller antenna structures, and lower loss tangent values prevent high-frequency losses. The impedance bandwidth can be increased by the thicker substrate, but increasing the cost of manufacturing and difficulty in fabrication. The commonly used substrates are FR4, Rogers, ceramic, and organic sheets.

1. **FR4.** It is flame-retardant. It is the composite of glass with epoxy resin bound together. The cost of FR4 is very cheap, and it is most commonly used for antenna design. It is available with dielectric values between 3.8 and 4.7 and loss tangent values between 0.02 and 0.03. Due to moisture absorption and electric dispersion, it is not suitable for mmWave band over 24 GHz.

TABLE 15.2
Performance Comparison between Various Organic Sheets

Material	Transmission loss	Flexibility	Stability	Moisture absorption	Heat resistance	Cost
Polyimide	Poor	Poor	Poor	Better	Better	High
Modified polyimide	General	General	General	General	General	High
Liquid crystal polymer	Better	Better	Better	Poor	Poor	Very high

2. **Rogers substrate.** It is expensive and has low loss substrate compared to FR4. The dielectric value ranges between 2.2 and 10.2. The different varieties of Rogers are available, such as RT5880, RO4003, and RO4350.

   RT 5880 has dielectric constant of 2.2 and loss tangent of 0.0009 at 10 GHz. The low absorption loss makes it suitable for 5G applications. However, the cost of the material is high. RO4003 and RO4350 have similar characteristics and have low loss behaviour. These substrates are popularly used in designing the 5G antenna.
3. **Ceramic substrate.** The important and popular ceramic substrate is low-temperature co-fired ceramic (LTCC), which has high dielectric constant between 5 and 20 and low loss tangent of less than 0.002. Due to this nature, the size of the antenna can be reduced, and it can be easily integrated into the 5G module. It can withstand high amount of current and temperature and has high conductivity as compared to other PCBs. Since the moisture absorption of LTCC is very low, it leads to the solution of antenna in package.
4. **Organic sheets.** Different forms of organic materials, such as polyimide, polytetrafluoroethylene (PTFE), and liquid crystal polymer (LCP), are commonly used. It has features such as flexibility, high temperature resistance, light weight, very thin, and high thermal stability. This makes suitable for integrating inside the mobile phones and in other electronic devices. However, the cost of the material is high compared to other substrates.

The performance comparison of various organic sheets is given in Table 15.2.

As seen in Table 15.2, liquid crystal polymer has high stability compared to polyimides. Due to the cost of the LCP, it cannot replace polyimide substrate.

15.6 FREQUENCY-TUNABLE MILLIMETRE WAVE 5G ANTENNA

As the technology improves, the communication capacity also needs to be improved to support the development. The multiband antenna design is used for transmission/reception of signals at various frequency bands. However, to further improve the simplification in size, to change the characteristics, such as frequency, pattern, polarization, reconfigurable antenna is preferred in 5G bands. The frequency-reconfigurable antenna can change the frequency characteristics in real time. The reconfiguration is based on various components, such as pin diodes, varactor diodes, MEMS, etc. The pin diode is more cost-effective compared to other components and has high power handling capability.

The dipole antenna proposed in [1] has dipole antenna arms and four-pin diode BAR64–02 V connected to it. When all the switches are connected, the antenna resonates at 3.5 GHz. Similarly, when all the switches are disconnected, the resonant frequency is 5.5 GHz. The electrical length in the first state of operation (at 3.5 GHz) is high compared to the second state of operation. The differentially fed proximity coupling feeding [2]–based antenna is designed for sub-6 GHz 5G application bands. It consists of two layers of substrate. Two pin diodes are placed in the top layer, and another two diodes are placed in the bottom substrate layer near the feed line. The radiating elements consist of slot and the patch. Two diodes control the current flows on the patch and slot, which give rise to resonant frequency at 2.4 GHz (when it is OFF) and resonance at 3.5GHz (when it is ON). As the non-linear elements are used in the communication systems, it will give rise to harmonics in the frequency bands. The suppression of harmonics affects the impedance matching of the system. The filtering branches are used in [3] to eliminate the harmonics, and along with this, frequency reconfiguration is achieved by tuning the pin diodes in triangular-shaped dipole antenna.

15.7 HIGH-GAIN ANTENNAS FOR 5G APPLICATIONS

The amount of power transmitted or received by the antenna is defined by the term *gain*. There exists a relationship between the opening size of the aperture and the gain, as shown in Figure 15.3. If the aperture of the antenna is larger, then the gain of the antenna increases. The directivity also increases with increase in size of the aperture. Another method for increasing the gain is to increase the number of antenna element, which is called as arrays. However, implementation of arrays suffers from other effects, such as feed-line losses due to blending of feeds and complex network. In the 5G scenario, increasing the size of the aperture is not a reliable solution. Other approaches include resonant cavity antennas, which include cavity structures, reflecting surfaces, and other forms of resonant structures. The partial reflecting surface used in [4] has limitation on its bandwidth. The gap between the antenna and the reflecting surfaces determines the impedance bandwidth. For enhancing gain over the broad band, the dielectric slab structures with holes are used.

By employing epsilon near zero/mu near zero materials, the gain of the antenna is improved along the orthogonal planes (either in H-plane or in E-plane) [5]. This technique is limited by low radiation bandwidth and natural properties of metamaterials. The near-field phase correction technique implemented in [6] consists of arrays of unit cells with phase shifting. This modifies the phase of the radiation; however, it results in large size of the antenna and is suitable for narrow band of operation. For a wide band of operation, the phase-correcting technique proposed by [7] improves the bandwidth and efficiency.

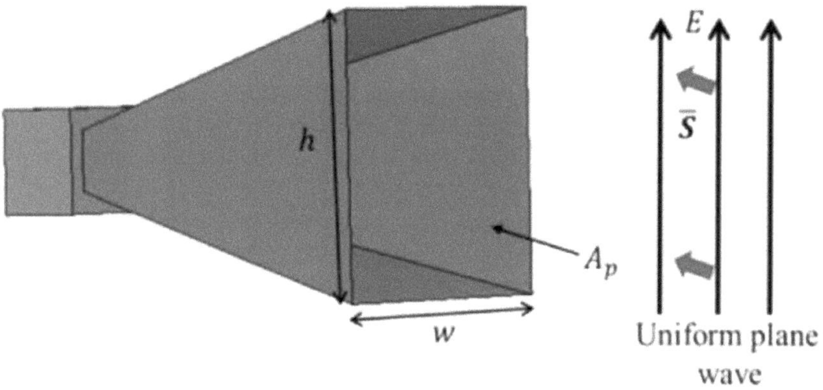

FIGURE 15.3 Plane wave front on the horn antenna.

The dielectric slab is stacked on top of the antenna to enhance the gain of the antenna in 5G applications. This design of superstrate increases the fabrication complexity, and it is sensitive to the operating frequency bands. The stacked patch is combined with substrate-integrated suspended line. This method comprises of five layers of substrates, ten metal layers with cavity [8]. The TM modes are generated for the suppression of the beam split and to improve the gain of the antenna at 3.2 GHz, 5.4 GHz. Whenever nulls are created in the radiation pattern, it further reduces the gain of the structure. So the researchers concentrate on reducing the nulls of the pattern.

In the feeding network, slot vias are combined to improve the gain at 38 GHz applications. The via pillars used in [9] reduce the loss of the substrate. The structure consists of both metallic vias and air cavity. The radiator printed on substrates 1 and 2 is given with orthogonal feed for polarization. The pillars are symmetrically supported by the aluminium layer. The array of these elements is increased to 64 to achieve a gain of 24 dBi. The Vivaldi antenna is connected as an array for enhancing the gain at 24 GHz. The superstrate can be kept at the optimal distance from the radiator, and it has advantages such as improved gain and bandwidth. The distance between the superstrate and the radiator [10] is maintained by another Rogers substrate placed between them. SIW-based cavity antenna is proposed in [11] for gain enhancement without compromising the size of the antenna. The dipole antenna is printed on a substrate, and it is separated from the slot cavity by means of air gap. In order to reduce the complexity in stacking structures of patch design, the patch antenna can be segmented. This segmentation provides 3 dB enhancement [12] in the gain of the antenna without using any superstrates. The diffraction grating is the principle [13] used in dielectric lens, and it is combined with yagi antenna. The perturbation is introduced to improve the gain of the antenna. By adjusting the perturbation, the wave propagation in the antenna can be altered. So compared to the planar lens antenna, the performance of the yagi antenna with perturbed lens antenna is higher.

15.8 MILLIMETRE WAVE MIMO AND BEAMFORMING

The millimetre wave antenna suffers from large path loss, fading, and large penetration loss. So various techniques such as MIMO (multiple-input, multiple-output; arrays are present both at the transmitter and receiver), MU-MISO (multiuser multiple-input, single-output; antenna arrays are present only at the base station), massive MIMO, and beamforming are used for transferring a large amount of data.

The metamaterial slab array is combined with the dipole antenna [14], and the resulting radiated beam is shifted by 35°. The mu near zero property of metamaterial shifts the incoming beam, and hence, the desired beam scanning can be obtained. The butler matrix is derived from SIW-based slot antenna presented in [15]. The slot antenna provides beam angle of 90°. The folded monopole antenna is arranged in a 360° fashion, and it is connected to the pin diode [16]. By controlling the voltage given to the pin diode, the corresponding folded monopole antenna is connected. Similarly, the current of the antenna is controlled and, in turn, controls the beam pattern. Thus, the electronically beam-steerable antenna has low cost and eliminates the need for superstrate. The pin diode has 12 possible states for the frequencies 1.8 GHz to 2.2 GHz and 2.85 GHz to 3.15 GHz. It can switch between any of the 360° beam direction, depending on the states of the pin diode.

The large number of arrays is required at the base station. As the array increases, the spacing between the element decreases, and it is highly required in case of massive MIMO arrays. The triple metasurface is introduced in [17], and it consists of identical elements in the top and bottom layers. The amplitude and phase of the waves coming from the bottom and top layer get cancelled due to same phase and amplitude. This principle is used as the decoupling mechanism between the radiators. The passive and active patch radiator is printed in the substrate for operating at 5.1–5.9 GHz [18]. The spacing between the elements is analyzed, and the optimum value is chosen. The uniform patch array is shifted to different patterns by changing the distance between the elements [19]. The radiation pattern of uniform and shifted array is compared. In the horizontal plane, the radiation patterns in both the cases look identical. However, in the vertical plane, the pattern of shifted array is widened compared to the uniform array. The gain of the shifted pattern array is less compared to the uniform array. But isolation is improved in shifted array pattern, and this method can be applied by comprising the gain value.

The lens antenna which is acting as metasurface is fed by the stacked array of antenna [20]. The lens antenna is used to adjust the phase shift of the incident waves, and it has low cost, weight, compactness, etc. The data rate can be further improved by incorporating techniques such as space division multiplexing and frequency-division multiplexing. The lens antenna supports these techniques and changes the phase of the radiation beam. When the beams are isolated, the lens antenna helps in achieving the frequency reuse in massive MIMO applications.

15.9 INTEGRATION OF 4G AND MMWAVE 5G ANTENNAS FOR SMARTPHONE APPLICATIONS

The wireless devices have been increased in recent years to a great extent. Due to this rapid development of devices, there is greater demand for wider bandwidth and data rates. High throughput and multiple reception can be achieved through MIMO technology. To meet this data requirements, antennas that operate for 4G and 5G applications need to be integrated. Several designs of antenna are available in the literature for integrating the multiple antennas. The 4G antenna operates at 2.2–4.9 GHz, and the 5G antenna that covers 26–28.8 GHz was integrated in the smartphones [21]. The two configurations of the antennas are isolated by the distance of 8 mm. The same footprint is used by the two antennas. The slot area is used as Vivaldi antenna [22], which operates at 28 GHz (5G applications) and four-element planar array used for 4G applications.

The monopole antenna and Vivaldi antenna used in [23] are isolated by the slots in between them. The two-monopole antenna which operates for 4G applications is separated by the connected slots. The connected slots are excited by the array feed printed on another side of the substrate. The inverted L-shaped 4G antenna is printed in the ground plane, and 5G antenna is used as beam scanning using phase shifter [24]. Similarly, mmWave antenna arrays are fixed at the corners of the smartphones, and 4G MIMO antennas are placed at the edges of the gadgets for LTE applications.

15.10 FABRICATION CHALLENGES OF MILLIMETRE WAVES

Even though the 5G mmWave designs bring lots of benefits in the current scenario, there are challenges to the PCB fabrication at high frequencies. At lower frequencies, generally, FR4 substrates are used for the design. It is very cheap and has low dielectric losses below 1 GHz. At higher frequencies, using this standard of FR4 becomes a challenge due to increase in loss factor.

The next problem is, increase in frequencies brings uncertainty in the tolerance limit of dielectric thickness and dielectric constant. The dimensions are reduced at higher frequencies, which, in turn, reduces the gap between two copper lines. While manufacturing these designs, difficulty arises to meet the requirements of too small dimensions. The components such as SMD device properties get changed after 6 GHz. After 6 GHz, the SMD components fabricated along with the antenna design produces different output characteristics compared to simulation. Next issues are related to the connectors used in the higher frequencies. These connectors are costly, and it has variation in the real-time measurement results. Therefore, in most of the cases, expert manufacturing process, materials, and tolerance limits are required.

REFERENCES

[1] G. Jin, C. Deng, Y. Xu, J. Yang and S. Liao, "Differential Frequency-Reconfigurable Antenna Based on Dipoles for Sub-6 GHz 5G and WLAN Applications," in *IEEE Antennas and Wireless Propagation Letters*, vol. 19, no. 3, pp. 472–476, 2020.

[2] G. Jin, C. Deng, J. Yang, Y. Xu and S. Liao, "A New Differentially-Fed Frequency Reconfigurable Antenna for WLAN and Sub-6GHz 5G Applications," in *IEEE Access*, p. 1, 2019.

[3] Z. Nie, H. Zhai, L. Liu, J. Li, D. Hu and J. Shi, "A Dual-Polarized Frequency-Reconfigurable Low-Profile Antenna with Harmonic Suppression for 5G Application," in *IEEE Antennas and Wireless Propagation Letters*, p. 1, 2019.

[4] J. Hao, N. Yan, Y. Luo, H. Fu and K. Ma, "A Low-Cost Dual-Band Multimode High-Gain Stacked-Patch Antenna Based on SISL for 5G Applications," in *IEEE Antennas and Wireless Propagation Letters*, vol. 21, no. 1, pp. 4–8, 2022.
[5] Q. Gu, B. Wang and P. Zhao, "Broadband Dual-Polarized Antenna Array with Enhanced Radiation Pattern Stability," in *Electromagnetics*, vol. 43, no. 7, pp. 506–522, 2023.
[6] S. Zhu, H. Liu, P. Wen and Z. Chen, "A Compact Gain-Enhanced Vivaldi Antenna Array With Suppressed Mutual Coupling for 5G mmWave Application," in *IEEE Antennas and Wireless Propagation Letters*, p. 1, 2018.
[7] Y. Al-Alem and A. Kishk, "Low-Cost High Gain Superstrate Antenna Array for 5G Applications," in *IEEE Antennas and Wireless Propagation Letters*, p. 1, 2020.
[8] M. Asaadi and A. Sebak, "High-Gain Low-Profile Circularly Polarized Slotted SIW Cavity Antenna for MMW Applications," in *IEEE Antennas and Wireless Propagation Letters*, vol. 16, p. 1, 2016.
[9] J. Kim and H. L. Lee, "High Gain Planar Segmented Antenna for mmWave Phased Array Applications," in *IEEE Transactions on Antennas and Propagation*, vol. 70, no. 7, pp. 5918–5922, 2022.
[10] E. H. Mujammami and A. B. Sebak, "Wideband High Gain Printed Quasi-Yagi Diffraction Gratings-Based Antenna for 5G Applications," in *IEEE Access*, vol. 7, pp. 18089–18100, 2019.
[11] A. Dadgarpour, B. Zarghooni, B. S. Virdee and T. A. Denidni, "One- and Two-Dimensional Beam-Switching Antenna for Millimeter-Wave MIMO Applications," in *IEEE Transactions on Antennas and Propagation*, vol. 64, no. 2, pp. 564–573, 2016.
[12] G. -g. Namgung, C. Lee, H. Park, Y. Seo, S. Kahng and D. Lim, "Design of Wideband SIW Beamforming Circularly Polarized Antennas for 5G-band," in *2019 8th Asia-Pacific Conference on Antennas and Propagation (APCAP)*, Incheon, Korea (South), pp. 297–298, 2019.
[13] H. Liu, S. Gao and T. H. Loh, "Compact Dual-Band Antenna with Electronic Beam-Steering and Beamforming Capability," in *IEEE Antennas and Wireless Propagation Letters*, vol. 10, pp. 1349–1352, 2011.
[14] S. Luo, Y. Zhang, G. F. Pedersen and S. Zhang, "Mutual Decoupling for Massive MIMO Antenna Arrays by Using Triple-Layer Meta-Surface," in *IEEE Open Journal of Antennas and Propagation*, vol. 3, pp. 1079–1089, 2022.
[15] J. Zhang, K. Yang, J. Meng, W. Li, E. Eide and G. A. E. Vandenbosch, "Dual-Band Dual-Polarized Massive MIMO Array for Maritime Applications," in *IEEE Sensors Letters*, vol. 6, no. 1, pp. 1–4, 2022.
[16] M. Temiz, E. Alsusa, L. Danoon and Y. Zhang, "On the Impact of Antenna Array Geometry on Indoor Wideband Massive MIMO Networks," in *IEEE Transactions on Antennas and Propagation*, p. 1, 2020.
[17] M. Khalid, S. I. Naqvi, Y. Amin and H. Tenhunen, "An Integrated 4G Array With mm-Wave 5G MIMO Antenna for Future Mobile Applications," in *2020 3rd International Conference on Computing, Mathematics and Engineering Technologies (iCoMET)*, Sukkur, pp. 1–6, 2020.
[18] M. Ikram, N. Nguyen-Trong and A. Abbosh, "Hybrid Antenna Using Open-Ended Slot for Integrated 4G/5G Mobile Application," in *IEEE Antennas and Wireless Propagation Letters*, vol. 19, no. 4, pp. 710–714, April 2020.
[19] N. Nguyen-Trong and M. Ikram, "Multi-Functional Structures for 4G/5G Antennas Utilizing Slot Geometry," in *2020 4th Australian Microwave Symposium (AMS)*, Sydney, NSW, pp. 1–2, 2020.
[20] T. Yamagajo, Y. Koga, M. Kai, T. Tonooka, H. Sumi and M. Hoshino, "A Nobel 4G and 5G Antenna Solution for Future Smartphones," in *2018 IEEE International Symposium on Antennas and Propagation & USNC/URSI National Radio Science Meeting*, Boston, MA, pp. 1785–1786, 2018.

[21] J. Park, J. Kim, J. Choi, D. Choi and W. Hong, "Concept of Integrating 4G LTE and Millimeter-Wave 5G Antennas within Zero-Bezel Cellular Devices," in *2020 IEEE International Symposium on Antennas and Propagation and North American Radio Science Meeting*, Montreal, QC, pp. 1675–1676, 2020.

[22] M. Ikram, N. Nguyen Trong and A. Abbosh, "Hybrid Antenna Using Open-Ended Slot for Integrated 4G/5G Mobile Application," in *IEEE Antennas and Wireless Propagation Letters*, vol. 19, no. 4, pp. 710–714, 2020.

[23] R. Golla and S. Nelaturi, "Compact 4-Port Vivaldi MIMO Antenna for 5G Wireless Devices," in *Progress in Electromagnetics Research C*, vol. 131, pp. 13–24, 2023.

[24] S. P. Biswal, S. K. Sharma and S. Das, "Collocated Microstrip Slot MIMO Antennas for Cellular Bands Along with 5G Phased Array Antenna for User Equipments (UEs)," in *IEEE Access*, vol. 8, pp. 209138–209152, 2020.

16 Machine Learning Algorithms for Optimization and Intelligence in Wireless Networks

WSNs, MANETs, VANETs, and USNs

Lopamundra Hota, Biraja Prasad Nayak, and Arun Kumar

16.1 INTRODUCTION

In recent years, there has been a growing convergence between the domains of networks and machine learning techniques. The management, optimization, and security of networks might evolve as a result of ML integration. To make networks more adaptable, efficient, safe, and capable of providing a better user experience, ML is being increasingly integrated into networking [1]. The use of ML in networking is projected to increase as networking technologies develop further. Network resource allocation can be optimized using ML. ML algorithms, for instance, can optimise performance and energy efficiency in cloud networks by dynamically allocating server resources depending on traffic patterns and demand. Critical network services can be guaranteed to satisfy performance requirements by using ML to forecast and adjust QoS parameters in real time. By studying historical traffic patterns and dynamically modifying load-balancing algorithms, ML can assist in more effectively distributing network traffic. Network anomalies suggestive of security vulnerabilities can be found using ML models. For instance, ML can be used by intrusion detection systems to identify odd network traffic patterns and issue alerts. Large datasets can be analyzed using ML to find new vulnerabilities and threats. By spotting fresh attack patterns, it can assist security teams in staying one fig step ahead of attackers. With the use of ML, it is possible to create models of typical user and object behaviour, which makes it simpler to identify deviations that might point to a security breach [2].

DOI: 10.1201/9781032625034-16

By examining network characteristics and traffic patterns, ML can enhance routing decisions. In instances like traffic engineering in expansive networks, this is especially helpful. Network administrators can more effectively allocate resources when using ML models to predict network traffic loads. Machine learning is used by content producers and content delivery networks (CDNs) to personalize recommendations, improve user experience, and optimize content delivery. Machine learning can enhance beamforming methods in wireless communication to boost network coverage and signal quality [3]. In wireless networks, ML can be used to optimize frequency allocation and lessen interference [4]. For intelligent network control and service orchestration, ML can improve software-defined networking (SDN) and network function virtualization (NFV) [5]. By foreseeing network equipment failures, ML enables preventive maintenance to reduce downtime. The idea of autonomous networks, in which networks employ ML to make decisions in real time without human involvement, is being researched. For instance, based on traffic conditions, an autonomous network might dynamically modify its routing algorithms. Machine learning is anticipated to be important in the era of 5G and future wireless technologies, managing vast IoT devices, and assuring ultra-low latency for critical applications like autonomous vehicles and augmented reality [6].

The computing procedures are becoming more reliable, cost-effective, and efficient due to ML applications. By automatically, more quickly, and with greater accuracy, analyzing even more complex data, ML generates various models. The advantage of ML is its capacity to deliver generic solutions via an architecture that can adapt and perform better. Its interdisciplinary nature makes it essential in a number of disciplines, including engineering, medicine, and computing [7].

The diverse and numerous ML applications have a significant impact on WSNs, VANETs, MANETs, and USNs. By modifying the operational behaviour of sensor nodes in response to data patterns and network conditions, ML can reduce the energy consumption in WSNs. The lifespan of the network is thereby increased. By learning from sensor data to judiciously combine and process data from various sensors, machine learning can enhance data aggregation. As a result, redundant transmissions are decreased, and energy is saved [8]. Sensor data anomalies can point to system errors or unusual environmental conditions, which ML algorithms can detect. The accuracy of sensor node localization, which is essential for applications like environmental monitoring and asset tracking, can be improved by ML techniques [9]. Through proactive maintenance and reduced network downtime, ML can forecast when sensor nodes or other equipment are likely to fail. By examining network traffic patterns, ML can help with intrusion detection and maintain data security and integrity by identifying threats.

Based on real-time vehicle data, ML can predict congestion, improve routing, and optimize traffic flow. This is essential for lowering traffic congestion and raising driving standards. ML algorithms can identify potential collisions and start automatic braking or warning systems. By providing real-time traffic data, enhancing traffic signal timings, and enhancing traffic management, ML can improve ITS. By

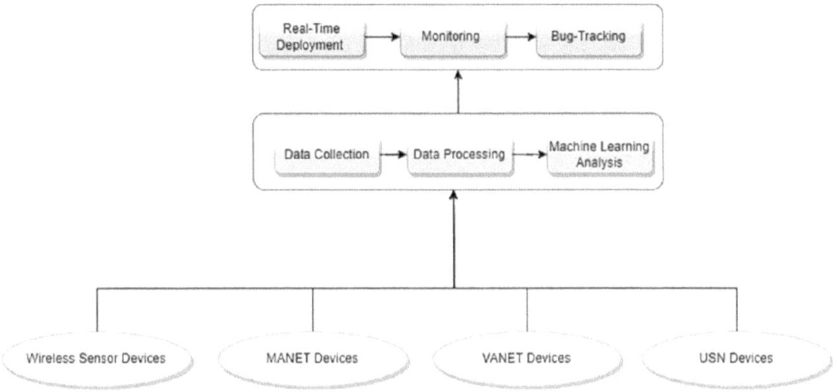

FIGURE 16.1 ML implementation in network scenarios.

examining driving patterns and recommending effective routes, ML models can aid in improving vehicle fuel efficiency. In VANETs, ML can identify and stop security threats, like false information injection or cyberattacks [10].

In MANETs, ML can improve routing protocols by learning from the state of the network to make wise routing decisions and adjust to changes. In MANETs, ML can optimize the distribution of resources like bandwidth or power, increasing network effectiveness. To prioritize important data, ML can predict the QoS requirements for various applications and allocate network resources accordingly. Network security in MANETs is improved by ML algorithms' ability to spot anomalies in network traffic and identify intrusion attempts [11].

ML can examine sensor data to track temperature, salinity, and pollution levels in underwater environments. Underwater networks can benefit from ML's ability to adapt to changing conditions and improve the reliability of data transmission. For purposes like marine biology research and underwater exploration, ML can improve underwater node localization. By examining data from underwater sensors, ML can forecast seismic activity, assisting in the development of tsunami and earthquake early warning systems [12]. A pictorial representation of ML implementation in various network scenarios is depicted in Figure 16.1. ML can enhance the capabilities of networks in various domains, from environmental monitoring to transportation and beyond. As these techniques continue to advance, their impact on networks is likely to expand.

16.2 MACHINE LEARNING PRELIMINARIES

This section introduces various ML algorithms used in various fields surveyed in this chapter, including supervised learning, unsupervised learning, deep learning, transfer learning, and reinforcement learning. A schematic diagram of various ML algorithms is depicted in Figure 16.2.

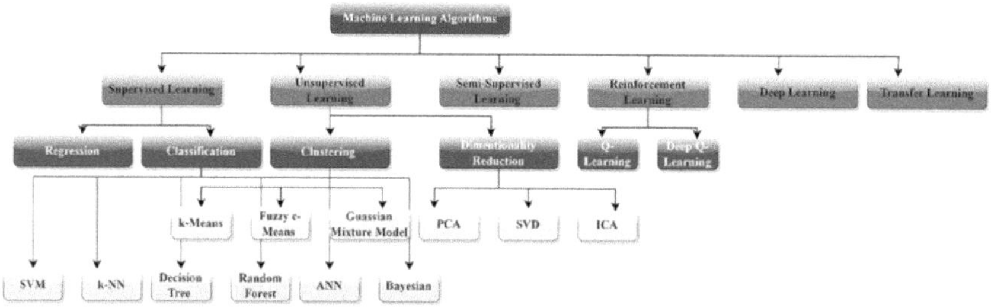

FIGURE 16.2 Classification of machine learning algorithms.

16.2.1 Supervised Learning

Given a labelled dataset, supervised learning is a machine learning task that aims to learn a mapping function from the input to the output. Based on the consistency of the output, supervised learning can specifically be further divided into regression and classification. One of the most crucial methods for processing data in ML is supervised learning. In supervised learning, the system is provided with a set of inputs and outputs (datasets with labels), and the system learns the relationship between them. The function is identified from an input x with the best estimate of an output y (f: x y) after the training process. One of the main tasks of supervised learning algorithms is to create a model that illustrates the connections and dependencies between input features and forecasted objective outcomes. Anomaly detection, localization, fault detection, routing, MAC, data aggregation, synchronization, congestion control, target tracking, event detection, and energy harvesting are some of the problems that supervised learning addresses in WSNs. Regression and classification are two categories of supervised learning. There are several different classification algorithms, including instance-based k-nearest neighbours (KNN), logic-based (decision tree and random forest), perceptron-based [Artificial Neural Networks (ANN) and deep learning], statistical learning (Bayesian and SVM), and decision tree and random forest-based.

In support vector machine (SVM), a linear classifier, the model seeks to divide data points that are p-dimensional vectors using a p-1 hyperplane. The best hyperplane, also known as the maximum-margin hyperplane, is the one that results in the greatest separation or margin between the two specified classes. The dataset, however, frequently cannot be linearly separated in the original space. The dataset, however, frequently cannot be linearly separated in the original space. By using kernel functions, such as polynomial or Gaussian kernels, in this situation, the original space can be mapped to a much higher dimensional space, producing a non-linear classifier [13].

For classification and regression, KNN is a non-parametric lazy learning algorithm that does not require any assumptions about the distribution of the data. The fundamental idea of KNN is to classify a test point in the form of a feature vector based on the majority voting of its KNN, using the classification task. A weighted method can be used by assigning each neighbour a weight that is proportional to the inverse of its distance from the test point, taking into account that training data from very frequent classes can significantly influence the prediction of test data. The tradeoff of the parameter *K* is also one of the keys to using KNN; for more information on value selection, see [13].

16.2.2 Unsupervised Learning

In unsupervised learning, there is no output (unlabeled) connected to the inputs; the model tries to infer the connections from the data. Dimensionality reduction, anomaly detection, and clustering of sets of related patterns are all examples of unsupervised learning techniques. Unsupervised learning makes its greatest contributions to WSNs by addressing a number of problems, including connectivity problems, anomaly detection, routing, and data aggregation. Unsupervised learning is further divided into dimensionality reduction (PCA, ICA, and SVD) and clustering (k-means, hierarchical, and fuzzy c-means). The k-means clustering algorithm divides n data points into k-clusters, and each data point is assigned to the cluster with the closest mean. *Iterative refinement* is the foundation of the k-means algorithm that is used most frequently. K-means are initially initialized at random. The mean of each cluster is then updated after assigning each data point to the exact cluster whose mean is closest to it in Euclidean space for each iteration. The algorithm iterates until no new members are added to any clusters [14].

PCA is an orthogonal linear transformation that transforms the data into a new coordinate system. It is a dimension-reduction tool that can be used to condense a large set of variables into a smaller set while retaining the majority of the information in the larger set [14].

16.2.3 Reinforcement Learning (RL)

To take specific actions, the RL algorithm continuously learns by interacting with the environment. RL maximizes performance by extracting the best outcome from the surroundings. One of the model-free reinforcement learning approaches is Q-learning. Each agent in Q-learning interacts with the environment and produces a series of observations known as state-action rewards (for instance: $s_0, a_0, r_1, s_1, a_1, r_2, s_2, a_2, r_3, \ldots >$), the reward matrix R(S, A), where *S* stands for *states* and *A* for *actions*, respectively. The actions of the agent in Q-learning are represented by the matrix Q(S, A), which has initial values of zero and has a size equal to the size of *R*. The current agent state and potential next state are represented, respectively, by the rows and columns of the matrix Q. The transaction rule, as shown in the following equation, updates each entry of the matrix *Q* with the sum of the value for that entry in the matrix *R* and the learning parameter, multiplied by the maximum value of *Q* for all actions that could be taken in the following state. Some of the RL mechanisms include Q-learning, multi-armed bandit learning, actor-critic learning, deep RL, and deep neural networks [15].

16.3 APPLICATIONS OF ML IN WSN

ML has been successfully used in many fields, including computer vision, medical diagnosis, search engines, and speech recognition, as a key enabling technology for artificial intelligence. Computers can now learn without explicit programming due to ML applications. ML is used to separate the faulty sensor nodes from the working sensor nodes and increase the network's effectiveness. For target area coverage problems, ML techniques make it simple to determine the ideal number of sensor nodes to cover the area. Sensor nodes may relocate as a result of both internal and external factors. With the aid of ML algorithms, accurate localization is effortless and quick. Energy harvesting offers the WSNs deployed in the harsh environment self-powered and long-lasting maintenance. WSNs perform better when forecasting how much energy will be harvested during a specific time period, thanks to ML algorithms. Data routing plays a significant role in extending the network lifetime. Dynamic routing is necessary due to the sensor network's dynamic behaviour. ML mechanisms can improve the efficiency of the overall system [16].

Some of the application of ML in WSN includes:

1. **Localization.** Recognizing a sensor node's geographic or physical location is referred to as *localization* in WSNs. Many WSN applications deploy sensor nodes into the field without first knowing their locations or having the necessary infrastructure to find them later. However, finding the sensor node's location is a crucial task. The location of sensor nodes can be determined manually or by using a GPS system; some specialized sensor nodes are known as anchor nodes or beacon nodes. Localization is primarily divided into proximity-based, range-based, angle- and distance-based, and known location–based categories [17]. Due to numerous external factors, the location of sensor nodes in the environment can change dynamically. The authors in [18] introduce a technique to increase accuracy of localization using the fuzzy c-means technique and the k-means algorithm. This system needs RSSI data to be trained. In [19], neural networks were used to create a new PSO-based neural network (LPSONN) scheme for WSNs. For adoptive localization in WSNs, a Bayesian compressive sensing (BCS) framework has been presented in [20]. The target locations and ambient noise are estimated by the BCS framework. The accuracy offered by this method is higher than that of the conventional ones. A semi-supervised hidden Markov model (HMM)–based localization method for mobile sensor nodes in WSNs is presented by the authors in [21]. Even with fewer training data, this model can be used in a variety of situations. This algorithm operates in both an indoor and an outdoor setting and has an O(n) time complexity. ML-based localization in wireless networks is a new and significant field of study and application. It entails using data gathered from wireless signals, such as Wi-Fi, Bluetooth, cellular, or RFID, to pinpoint the precise location of wireless devices or sensors. Wireless localization systems can be made more reliable and accurate by utilizing ML techniques.

2. **Anomaly detection.** An *anomaly* is defined as an inconsistent observation that significantly differs from a given set of data readings [22]. Misbehaviour in WSNs is detected through either sensor data analysis or traffic-related attributes. In the majority of WSN applications, sensors continuously collect data from their surroundings and transmit it to the base station via relay nodes. Sensing data must be protected because abnormal attacks during data transmission have the potential to result in data loss. In WSNs, anomaly detection reduces communication overhead while distributing data among the sensor nodes.

   To safeguard against various attacks and misunderstandings, ML approaches have been used to detect anomalies in WSNs. A variety of algorithms have been used in WSNs to guard against anomalies, including self-learning threshold and sleep scheduling approaches [23] for a heterogeneous sensor network based on geological sensor nodes, [23] for sinkhole attacks and denial-of-service (DoS), which offer resilience and accurate information. A hybrid anomaly is the result of several attacks combined, allowing for the identification of the node where an anomaly is occurring and its effects. Utilizing clustering algorithms reduces the problem's complexity and communication overhead. ML approaches promise to handle errors, attacks, and outliers in WSNs for detecting anomalies in non-stationary environments [24]. It is possible to use historical data to dynamically adjust the parameters for online anomaly detection.
3. **Fault detection.** WSNs have been installed in a range of application-specific environments, including hostile, harsh, and unattended conditions. Therefore, WSNs may experience faults, like communication problems, battery problems, hardware problems, software problems, ineffective base stations, or topological changes. Due to several factors, including resource constraints, a variety of environments (forest, indoor), changes in deployment, and detection accuracy between normal and faulty nodes, fault detection in WSNs is difficult [25]. The authors of [26] have presented a fault detection scheme (FDS) that uses the batteries and sensing data from the nodes to identify the faulty ones in WSNs. Faulty nodes can be discovered in two-level verification in FDS. A naive Bayesian classifier was used in the first phase to identify the fault in the sensor node, and in the second phase, the cluster head or gateway assessed fault detection. Through simulation results, this method demonstrates a 100% accuracy rate. A fault node distribution and management approach based on the fuzzy rules in WSNs has been presented in [27]. The main focus of this algorithm was to make faulty nodes usable again by implementing an effective route to the base station. It offers better network lifetime and QoS.
4. **Routing.** Due to limited power supply, low transmission bandwidth, low memory and processing capacity, routing is one of the main challenges in WSNs. Each sensor node in a WSN gathers data from the environment and transmits it to the base station for processing. Sensors are distributed randomly throughout the environment. Due to their role as relay nodes, nodes close to the base station typically use more energy. The design of the

routing protocol aims to prolong the network lifetime and lower the energy consumption of sensor nodes. The researcher recently developed several routing strategies [28] for WSNs using various methodologies. The ML algorithm adapts to changes in the environment without re-programming to choose an ideal set of CHs for routing in WSNs. ML is used in WSNs for a variety of purposes, including optimal routing, reducing communication overhead, and delay-aware [29]. Authors have created an ELDC approach based on artificial neural networks (ANN) in [30] to decide on reliable and energy-efficient routing for WSNs. ANN is used in ELDC to train the protocol. WSN parameters like residual energy and the separations between nodes, cluster heads (CH), border nodes, and sink or base stations are used in ANN training to train protocols. The dynamic 3D fuzzy routing based on traffic probability (DFRTP) routing protocol was presented by the authors in [31] using a fuzzy system. The fuzzy system generates routing probability as an output using distances and neighbour nodes as inputs. As the data forwarding node towards the sink, one of each node's neighbours that offers the best (lowest) traffic probability is selected. DFRTP lowers the energy consumption of the sensor nodes while increasing the data delivery ratio. To execute real-time energy-efficient routing policies for WSNs, the authors present a new scheme, MeFi, based on mean field reinforcement learning. They characterize the network lifespan maximization problem as a decentralized partially observable Markov decision process. The training of policies is guided by the use of mean field theory, which successfully simplifies the complex interactions among several agents. Furthermore, a prioritized-sampling loop-free method is created to get rid of routing loops and steer clear of energy-intensive routing policies [32].

5. **MAC.** By creating medium access control (MAC) protocols that are energy-efficient, network lifetime in WSNs can be maximized. A sublayer of the data link layer, the MAC layer's main responsibilities include addressing, data transfer from higher layers, channel allocation, error prediction, and frame recognition. Because of the dynamic behaviour of adding or removing (possibly dying) sensor nodes, noise data transmission, and other factors, designing energy-efficient MAC protocols for WSNs is difficult. The two main types of MAC protocols are contention-based (there is no need for centralized coordination) and schedule-based (each node communicates only during designated time slots) protocols [33]. The energy efficiency and avoidance of latency are guaranteed by ML approaches for MAC layer protocol design. It lessens the additional work required to reconfigure any nodes that have recently joined or died in the network. It increases the effectiveness of the network's self-learning process and decreases end-to-end latency. Cooperative approaches for WSNs based on reinforcement learning have been proposed in [34]. Both distributed and centralized sensor networks can use this technique. Depending on the network situation, the system's parameters are dynamically activated or deactivated. Dynamic activation or deactivation policy may have a positive or negative impact on the network's performance. Self-learning techniques based

on reinforcement learning for the best possible set of services have been presented in [35]. This approach is extremely dynamic and offers a set of network services that are both effective and compelling. This algorithm increases network dependability by decreasing the end-to-end delay of multi-hop transmission.

6. **Data collection and aggregation.** Data aggregation is the process of gathering and combining data from sensor nodes. In WSNs, data aggregation has an impact on a number of variables, including power, memory, communication overhead, and computational units. Data aggregation plays a significant role in WSNs in order to lower the amount of transmission and communication overhead. The network lifetime is increased, and the energy consumption of the sensor nodes is balanced by an effective data aggregation method. Depending on the network structure, different types of data aggregation techniques exist, including cluster-based, tree-based, in-network, and centralized data aggregation. In [36], several methods for data aggregation in WSNs have been put forth. For data aggregation, ML methods choose effective CHs in the network, which will help to significantly balance the energy used by the sensor nodes [37]. ML algorithms are useful for dimensionality reduction of the data at the level of the sensor node, which lowers network communication overhead. To reduce data transmission delays, the reduction may be carried out at sensor nodes or cluster heads. In the context of data aggregation, ML adapts to the environment and works in accordance with it without reconfiguring or reprogramming. Temporal brevity for WSNs, a method for estimating sensor drift based on Bayesian learning, has been proposed in [38]. With this approach, the drifting sensor can be found without dense deployment or prior understanding of data models. A brand-new PCA-based framework for data recovery, prediction, and compression has been put forth in [39]. This strategy's main objective is to cut down on communication overhead. This algorithm achieves lower communication overhead by compressing the data at CHs. For extending the network lifetime, this method offers an accurate prediction, recovery, and efficient compression mechanism. An innovative power-aware hybrid data aggregation method for large-scale WSNs was investigated in [40] using compression sensing and hierarchical clustering. With the help of various threshold values, this method varies cluster sizes at each leave to maximize the number of data transmissions throughout the network.
7. **Energy harvesting.** The lifespan of the network depends on the energy used by the sensor nodes, which is one of the main current energy sources for sensor nodes in WSNs. Most WSN applications demand longer network lifetimes, which can range from months to years. Either use energy-efficient protocols or give sensor nodes energy harvesting strategies to extend the lifespan of WSNs. For energy-efficiency, a number of routing, sleep scheduling, mobile sink, mobile charger, etc. protocols have been introduced. However, due to high computational resource requirements, inaccessible sensor nodes, and additional maintenance, the energy requirement is not met. Energy harvesting includes long-lasting, self-powered,

and maintenance-free WSNs. Energy harvesting provides continuous power for sensor nodes from ambient sources like radio frequency, wind, solar, thermal, mechanical, and vibration energy. It has been possible to track efficient energy harvesting techniques for WSNs using a number of ML-based models [41]. The performance of WSNs to predict how much energy will be harvested during a specific time period is improved by ML algorithms [42]. To calculate the amount of energy harvested and maintain it appropriately by balancing the energy consumption, ML approaches reduce computational complexity. In [43], a fault prediction and diagnosis method based on deep learning for wind power generation was developed. Without human interaction, it produces high accuracy rates and low error rates of fault prediction under any circumstances. The authors of [44] have concentrated on energy harvesting for heterogeneous WSNs using a hierarchical clustering algorithm. The algorithm identified static cluster heads for data collection, and all other sensor nodes except for these cluster heads are non-renewable. This method's goal is to locate the cluster heads in the most effective locations possible so that the sensor nodes use less energy.

8. **Quality of service (QoS).** The QoS a WSN provides to its users determines the level of service it provides. The two main subcategories of QoS are network-specific and application-specific. The energy consumption rate of sensor nodes and bandwidth are the network-specific QoS parameters. The measurements of sensor nodes, their deployment, and how many sensor nodes are currently active in the network are the application-specific parameters. Severe resource constraints, unbalanced traffic, data redundancy, dynamic networks, energy balancing, scalability, multiple sinks, and the variety of traffic types are some of the challenges to maintaining the QoS for WSNs [45]. To achieve QoS in WSNs, authors in [46] proposed an energy-efficient method for data fusion based on fuzzy logic. Instead of gathering complete data from the WSNs, this method aggregates only accurate information. [47] describes the development of multi-agent RL, energy-aware topology control, and data dissemination protocols for WSNs to effectively self-organize. To keep the trustworthy topology, the multi-agent RL algorithm was used to choose the active neighbour nodes. A convex-hull algorithm maintained the boundary nodes' connectivity and coverage. The performance of the sensor network is enhanced by using this method, which offers the best QoS. In order to address the uncertainty of QoS and the dynamic behaviour of a service for WSNs, a Q-leaning-based mechanism has been used in [48]. The safety service is achieved using this method at a global optimal level. The network's best relay nodes are chosen by the reinforcement learning algorithm, and each node's transmission mode is selected to maximize reliability.

16.4 APPLICATIONS OF ML IN MANET

Without the need for any fixed infrastructure, mobile nodes form dynamic, decentralized networks known as mobile ad hoc networks, or MANETs. MANETs are used for various applications, including military operations, emergency relief, and automobile communication. Nevertheless, MANETs are also susceptible to challenges, including energy limitations, limited bandwidth, and node mobility. Many of the problems MANETs have can be solved with ML. ML, for instance, can be utilized in the dynamic MANET environment to create routing protocols that are more reliable and efficient. ML can be used to create defences against assaults on MANETs that are more adept at identifying and thwarting them. ML can be used to create energy-saving plans that increase MANET node battery life. ML can be used to create QoS methods that guarantee the dependable delivery of real-time traffic in MANETs [49].

Some of the ML applications as per literature include:

1. **Routing.** Several ML-based routing protocols, including routing based on genetic algorithm (GA), reinforcement learning (RL), and ant colony optimization (ACO), have been developed for massively adaptive MANETs. It has been demonstrated that these protocols perform better than conventional routing protocols in terms of metrics like throughput, packet delivery ratio, and end-to-end delay. An effective, trust-based intrusion-detection technique that uses local information for data transmission in a highly dynamic MANET was presented by Sargunavathi, S., et al. in [50]. The defence against both blackhole and greyhole attacks involved the selection of trusted nodes to maintain security. Additionally, the suggested method is contrasted with the conventional AODV methodology using a range of performance measures. In [51], RL is defined formally in terms of a learning algorithm and a distributed environment. It provides a survey on techniques for reinforcement learning created specifically to address the point-to-point routing issue in MANETs. Collaborative RL (CRL) makes use of a decay function (which is akin to pheromone evaporation in ant colony optimization [ACO]) in order to better accommodate the characteristics of ad hoc networks [52]. The strategy is workable for WSNs as well because it necessitates little additional expense.
2. **Security.** ML can be utilized to create a range of intrusion detection and prevention systems (IDSs), and other security measures for MANETs include DoS, wormhole attacks, and black hole attacks, which can all be identified and stopped with the use of these technologies. Anomaly detection systems and signature-based detection systems are the general categories of intrusion detection systems [53]. Through the use of regular comparisons between node actions and regular normal patterns, anomaly detection systems can filter outlier or unauthorized nodes. A node is classified as intrusive if it is discovered to exhibit outlier behaviour. On the other hand, because signature-based and misuse detection systems rely on existing signatures or behaviour patterns, they are unable to identify new threats [54]. Because anomaly detection systems operate under the assumption that an intruder would never follow a normal attack pattern, they perform better in finding new vulnerabilities; however, a significant number of false

alarms are the major drawback. To implement techniques that aid in lowering true negatives and false negatives in the systems and improving security in MANETs, ML-based intrusion detection systems have emerged as an intriguing option for researchers. Different machine learning classification techniques can be used to categorize regular and invader nodes. There are a lot of IDSs out there, but the majority are for wired networks, which require more memory and have comparatively higher computational complexity. Additionally, a few low-tech methods for intrusion detection are presented. To identify malicious nodes, the majority of them rely on thresholds or the observation of certain parameters [55]. It determines the behaviour of nodes by analyzing important features or a set of 20 features. A methodology for evaluating the performance reliability of intrusion detection techniques has also been provided in this work. It makes use of the fuzzy logic system (FLS) [56] and scales the effectiveness of techniques based on memory and energy consumption input factors.

3. **Energy management.** ML can be used to create energy-saving plans that increase MANET node battery life. By optimizing a node's power control, sleep scheduling, and routing behaviour, these approaches can lower the node's energy usage. ML-based energy management schemes offer several advantages when it comes to energy management in MANETs, which are more flexible than conventional energy management strategies, in dealing with the changeable topology of MANETs able to account for a greater variety of parameters, including network traffic load, node energy levels, and node distance, which can be used to coordinately optimize a node's power control, sleep scheduling, and routing behaviour [57]. To increase MANET energy efficiency, a machine learning–based routing protocol is presented in this study. The authors provide a technique that considers network characteristics and node energy status when making routing decisions using machine learning [58]. This paper discusses the use of machine learning for energy management in MANETs, particularly for Internet of Things (IoT) applications. It provides insights into energy-efficient data forwarding and routing techniques that utilize machine learning models. Specifically, the use of machine learning for IoT applications in MANETs for energy management is covered in this study. It offers information on machine learning model–based energy-efficient data forwarding and routing strategies.
4. **QoS.** ML can be used to create QoS systems that guarantee the dependable delivery of real-time traffic in MANETs. These techniques can be applied to effectively distribute bandwidth resources and give priority to real-time traffic over non-real-time traffic. This study provides a machine learning–based MANET routing method that takes packet loss and latency into account, along with other QoS measures. The suggested algorithm determines routing in real time based on historical network data. [59] provides a machine learning–based MANET routing method that takes packet loss and latency into account, along with other QoS measures. The suggested

algorithm determines routing in real time based on historical network data. The potential and difficulties of applying machine learning to improve QoS in MANETs are examined in this paper. It discusses how dynamic MANETs are and how adaptive QoS solutions are necessary. A QoS-aware routing method based on network conditions and QoS needs is introduced, which uses fuzzy logic and machine learning to determine routing decisions in MANETs [60]. ML is a promising technique that can enhance MANET security and performance. Future ML applications in MANETs should be even more innovative as ML algorithms continue to advance.

5. **Load balancing.** Load balancing, or the distribution of network traffic and computational jobs across network nodes to maximize resource utilization and network performance, is one use case for ML in MANETs. By reallocating resources and optimizing routes ahead of time in anticipation of changes in traffic loads, ML algorithms can forecast future patterns of network traffic and aid in proactive load balancing. ML models have the ability to dynamically modify the degree of data compression in response to the present network load [61]. This guarantees effective data transfer while lessening the load on nodes. With the aid of machine learning, tasks or data transmissions can be scheduled to reduce congestion and optimize the effective use of network resources. Because of their dynamic and self-organizing characteristics, MANETs are especially well-suited for ML-driven strategies for load balancing. Load balancing in cognitive radio MANETs, where ML techniques are employed for efficient spectrum utilization and load distribution, considers dynamic spectrum access and its impact on load balancing [62]. This study investigates load balancing in cognitive radio MANETs, where load distribution and efficient spectrum utilization are achieved through the use of machine learning algorithms.
6. **Fault detection and recovery.** To maintain the dependability and resilience of the network, ML applications for fault detection and recovery in MANETs are crucial. Based on previous data and the state of the network, machine learning models are able to forecast possible errors. Network administrators are able to avert disturbances by detecting patterns that lead to errors. Based on previous data and the state of the network, machine learning models are able to forecast possible errors. Network administrators are able to avert disturbances by detecting patterns that lead to errors. Based on previous data and the state of the network, machine learning models are able to forecast possible errors. Data-driven insights facilitate the creation of effective recovery strategies and minimize downtime. Collaborative methods based on machine learning allow nodes to exchange fault and recovery data. MANETs can enhance the overall recovery process by implementing coordinated recovery measures through the utilization of collective intelligence. To improve recovery tactics, this research highlights proactive fault identification by machine learning–based predictive fault detection in MANETs [63]. MANETs can be made more resilient and adaptive by utilizing ML approaches, which will guarantee constant network connectivity and better fault management.

16.5 APPLICATIONS OF ML IN VANET

There are new requirements for networking algorithms that deal with network control and resource allocation because of the dynamic nature of vehicular networks and their underlying heterogeneous structure. While network resource allocation deals with the spectrum, transmission power, and computing resources, network control deals with the handover, routing, and offloading of network traffic. The use of ML methods in this area of vehicular networking, as well as a combination of enabling technologies, is presented in this section because traditional methods cannot capture the underlying patterns in vehicular networks.

The safety of every mobile node is the main priority in VANET, along with lowering accident rates and data transfer delays. Routing problems result from the network structure's volatility. Data congestion on RSUs results from the VANET system's data-rich environment. Traditional prediction-based protocols have been developed for a variety of specialized applications, including routing, traffic control, safety, and others. The failure of any of them to incorporate an ML technique for additional data analytics drives the integration of ML models with traditional models to enhance performance [64]. By examining the data flow in the VANET system, various machine learning approaches, including SVM, KNN, k-means clustering, naive Bayes (NB), convolutional neural network (CNN), and artificial neural network (ANN), reveal these issues. Additionally, one of the main goals of the intelligent transportation system is safety.

Millimetre wave bands are most suited for Gbps lines, which are needed for 5G-VANET data transport. By taking into account the surrounding environmental factors, a suitable beam selection employing machine algorithms can be created [65]. By utilizing the Bayesian coalition game theory, the integration of ML into VANET can help decide trustworthy data processing and forwarding from one node to another or between various ITS system levels. The processing of the data by the intermediate node can be cooperative or non-cooperative. Reliability of nodes for processing and conveying spatiotemporal data can be assessed using ML [66]. At various VANET components, numerous types of data are produced. RSUs, vehic ular nodes, and the central station all have access to information like the specifics of nearby vehicular nodes, routing details, congestion-related data, weather-related data, and information necessary for communication between VANET components. Due to the dynamic nature of VANET, this tremendous volume of data is constantly generated. Machine learning methods can be combined with VANET to effectively use this data for a variety of VANET applications.

1. **Driver assistance.** The various sensors that are integrated into the cars produce a lot of data. Additionally, information gathered from the environment can be processed to comprehend the behaviour and thinking of the driver, which helps give them intelligent assistance. Eldar et al.'s [67] method uses the front camera as the source to create a road-type categorization solution. The performance of a vehicle's steering, acceleration, and braking can all be influenced by accurate and reliable information on the type of road. Six different types of road pavement are classified by the DNN algorithm,

which provides an average accuracy of 88.8% for road type and condition. Additionally, by forecasting friction coefficient from the given data, they have created a mathematical model for enhancing vehicle dynamics management via the anti-lock braking system (ABS) [68]. The proposed work reduces stopping distance for all analyzed road conditions and kinds. Accidents can occur with a slight amount of neglect. With fivefold cross-validation, Nguyen et al.'s [69] classification algorithms (NB, KNN, SVM, and decision tree) outperformed the others with a 90% accuracy rate. After removing outliers, the dataset was trained for further improvement, yielding an additional improvement of 3.4%. Autonomous vehicles will be crucial to ITS in the future.

2. **Collision alert.** By warning the driver of a serious accident before it occurs, safety can be assured to a vehicular node. It comprises the identification of potential obstructions to the typical course, vehicle behaviour, and crash probability. A real-time traffic accident detection method merged several ML techniques with V2V (vehicle-to-vehicle) communication, which can offer collision alarms by examining the behaviour of the vehicles. Data was gathered from the vehicular network based on the speeds and coordinates of the vehicles [70]. Although driving safety prediction algorithms have made great strides, they still do not take three distinct factors – driving information, traffic information, and driver information – into account.
3. **Safety.** *Vehicle safety* is the term used to describe the information that on-board units give the driver about the internal data of the car, such as its structural integrity, and other external aspects, such as lane detection and mode of transportation. This data aids in evaluating the vehicle's safety. The work done in this direction is covered in this section. Routing is crucial for communication between vehicles and RSUs. In their work, Peng et al. [71] first sought to gather information through mobile data sensing in VANET and afterwards recognized two important features of the vehicle, namely, road and driving safety analyses. For real-time road safety prediction, DeepRSI, a unique deep learning technique, is suggested. The study takes into account how GPS trajectories and other environmental elements relate spatially and temporally. With a mean precision of 83.2%, a maximum precision of 90.7%, and a minimum precision of 74.1%, DeepRSI outperforms other machine learning methods, including decision tree, KNN, and SVM technique.
4. **Communication.** The ITS's most crucial component is communication. All VANET applications rely on effective communication between vehicles and infrastructure(s), either directly or indirectly. The VANET system generates vast amounts of data for many uses, including safety alarms, traffic detection, and traffic forecast, which must be evaluated appropriately. In light of the network's frequent changes brought on by the high mobility of vehicular nodes, managing this data is an essential responsibility. Broadcasting urgent priority messages to all adjacent affected vehicle nodes is essential in the VANET [72]. Routing is crucial for communication between vehicles and RSUs. Due to its changeable architecture, routing in VANET

is a difficult task. The network is dynamic; thus, the routing decision must be made quickly and precisely. The path selection for transmission in the network is impacted by network anomalies, variations in packet delivery ratio (PDR), connection timing between nodes, secured packet transmission, etc. Machine learning–assisted route selection (MARS) technology is utilized to stabilize transmissions, make effective routing decisions, and minimize transmission delays for the vehicular nodes in VANET. Based on the expected connectivity degree (ECD), route selection for packet transfer benefits from real-time information on vehicular traffic [73]. Utilizing a variety of machine learning techniques, these dynamic routing decisions have dramatically reduced delay and packet loss rates while raising the trust value for the chosen path. ML has proved to increase the routing protocol's effectiveness.

5. **Misbehaviour detection.** In the VANET communication system, the term 'misbehaviour detection' refers to the identification of unusual activities. This undesirable behaviour may be caused by a bug in the vehicle nodes, which causes data packet dropping, erroneous resource requests, DDoS attacks, etc. Communication errors can lead to decreased PDR, limited throughput, end-to-end delay (EED), etc. One method for preventing malicious attacks is an ML-based intrusion detection system (IDS) with SVM. To find malicious nodes, Sharma et al. [74] suggested a multi-cluster head IDS technique. This method begins with the formation of a cluster of the vehicle and the assignment of a cluster head. After that, the cluster head uses the improved SVM algorithm to identify malicious nodes in the relevant cluster. Using the hybrid fuzzy multi-criteria decision-making (HF-MCDM) technique, other cluster chiefs are chosen after harmful nodes have been removed from the cluster. Average detection rate, average false positive rate, and average detection time for the suggested model were determined to be 99%, 0.87%, and 44 ms, respectively.
6. **Link management.** *Link management* is the term for the control of connections between RSUs and vehicular nodes. The link connection time with the surrounding nodes aids in the data's proper routing. This link management aids in the evaluation of the node's reliability for future data transmission. The continual movement of the cars and the frequent changes in the connected RSUs make vehicular networks unstable. Therefore, it is difficult to predict in advance how long a link will last between two vehicular nodes. Link duration between cars is a crucial metric for clustering and data distribution. A concept was put up by Gao et al. [75] to increase the dependability of data transfer between various RSUs or data nodes. The Manhattan mobility model was utilized to examine the movement patterns of the vehicles. In order to improve routing protocols, a hybrid network of wired and wireless links is suitable. A GPSR wireless routing strategy is created for the wireless transmission scheme. By leveraging RSUs as supporting infrastructure, they created a hybrid VANET routing architecture that combines wired and wireless transmission.

7. **Routing.** *Routing* is the process of determining the best route for sending data from a source to a destination. Data should be routed so that it consumes less computation and gets to its destination within the appropriate time frame. The VANET's routing algorithms do not take into account their prior knowledge of network conditions, like congestion, bad behaviour, and so forth. In order to address this, Tang et al. [76] created a deep learning–based method for effective routing. At first, routers determine every route that could connect two nodes. Depending on factors like hop number, distance, and priority, these paths are saved in a 3D path matrix. There has been an increase of crashes during the last few decades. The high speed on a motorway, which causes significant congestion and may result in several damages, is one of the causes. For the best decision-making to regulate traffic lights and dynamic route planning with real-time traffic predictions, reinforcement learning, deep learning, and numerous other algorithms are helpful.
8. **Traffic flow management.** Numerous factors, including the weather, uneven traffic distribution, accidents, and many others, have an impact on traffic flow. This element influences the flow of traffic by introducing uncertainty into it and by making it more difficult to collect data. Numerous studies that take into account things like the weather, previous traffic patterns, the kinds of vehicles, speeds, traffic volume, and density have been published. Additionally, efforts have been made to address the issues of data and traffic uncertainty. There is a lot of work being done to improve forecast accuracy [77]. Fuzzy deep convolutional network (FDCN), a method based on fuzzy deep learning, was proposed by Chen et al. [78] for analyzing and forecasting traffic flow. The FDCN presents certain fuzzy rules utilizing the learning method by fusing the fuzzy theory with DBN. The uncertainty issue is somewhat resolved by the suggested approach. The spatial and temporal characteristics of the traffic flow data are examined by the FDCN technique. For optimizing the supplied model, various layers and regression functions were taken into account. According to the test results, the proposed model is more accurate and capable of better representation than the practises already in use.

16.6 APPLICATIONS OF ML IN UWSNs

Underwater monitoring and communication systems can be made more effective, reliable, and functional with the help of machine learning technologies in underwater sensor networks (USNs). Underwater exploration, research, and monitoring could be revolutionized by machine learning in USNs. It improves the capabilities of these underwater sensor networks in applications ranging from marine biology to offshore industry and disaster monitoring by offering intelligent data analysis, resource management, and network optimization.

Some of the applications of ML in USN include the following.

In underwater sensor networks, which are used to monitor and gather data in aquatic environments, machine learning has numerous applications. These networks are usually made up of sensors that are placed in submerged environments, including lakes, rivers, and seas. Examples of these sensors are temperature, pressure, and sound sensors. These networks can be made more capable by machine learning, which offers intelligent data processing, boosts sensor performance, and permits real-time decision-making [79].

1. **Environmental monitoring.** To track changes in temperature, salinity, and other environmental characteristics, as well as to monitor pollution and water quality, machine learning can assist in the analysis of data from a variety of sensors. Events like oil spills and algae blooms can also be predicted by it. A number of sensors, including two underwater cameras that are streamed live on the Project Natick homepage, are used in the data centre to keep an eye on environmental factors and server settings. Cameras are used to keep an eye on the surrounding area in real time from two permanent positions outside the data centre thanks to these cameras, to tally the aquatic creatures captured by the cameras. It takes a lot of work to manually count all the marine creatures in every frame of the video stream. Object detection is implemented to automate the counting and observation of marine creatures in order to address this [80]. A probabilistic model–based approach for entity detection and localization is presented in [81]. A range of models were presented to evaluate their degree of adaptability and usefulness. The results showed that a probabilistic sensing paradigm could offer accurate entity observation and deliciousness in a range of settings.
2. **Energy consumption.** By modifying duty cycles, sensor activation schedules, and data transmission rates, algorithms can optimize power consumption. For the sensors in a UASN to function, they must have energy stored in their batteries. Analyzing the energy used for various sensor node operations, such as transmission, reception, and encryption, is crucial when building a sensor network architecture. UASNs typically use many hops to send their sensing data, with nearby nodes serving as relays to deliver the data to the sink node. In most cases, researchers employ optimization and data analysis to look at the nodes' energy use. On the other hand, machine learning techniques can be used to forecast various network parameters in place of executing laborious optimizations for various network architectures. A low-energy routing protocol with deep control was presented by Lilhore et al. [82]. This protocol allows for exact control over routing and efficient energy management by continuously monitoring the depth of underwater sensor nodes. This allows for the regulation of the number of nodes, depth spanning, and movement speed. When compared to conventional routing protocols, the research findings show that this protocol dramatically lowers node energy consumption while improving network efficiency. In the domains of environmental monitoring, underwater communication, and ocean observation, UWSNs have become an indispensable technology. However, optimizing node deployment and guaranteeing network performance are difficult due to the complicated underwater environment and

the restricted energy and communication range of sensor nodes. To tackle this problem, researchers have put forth several solutions, including node placement techniques based on approaches like adaptive mobility, energy balance, and direction optimization. The main goals of current research endeavours are to maximize network coverage, optimize energy utilization, and improve communication quality in order to support a variety of application scenarios, including underwater detection and ocean monitoring [83].

3. **Navigation and autonomous underwater vehicles (AUVs).** Recent years have seen a significant increase in the use of autonomous underwater vehicles (AUVs), which are essential tools for ocean research. By collecting information from several relevant sensors, the navigation system could ascertain the AUV's position, velocity, and other navigational parameters. Ocean research, therefore, requires AUV navigation technology as a prerequisite. Given that the global positioning system (GPS) is useless when the AUV is operating underwater, underwater autonomous navigation technology is necessary. The main navigation approaches at the moment are inertial navigation, acoustic navigation, simultaneous localization and mapping (SLAM) [84], map-based navigation, and integrated navigation. From the aforementioned navigation techniques, integrated navigation is the most often used method for AUVs. The Doppler velocity log (DVL) and inertial measurement unit (IMU) of the AUV were employed by the authors to obtain linear and angular velocity data, and they used a deep RL actor-critic architecture to produce direct thruster orders. The generative adversarial imitation learning technique for AUV navigation was put into practise by the authors in [85]. In the path-following task, the technique performs at the same level as a reinforcement learning algorithm.
4. **Fault detection and diagnostics.** To identify malfunctioning sensors or communication problems, machine learning models can be employed to detect errors or anomalies in the sensor network. This guarantees the accuracy of the information gathered. Underwater environments are susceptible to defects induced by ambient flowing water generated by surface waves or by other factors that jostle the sensor network components over time. Furthermore, defects may be brought about by aquatic or human species. Numerous factors impede communication underwater, and many communication methods are not available [86]. The physical integrity of sensor nodes needs to be given additional consideration and resources, since water leaks could flood the hardware. However, underwater, overheating issues ought to be uncommon and preventable. Fault tolerance has so far been addressed in the underwater context for space localization, underwater pipeline monitoring, and dependant UWSN networking.
5. **Underwater fault tolerance concepts.** The environment is harsher and more demanding, and faults are more expensive. Implementing certain more demanding techniques – like cloud computing – in USNs might not be the best idea. In the undersea realm, the authors do not perceive any barriers to the application of fault-tolerant methodologies that result in suitable

communication methods, low network bandwidths, and power requirements [87].

6. **Underwater target tracking.** Underwater targets, such as ships or marine animals, can be tracked, and their movements predicted using machine learning. Applications like fishery management and tracking the movement of potentially dangerous undersea objects depend on this. SVM is one of the statistical learning techniques frequently utilized in underwater target recognition [88]. The main concepts involved are determining the decision surface that separates two classes of data, aligning the samples such that they fall on opposite sides of the decision surface, and ensuring that the samples are sufficiently distant from the decision surface. Single-class SVM is used to detect targets in passive sonar systems. Due to the advancements in computer power, deep learning (DL) research based on neural networks has expanded quickly in the last few years [89]. Multiple processing layer network modules can make up a deep learning model. These layers extract features at varying degrees of abstraction and use backpropagation to automatically modify settings until the right data features are extracted for jobs that come after. Because of their strong feature extraction capabilities, deep learning models are frequently utilized in intelligent control, image processing, speech recognition, expert systems, and other domains. Based on deep learning, underwater target detection models are typically supervised; they use labelled datasets to train deep neural networks so that the networks can predict the kind of unknown samples. Numerous research have demonstrated the effectiveness of data augmentation techniques, including transfer learning (TL), in resolving the issue of model training when there is inadequate data [90]. These techniques not only lessen the training load on a small dataset but also shorten the target source's training period and provide reliable models. Additionally, in the event of insufficient data, data augmentation techniques like generative adversarial networks (GANs) [91] offer a solution for model training by expanding the dataset by producing new examples. In order to train the underwater network, it is challenging to build a standard dataset large enough in scale. As a result, for tracking of objects in underwater, several researchers employ transfer learning or data augmentation strategies.
7. **Security and intrusion detection.** UWSN sensor nodes are frequently sporadically placed in hazardous and challenging environments. They are hence vulnerable to network attacks. Security is one of the most crucial components of UWSNs in order to guarantee that an application runs well and produces secure data. Because of their limitations in terms of computational capacity, bit error rates, high propagation delay, and bandwidth, UWSNs have been the target of both internal and external attacks. The researchers claim that DoS, impersonation, replication, and data transfer in physical node attacks were the methods used to infiltrate UWSNs. DoS assaults are frequent in UWSNs because of their difficulties, cheap operating costs, and great efficacy, according to [92]. Authors in [93] have carried out a deeper investigation into DoS attacks on the physical layer. Within

an actual setting, they test the outcomes. Another typical security risk with UWSNs is a data assault. Information-centric architecture data management is one technique to safeguard data against DoS attack. By implementing a cloud paradigm to provide a single point of trust management in underwater sensor networks, the authors in [94] improved the application of the previously discussed trust model. The aforementioned management approach uses a mathematical way to compile trust proof in order to regulate each sensor node's reliability.

16.7 CONCLUSION AND FUTURE SCOPE

A comprehensive analysis of ML applications in WSN, MANET, VANET, and USN is presented in this chapter. According to the study, ML is a critical component in improving these various wireless network paradigms' intelligence, adaptability, and efficiency. ML algorithms have demonstrated their adaptability and versatility in various networks. Numerous optimization and intelligence problems, including routing, data aggregation and control, fault detection, and improving QoS, can be addressed using efficient ML algorithms. It has been shown that ML-driven methods can greatly improve network performance by maximizing resource use, optimizing energy use, and addressing a variety of network issues. Networks can now adapt and make decisions in real time in response to changing conditions and network dynamics due to the application of ML models. ML is useful for handling complicated issues like traffic management in VANETs and route optimization in MANETs that might be challenging to resolve with conventional algorithms. By assisting with intrusion detection, anomaly recognition, and threat mitigation in wireless networks, ML applications support security. This chapter highlighted the significance of ML algorithms in optimizing and adding intelligence to various network scenarios.

The future scope of this technology includes data privacy using federated learning for cooperative model training. This strategy is pertinent in networks because sensitive data may be involved. To expedite decision-making and lower latency, ML models can be pushed to the edge and fog layers of wireless networks. It is mostly essential for time-sensitive applications, like autonomous vehicles in VANETs. Examining methods for coordinating ML algorithms among various wireless networks, particularly in situations involving many domains, might enhance network effectiveness and administration. Energy-efficient ML model development is essential, particularly in resource-constrained networks and energy-harvesting contexts. A promising field of study includes applying machine learning to nodes with low resources. More research into dynamic network optimization using reinforcement learning approaches may result in networks that are self-healing and self-optimizing. It is crucial to make advances in ML-driven trust management and security methods to protect the confidentiality and integrity of data in networks. As these networks get more integrated, standardization initiatives that guarantee compatibility across various wireless network types and ML models will become more crucial. Promising research prospects exist in the future for improving ML methods as well as for tackling the particular difficulties and needs of various kinds of wireless networks.

REFERENCES

[1] Klaine, P. V., Imran, M. A., Onireti, O., & Souza, R. D. (2017). A survey of machine learning techniques applied to self-organizing cellular networks. *IEEE Communications Surveys & Tutorials*, *19*(4), 2392–2431.
[2] Challita, U., Ryden, H., & Tullberg, H. (2020). When machine learning meets wireless cellular networks: Deployment, challenges, and applications. *IEEE Communications Magazine*, *58*(6), 12–18.
[3] Chemouil, P., Hui, P., Kellerer, W., Limam, N., Stadler, R., & Wen, Y. (2020). Guest editorial special issue on advances in artificial intelligence and machine learning for networking. *IEEE Journal on Selected Areas in Communications*, *38*(10), 2229–2233.
[4] Kulin, M., Kazaz, T., De Poorter, E., & Moerman, I. (2021). A survey on machine learning-based performance improvement of wireless networks: PHY, MAC and network layer. *Electronics*, *10*(3), 318.
[5] Gebremariam, A. A., Usman, M., & Qaraqe, M. (2019, March). Applications of artificial intelligence and machine learning in the area of SDN and NFV: A survey. In *2019 16th International Multi-Conference on Systems, Signals & Devices (SSD)* (pp. 545–549). IEEE.
[6] Lei, L., Tan, Y., Zheng, K., Liu, S., Zhang, K., & Shen, X. (2020). Deep reinforcement learning for autonomous internet of things: Model, applications and challenges. *IEEE Communications Surveys & Tutorials*, *22*(3), 1722–1760.
[7] Boutaba, R., Salahuddin, M. A., Limam, N., Ayoubi, S., Shahriar, N., Estrada-Solano, F., & Caicedo, O. M. (2018). A comprehensive survey on machine learning for networking: evolution, applications and research opportunities. *Journal of Internet Services and Applications*, *9*(1), 1–99.
[8] Zhang, C., Patras, P., & Haddadi, H. (2019). Deep learning in mobile and wireless networking: A survey. *IEEE Communications Surveys & Tutorials*, *21*(3), 2224–2287.
[9] Jagwani, N., & Poornima, G. (2023, February). A survey on detecting location-based faults in wireless sensor networks using machine learning and deep learning techniques. In *Proceedings of 3rd International Conference on Recent Trends in Machine Learning, IoT, Smart Cities and Applications: ICMISC 2022* (pp. 493–507). Springer Nature.
[10] Gillani, M., Niaz, H. A., & Tayyab, M. (2021). Role of machine learning in WSN and VANETs. *International Journal of Electrical and Computer Engineering Research*, *1*(1), 15–20.
[11] Szott, S., Kosek-Szott, K., Gawłowicz, P., Gómez, J. T., Bellalta, B., Zubow, A., & Dressler, F. (2022). Wi-Fi meets ML: A survey on improving IEEE 802.11 performance with machine learning. *IEEE Communications Surveys & Tutorials*, *24*(3), 1843–1893.
[12] Niu, H., Li, X., Zhang, Y., & Xu, J. (2023). Advances and applications of machine learning in underwater acoustics. *Intelligent Marine Technology and Systems*, *1*(1), 8.
[13] Sun, Y., Peng, M., Zhou, Y., Huang, Y., & Mao, S. (2019). Application of machine learning in wireless networks: Key techniques and open issues. *IEEE Communications Surveys & Tutorials*, *21*(4), 3072–3108.
[14] Usama, M., Qadir, J., Raza, A., Arif, H., Yau, K. L. A., Elkhatib, Y., . . . & Al-Fuqaha, A. (2019). Unsupervised machine learning for networking: Techniques, applications and research challenges. *IEEE Access*, *7*, 65579–65615.
[15] Ladosz, P., Weng, L., Kim, M., & Oh, H. (2022). Exploration in deep reinforcement learning: A survey. *Information Fusion*, *85*, 1–22.
[16] Wang, M., Cui, Y., Wang, X., Xiao, S., & Jiang, J. (2017). Machine learning for networking: Workflow, advances and opportunities. *IEEE Network*, *32*(2), 92–99.
[17] Kuriakose, J., Joshi, S., Vikram Raju, R., & Kilaru, A. (2014). A review on localization in wireless sensor networks. In *Advances in signal processing and intelligent recognition systems* (pp. 599–610). Springer.

[18] Bernas, M., & Płaczek, B. (2015). Fully connected neural networks ensemble with signal strength clustering for indoor localization in wireless sensor networks. *International Journal of Distributed Sensor Networks*, *11*(12), 403242.
[19] Banihashemian, S. S., Adibnia, F., & Sarram, M. A. (2018). A new range-free and storage-efficient localization algorithm using neural networks in wireless sensor networks. *Wireless Personal Communications*, *98*, 1547–1568.
[20] Xiahou, Z., & Zhang, X. (2015). Adaptive localization in wireless sensor network through Bayesian compressive sensing. *International Journal of Distributed Sensor Networks*, *11*(8), 438638.
[21] Kumar, S., Tiwari, S. N., & Hegde, R. M. (2015). Sensor node tracking using semi-supervised Hidden Markov Models. *Ad Hoc Networks*, *33*, 55–70.
[22] De Paola, A., Gaglio, S., Re, G. L., Milazzo, F., & Ortolani, M. (2014). Adaptive distributed outlier detection for WSNs. *IEEE Transactions on Cybernetics*, *45*(5), 902–913.
[23] Wang, Y., Liu, Z., Wang, D., Li, Y., & Yan, J. (2017). Anomaly detection and visual perception for landslide monitoring based on a heterogeneous sensor network. *IEEE Sensors Journal*, *17*(13), 4248–4257.
[24] Yadav, P., & Sharma, S. C. (2023). A systematic review of localization in WSN: Machine learning and optimization-based approaches. *International Journal of Communication Systems*, *36*(4), e5397.
[25] Zidi, S., Moulahi, T., & Alaya, B. (2017). Fault detection in wireless sensor networks through SVM classifier. *IEEE Sensors Journal*, *18*(1), 340–347.
[26] Titouna, C., Aliouat, M., & Gueroui, M. (2016). FDS: Fault detection scheme for wireless sensor networks. *Wireless Personal Communications*, *86*, 549–562.
[27] Chanak, P., & Banerjee, I. (2016). Fuzzy rule-based faulty node classification and management scheme for large scale wireless sensor networks. *Expert Systems with Applications*, *45*, 307–321.
[28] Nayak, P., Swetha, G. K., Gupta, S., & Madhavi, K. (2021). Routing in wireless sensor networks using machine learning techniques: Challenges and opportunities. *Measurement*, *178*, 108974.
[29] Yau, K. L. A., Goh, H. G., Chieng, D., & Kwong, K. H. (2015). Application of reinforcement learning to wireless sensor networks: Models and algorithms. *Computing*, *97*, 1045–1075.
[30] Mehmood, A., Lv, Z., Lloret, J., & Umar, M. M. (2017). ELDC: An artificial neural network based energy-efficient and robust routing scheme for pollution monitoring in WSNs. *IEEE Transactions on Emerging Topics in Computing*, *8*(1), 106–114.
[31] Gharajeh, M. S., & Khanmohammadi, S. (2016). DFRTP: Dynamic 3D fuzzy routing based on traffic probability in wireless sensor networks. *IET Wireless Sensor Systems*, *6*(6), 211–219.
[32] Ren, J., Zheng, J., Guo, X., Song, T., Wang, X., Wang, S., & Zhang, W. (2023). MeFi: Mean field reinforcement learning for cooperative routing in wireless sensor network. *IEEE Internet of Things Journal, 11*(1), 995–10111.
[33] Hota, L., Nayak, B. P., Kumar, A., Ali, G. M. N., & Chong, P. H. J. (2021). An analysis on contemporary MAC layer protocols in vehicular networks: State-of-the-art and future directions. *Future Internet*, *13*(11), 287.
[34] Rovcanin, M., De Poorter, E., Moerman, I., & Demeester, P. (2014). A reinforcement learning based solution for cognitive network cooperation between co-located, heterogeneous wireless sensor networks. *Ad Hoc Networks*, *17*, 98–113.
[35] Rovcanin, M., De Poorter, E., van den Akker, D., Moerman, I., Demeester, P., & Blondia, C. (2015). Experimental validation of a reinforcement learning based approach for a service-wise optimisation of heterogeneous wireless sensor networks. *Wireless Networks*, *21*, 931–948.

[36] Osamy, W., Khedr, A. M., Salim, A., AlAli, A. I., & El-Sawy, A. A. (2022). Recent studies utilizing artificial intelligence techniques for solving data collection, aggregation and dissemination challenges in wireless sensor networks: A review. *Electronics*, *11*(3), 313.

[37] Raj, B., Ahmedy, I., Idris, M. Y. I., & Md. Noor, R. (2022). A survey on cluster head selection and cluster formation methods in wireless sensor networks. *Wireless Communications and Mobile Computing*, *2022*, 1–53.

[38] Wang, Y., Yang, A., Li, Z., Chen, X., Wang, P., & Yang, H. (2016). Blind drift calibration of sensor networks using sparse Bayesian learning. *IEEE Sensors Journal*, *16*(16), 6249–6260.

[39] Wu, M., Tan, L., & Xiong, N. (2016). Data prediction, compression, and recovery in clustered wireless sensor networks for environmental monitoring applications. *Information Sciences*, *329*, 800–818.

[40] Xu, X., Ansari, R., Khokhar, A., & Vasilakos, A. V. (2015). Hierarchical data aggregation using compressive sensing (HDACS) in WSNs. *ACM Transactions on Sensor Networks*, *11*(3), 1–25.

[41] Kim, T., Vecchietti, L. F., Choi, K., Lee, S., & Har, D. (2020). Machine learning for advanced wireless sensor networks: A review. *IEEE Sensors Journal*, *21*(11), 12379–12397.

[42] Kosunalp, S. (2016). A new energy prediction algorithm for energy-harvesting wireless sensor networks with Q-learning. *IEEE Access*, *4*, 5755–5763.

[43] Chen, F., Fu, Z., & Yang, Z. (2019). Wind power generation fault diagnosis based on deep learning model in internet of things (IoT) with clusters. *Cluster Computing*, *22*, 14013–14025.

[44] Awan, S. W., & Saleem, S. (2016, September). Hierarchical clustering algorithms for heterogeneous energy harvesting wireless sensor networks. In *2016 International Symposium on Wireless Communication Systems (ISWCS)* (pp. 270–274). IEEE.

[45] Asif, M., Khan, S., Ahmad, R., Sohail, M., & Singh, D. (2017). Quality of service of routing protocols in wireless sensor networks: A review. *IEEE Access*, *5*, 1846–1871.

[46] Collotta, M., Pau, G., & Bobovich, A. V. (2017). A fuzzy data fusion solution to enhance the QoS and the energy consumption in wireless sensor networks. *Wireless Communications and Mobile Computing*, *2017*.

[47] Renold, A. P., & Chandrakala, S. (2017). MRL-SCSO: Multi-agent reinforcement learning-based self-configuration and self-optimization protocol for unattended wireless sensor networks. *Wireless Personal Communications*, *96*, 5061–5079.

[48] Ren, L., Wang, W., & Xu, H. (2017). A reinforcement learning method for constraint-satisfied services composition. *IEEE Transactions on Services Computing*, *13*(5), 786–800.

[49] Gupta, A., Verma, P., & Sambyal, R. S. (2018). An overview of MANET: Features, challenges and applications. *International Journal of Scientific Research in Computer Science, Engineering and Information Technology*, *4*(1), 122–126.

[50] Sargunavathi, S., Manickam, J., & Leo, M. (2022). Enhanced trust based encroachment discovery system for Mobile Ad-hoc networks. *Cluster Computing, 22,* 4837–4847 (Retraction of 10.1007/S10586-022-03834-6, 2022; Retracted article. See DEC, 2022).

[51] Al-Rawi, H. A., Ng, M. A., & Yau, K. L. A. (2015). Application of reinforcement learning to routing in distributed wireless networks: A review. *Artificial Intelligence Review*, *43*, 381–416.

[52] Lanjewar, P. D., & Thakare, V. M. (2015). Performance and optimization of MANET routing protocols. *International Journal of Computer Applications*, *975*, 8887.

[53] Popli, R., Sethi, M., Kansal, I., Garg, A., & Goyal, N. (2021, August). Machine learning based security solutions in MANETs: State of the art approaches. *Journal of Physics: Conference Series*, *1950*(1), 012070.

[54] Nishani, L., & Biba, M. (2016). Machine learning for intrusion detection in MANET: A state-of-the-art survey. *Journal of Intelligent Information Systems*, *46*, 391–407.
[55] Hammamouche, A., Omar, M., Djebari, N., & Tari, A. (2018). Lightweight reputation-based approach against simple and cooperative black-hole attacks for MANET. *Journal of Information Security and Applications*, *43*, 12–20.
[56] Zhou, Y., & Wang, X. (2022). Adaptive fuzzy command filtering control for nonlinear MIMO systems with full state constraints and unknown control direction. *Neurocomputing*, *493*, 474–485.
[57] Sangeetha, A., & Rajendran, T. (2022). Supervised vector machine learning with brown boost energy efficient data delivery in MANET. *Sustainable Computing: Informatics and Systems*, *35*, 100761.
[58] Borole, Y. D., Dofe, J., & Dethe, C. G. (2021). Machine learning–enabled techniques for reducing energy consumption of IoT devices. In *Green internet of things and machine learning: Towards a smart sustainable world* (pp. 27–85). Willey.
[59] Kute, V. B., & Kharat, M. U. (2012). Survey on QoS for multi-path routing protocols in mobile ad-hoc networks. *International Journal of Machine Learning and Computing*, *2*(4), 458.
[60] Zafar, M. H., & Altalbe, A. (2021). Prediction of scenarios for routing in MANETs based on expanding ring search and random early detection parameters using machine learning techniques. *IEEE Access*, *9*, 47033–47047.
[61] Yao, H., Yuan, X., Zhang, P., Wang, J., Jiang, C., & Guizani, M. (2019). Machine learning aided load balance routing scheme considering queue utilization. *IEEE Transactions on Vehicular Technology*, *68*(8), 7987–7999.
[62] Duan, W., Tang, X., Zhou, J., Wang, J., & Zhou, G. (2018). Load balancing opportunistic routing for cognitive radio ad hoc networks. *Wireless Communications and Mobile Computing*, *2018*, 1–16.
[63] Sowah, R. A., Ofori-Amanfo, K. B., Mills, G. A., & Koumadi, K. M. (2019). Detection and prevention of man-in-the-middle spoofing attacks in MANETs using predictive techniques in Artificial Neural Networks (ANN). *Journal of Computer Networks and Communications*, *2019*.
[64] Gupta, R., Tanwar, S., Tyagi, S., & Kumar, N. (2020). Machine learning models for secure data analytics: A taxonomy and threat model. *Computer Communications*, *153*, 406–440.
[65] Nayak, B. P., Hota, L., Kumar, A., Turuk, A. K., & Chong, P. H. (2021). Autonomous vehicles: Resource allocation, security, and data privacy. *IEEE Transactions on Green Communications and Networking*, *6*(1), 117–131.
[66] Kumar, N., Rodrigues, J. J., & Chilamkurti, N. (2014). Bayesian coalition game as-a-service for content distribution in internet of vehicles. *IEEE Internet of Things Journal*, *1*(6), 544–555.
[67] Šabanovič, E., Žuraulis, V., Prentkovskis, O., & Skrickij, V. (2020). Identification of road-surface type using deep neural networks for friction coefficient estimation. *Sensors*, *20*(3), 612.
[68] Ulum, M., & Patriawan, D. A. (2022). Modeling and performance testing of anti-lock braking system (ABS) with variation of road friction coefficient to braking distance. *Rekayasa*, *15*(3), 340–345.
[69] Nguyen, H., Cai, C., & Chen, F. (2017). Automatic classification of traffic incident's severity using machine learning approaches. *IET Intelligent Transport Systems*, *11*(10), 615–623.
[70] Zhang, L., Yan, L., Fang, Y., Fang, X., & Huang, X. (2019). A machine learning-based defensive alerting system against reckless driving in vehicular networks. *IEEE Transactions on Vehicular Technology*, *68*(12), 12227–12238.

[71] Peng, Z., Gao, S., Li, Z., Xiao, B., & Qian, Y. (2018). Vehicle safety improvement through deep learning and mobile sensing. *IEEE Network*, *32*(4), 28–33.
[72] Taherkhani, N., & Pierre, S. (2016). Centralized and localized data congestion control strategy for vehicular ad hoc networks using a machine learning clustering algorithm. *IEEE Transactions on Intelligent Transportation Systems*, *17*(11), 3275–3285.
[73] Khan, S., Alam, M., Fränzle, M., Müllner, N., & Chen, Y. (2018). A traffic aware segment-based routing protocol for VANETs in urban scenarios. *Computers & Electrical Engineering*, *68*, 447–462.
[74] Sharma, S., & Kaul, A. (2018). Hybrid fuzzy multi-criteria decision making based multi cluster head dolphin swarm optimized IDS for VANET. *Vehicular Communications*, *12*, 23–38.
[75] Gao, H., Liu, C., Li, Y., & Yang, X. (2020). V2VR: Reliable hybrid-network-oriented V2V data transmission and routing considering RSUs and connectivity probability. *IEEE Transactions on Intelligent Transportation Systems*, *22*(6), 3533–3546.
[76] Tang, F., Mao, B., Fadlullah, Z. M., Kato, N., Akashi, O., Inoue, T., & Mizutani, K. (2017). On removing routing protocol from future wireless networks: A real-time deep learning approach for intelligent traffic control. *IEEE Wireless Communications*, *25*(1), 154–160.
[77] Chen, P., Li, K., & Sun, J. (2008, December). A method of traffic flow forecast and management. In *2008 International conference on information management, innovation management and industrial engineering* (Vol. 2, pp. 24–28). IEEE.
[78] Chen, W., An, J., Li, R., Fu, L., Xie, G., Bhuiyan, M. Z. A., & Li, K. (2018). A novel fuzzy deep-learning approach to traffic flow prediction with uncertain spatial–temporal data features. *Future Generation Computer Systems*, *89*, 78–88.
[79] Moniruzzaman, M., Islam, S. M. S., Bennamoun, M., & Lavery, P. (2017, September 18–21). Deep learning on underwater marine object detection: A survey. In *Advanced concepts for intelligent vision systems: 18th international conference, ACIVS 2017*, Antwerp, Proceedings 18 (pp. 150–160). Springer International Publishing.
[80] https://natick.research.microsoft.com/
[81] Nguyen, T. L. N., & Shin, Y. (2019). An efficient RSS localization for underwater wireless sensor networks. *Sensors*, *19*(14), 3105.
[82] Lilhore, U. K., Khalaf, O. I., Simaiya, S., Tavera Romero, C. A., Abdulsahib, G. M., & Kumar, D. (2022). A depth-controlled and energy-efficient routing protocol for underwater wireless sensor networks. *International Journal of Distributed Sensor Networks*, *18*(9), 15501329221117118.
[83] Krishnapriya, N., & Kumareshan, N. (2022). Machine learning based energy efficient high performance routing protocol for underwater communication. *Adhoc & Sensor Wireless Networks*, *54*.
[84] Hou, G., Shao, Q., Zou, B., Dai, L., Zhang, Z., Mu, Z., . . . & Zhai, J. (2019). A novel underwater simultaneous localization and mapping online algorithm based on neural network. *ISPRS International Journal of Geo-Information*, *9*(1), 5.
[85] Tal, A., Klein, I., & Katz, R. (2017). Inertial navigation system/Doppler velocity log (INS/DVL) fusion with partial DVL measurements. *Sensors*, *17*(2), 415.
[86] Tsai, C. M., Wang, C. S., Chung, Y. J., Sun, Y. D., & Perng, J. W. (2021). Multi-sensor fault diagnosis of underwater thruster propeller based on deep learning. *Sensors*, *21*(21), 7187.
[87] Subha, T. D., Subash, T. D., Jane, K. C., Devadharshini, D., & Francis, D. I. (2020). Autonomous under water vehicle based on extreme learning machine for sensor fault diagnosistics. *Materials Today: Proceedings*, *24*, 2394–2402.
[88] Szűcs, G., Papp, D., & Lovas, D. (2015, September). SVM classification of moving objects tracked by Kalman filter and Hungarian method. In *Working notes of CLEF 2015 conference*, Toulouse.

[89] Villon, S., Chaumont, M., Subsol, G., Villéger, S., Claverie, T., & Mouillot, D. (2016, October). Coral reef fish detection and recognition in underwater videos by supervised machine learning: Comparison between deep learning and HOG+ SVM methods. In *International conference on advanced concepts for intelligent vision systems* (pp. 160–171). Springer International Publishing.
[90] Moghimi, M. K., & Mohanna, F. (2022). Reliable object recognition using deep transfer learning for marine transportation systems with underwater surveillance. *IEEE Transactions on Intelligent Transportation Systems*, *24*(2), 2515–2524.
[91] Panetta, K., Kezebou, L., Oludare, V., & Agaian, S. (2021). Comprehensive underwater object tracking benchmark dataset and underwater image enhancement with GAN. *IEEE Journal of Oceanic Engineering*, *47*(1), 59–75.
[92] Cooperative its for mobility in European cities. Accessed 24 March 2019. https://cordis.europa.eu/project/rcn/196891/factsheet/en.
[93] Tanwar, S., Bhatia, Q., Patel, P., Kumari, A., Singh, P. K., & Hong, W. C. (2019). Machine learning adoption in blockchain-based smart applications: The challenges, and a way forward. *IEEE Access*, *8*, 474–488.
[94] Dogru, N., & Subasi, A. (2018, February). Traffic accident detection using random forest classifier. In *2018 15th learning and technology conference (L&T)* (pp. 40–45). IEEE.

17 An Efficient Localization for Resource Management in UAV-Assisted Wireless Networks Using Deep Learning

Visalakshi Annepu, Kalapraveen Bagadi, Naga Raju Challa, Vaegae Naveen Kumar, and Yamarthi Narasimha Rao

17.1 INTRODUCTION

Wireless sensor networks (WSNs) are crucial for real-time monitoring and control of physical environments, enabling efficient data collection and proactive decision-making [1]. WSNs offer cost-effective scalability, energy efficiency, and flexibility, making them valuable for diverse applications, such as environmental monitoring, industrial automation, and healthcare. Accurate node localization in WSNs is crucial for tasks such as target tracking, event detection, and data fusion. Traditional localization methods often rely on range-based or connectivity based techniques, which can be challenging to implement, especially in large-scale networks [2]. To address these challenges, researchers have explored the use of unmanned aerial vehicles (UAVs) to assist in node localization. This literature review provides an overview of the existing research in UAV-assisted node localization using neural networks [3–5]. UAV-assisted localization techniques have emerged as a promising solution to enhance the accuracy and scalability of node localization in WSNs. These techniques leverage the mobility and aerial perspective provided by UAVs to capture visual information from different angles. The captured images are then processed using neural network models to estimate the positions of the nodes. Various approaches have been proposed in the literature to achieve UAV-assisted node localization.

The study in [6] presents a comprehensive survey on the utilization of UAVs in WSNs. The focus of this survey is strongly on UAVs for data collection in WSNs. This chapter discusses various techniques to support UAVs in WSNs, considering factors like energy consumption and transportation costs. One approach is to use

DOI: 10.1201/9781032625034-17

traditional computer vision algorithms combined with UAV imagery. For instance, Li et al. [7] proposed a method that extracts features from UAV images using scale-invariant feature transform (SIFT) and performs node localization using an extended Kalman filter. Although effective, these methods heavily rely on handcrafted features and may not capture complex spatial relationships accurately. In another study, Al-Fuqaha et al. (2018) proposed a UAV-assisted localization system that utilized a hybrid approach of range-based and image-based localization [8]. The system employed a UAV equipped with both a camera and range sensors to collect data from the sensor nodes. The data was then processed using a combination of neural networks and range-based algorithms to estimate the node positions. This work presented in [9] addresses the challenge of finding and querying smart sensor nodes in industrial UAVs using a combination of ultrawideband (UWB) and wake-up radio (WUR) communication technologies. The UAV utilizes UWB range measurements for localizing sensor nodes and efficiently acquires data from them in terms of energy consumption. In another study, Wang et al. (2020) introduced a novel approach that combined UAV imagery and received signal strength (RSS) measurements for localization in large-scale WSNs [10]. They utilized a graph-based algorithm to estimate the positions of the sensor nodes based on UAV-captured images and RSS information. Furthermore, Yang et al. presented a comprehensive survey on UAV-assisted localization techniques in WSNs, highlighting different localization approaches, communication protocols, and energy-efficient mechanisms [11]. The survey discussed the challenges and opportunities in utilizing UAVs for node localization, providing insights into the advancements in this field.

In recent years, deep learning techniques, particularly NNs, have shown remarkable performance in computer vision tasks. Researchers have leveraged the power of neural networks to develop image-based node localization methods. In the engineering field, there is a growing trend in the use of soft computing approaches to address various challenges, indicating their increasing popularity and effectiveness [12–15]. DL techniques have become prominent in these days due to their wide range of applications [16–18]. As a family of DL techniques, the neural networks (NNs) are also used in many engineering applications [19–21]. The study in [22] proposed convolutional neural network (CNN) architecture to estimate node positions using UAV images. The network learned to extract discriminative features and achieved higher accuracy compared to traditional methods. Several studies have explored different aspects of UAV-assisted node localization. For instance, Li et al. (2019) proposed a localization approach using UAV-mounted cameras and deep learning techniques for accurate node localization in outdoor WSNs [23]. They achieved improved localization accuracy by leveraging visual information captured by the UAV. In a more recent study, Wang et al. proposed a UAV-assisted localization method that utilized a graph neural network (GNN) to estimate the node positions [24]. The method employed a UAV equipped with a camera and flew over the sensor network area, capturing images of the nodes from different angles. The images were then processed using the GNN to learn the spatial relationships between the nodes and estimate their positions. To address the non-linear distortion issue, the study in [25] proposes the use of an MLP as a robust tool for node localization in UAV-aided WSNs. MLPs can handle complex non-linear mappings between input (e.g. signal measurements) and output (e.g. node locations) data, resulting in improved localization accuracy and reduced deployment costs. In summary, UAV-assisted node localization using NNs has shown promising results in improving the accuracy and efficiency of node localization in WSNs. The proposed methods have utilized various deep learning techniques, including convolutional neural networks and graph neural networks, to capture complex spatial relationships between the sensor nodes. The use of UAVs for data collection has also provided advantages in coverage, scalability, and efficiency compared to traditional methods. On the other hand, RNNs are a type of NN that can effectively model sequential data by capturing temporal dependencies and contextual information [26]. They are well-suited for processing time series data, making them a natural choice for handling the dynamic and time-varying nature of node localization in WSNs. By leveraging the capabilities of RNNs, UAVs can intelligently analyze sensor data over time, improving the accuracy and reliability of node localization. The combination of UAVs and RNNs in node localization brings several advantages. Firstly, the mobility of UAVs allows for flexible positioning and collection of sensor data from multiple locations, enabling comprehensive coverage of the sensing area. Secondly, RNNs can capture long-term dependencies and patterns in the collected data, facilitating more accurate localization results. This is particularly beneficial in scenarios where nodes may undergo changes in their positions or exhibit complex movement patterns. In this work, we introduce a UAV-assisted node localization framework utilizing RNNs. We investigate the use of RNN architectures to effectively model the temporal dynamics of sensor data and predict node locations. We will also present experimental results and performance evaluations to demonstrate the effectiveness and accuracy of our approach compared to existing localization methods.

The chapter is structured as follows in the subsequent sections: Section 17.2 provides a comprehensive system model of related works in the field of node localization and UAV-assisted localization techniques. The MLP model for node localization is deliberated in Section 17.3. Section 17.4 presents the methodology and design of the proposed UAV-assisted node localization system using RNN. Section 17.5 describes the experimental set-up and presents the results and analysis. The chapter is concluded with a summary of the results in Section 17.6.

17.2 SYSTEM MODEL

Consider a three-dimensional (3D) sensor field as shown in Figure 17 1, where each of the P number of unmanned aerial vehicles (UAVs) carries an anchor node (AN). These UAVs with ANs are denoted as AUs. The UAVs are assumed to fly between h_{min} and h_{max}. These UAVs are used to localize randomly distributed Q number of 3D target nodes (TNs) in a given area of interest. We assume that the TNs are equipped with radio transmitters, such as Bluetooth, Wi-Fi, or ultrawideband (UWB) tags, which are linked to their respective IDs, for signal transmission to UAVs. Additionally, the UAVs can be equipped with RF sensors to measure the strength of the RF signal received from each TN. The communication technology between AUs

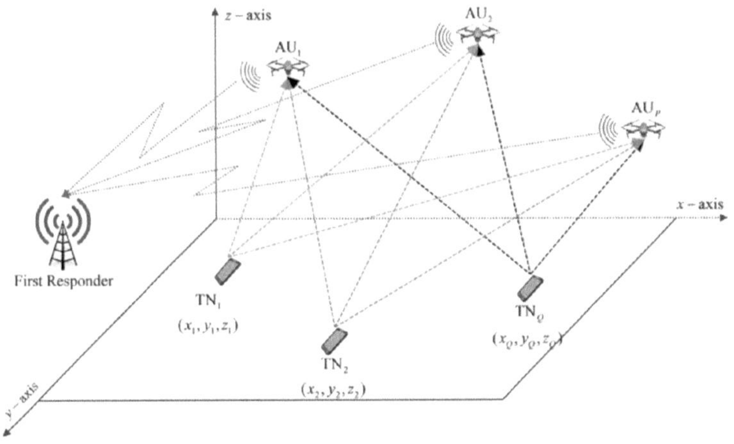

FIGURE 17.1 Localization of Q target modes using P UAVs.

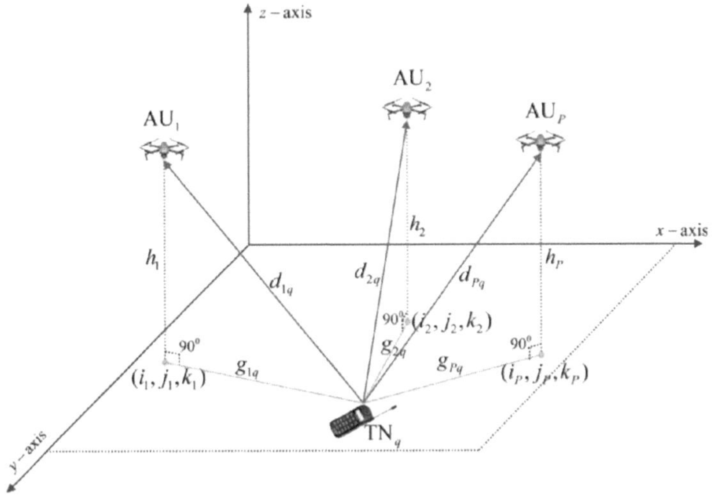

FIGURE 17.2 Localization of qth target node using P AUs.

and TNs could be LTE or Wi-Fi, or as simple as LPWAN [27]. The received signal strengths (RSSs) are used to compute distances between AUs and TNs [28]. Later, these distances are converted to coordinates of TNs using multilateration proves. The multilateration requires at least four UAVs to compute the coordinates.

The schematic diagram for localizing qth TN using P UAVs is depicted in Figure 17.2, where pth UAV flies at an altitude of h_p and P should be at least four. As illustrated in this figure, d_{pq} and g_{pq} refer to the direct and ground distances between pth AU and qth TN, where *ground distance* refers to the distance between TN and the perpendicular projection of UAV on the place of TN's existence. Let the 3D coordinate projection of pth AU on the plane of TN be (i_p, j_p, k_p), and (x_q, y_q, z_q) is the 3D coordinate of pth to the qth TN, where z_q is the altitude of TN from the ground field, and it becomes zero if TN is located on ground field. The log normal shadowing model (LNSM) is used to estimate the distances between AUs and the TN. Thus, the RSS measured in dBm at pth AU while receiving signal from qth TN is given by:

$$RSS[d_{pq}]dBm = P_0[d_0]dBm - 10\gamma \log_{10}\left[\frac{d_{pq}}{d_0}\right] - \chi_\sigma \tag{17.1}$$

The estimated distance between the qth TN and pth AU, based on the RSS measurement given in equation (17.1), is calculated using the parameters P_0 (RSS at reference distance d_0), d_{pq} (true distance between TN and AU), γ (path loss exponent), and χ_σ (representing log-normal shadowing effects). The value of χ_σ is a Gaussian distributed i.i.d. random variable with zero mean and variance σ^2.

$$\hat{d}_{pq} = d_0 10^{\left[\frac{P_0[d_0] - RSS[d_{pq}]}{10\alpha}\right]} \tag{17.2}$$

From the estimated direct distances, the square ground distances can be computed as: $\hat{g}_{pq}^2 = \hat{d}_{pq}^2 - h_p^2$. Let $(\hat{x}_q, \hat{y}_q, \hat{z}_q)$ be the estimated position of qth TN, then the set of square distances from N AUs to qth TN is mathematically written using multilateration as:

$$\begin{aligned}
(i_1 - \hat{x}_q)^2 + (j_1 - \hat{y}_q)^2 + (k_1 - \hat{z}_q)^2 &= \hat{g}_{1q}^2 \\
(i_2 - \hat{x}_q)^2 + (j_2 - \hat{y}_q)^2 + (k_2 - \hat{z}_q)^2 &= \hat{g}_{2q}^2 \\
&\vdots \\
(i_P - \hat{x}_q)^2 + (j_P - \hat{y}_q)^2 + (k_P - \hat{z}_q)^2 &= \hat{g}_{Pq}^2
\end{aligned} \tag{17.3}$$

For simplicity, the set of equation given in equation (17.3) can be written as a set of linear equations in terms of $\hat{x}_q$, $\hat{y}_q$, and $\hat{z}_q$. To accomplish this, remove the most recent equation from all earlier ones. Thus, equation (17.3) can be reformed as:

$$\begin{aligned}
&2(i_1 - i_P)\hat{x}_q + 2(j_1 - j_P)\hat{y}_q + 2(k_1 - k_P)\hat{z}_q \\
&= \hat{d}_{Pq}^2 - \hat{d}_{1q}^2 + i_1^2 - i_P^2 + j_1^2 - i_P^2 + k_1^2 - k_P^2 \\
&2(i_2 - i_P)\hat{x}_q + 2(j_2 - j_P)\hat{y}_q + 2(k_2 - k_P)\hat{z}_q \\
&= \hat{d}_{Pq}^2 - \hat{d}_{2q}^2 + i_2^2 - i_P^2 + j_2^2 - i_P^2 + k_2^2 - k_P^2 \\
&\vdots \\
&2(i_{P-1} - i_P)\hat{x}_q + 2(j_{P-1} - j_P)\hat{y}_q + 2(k_{P-1} - k_P)\hat{z}_q \\
&= \hat{d}_{Pq}^2 - \hat{d}_{(P-1)q}^2 + i_{(P-1)}^2 - i_P^2 + j_{(P-1)}^2 - i_P^2 + k_{(P-1)}^2 - k_P^2
\end{aligned} \tag{17.4}$$

Equation (17.4) is written in matrix form as

$$U \times L_q = V \tag{17.5}$$

where $L_q = [\hat{x}_q \ \hat{y}_q \ \hat{z}_q]^T$ is the position of qth TN:

$$U = 2\begin{bmatrix} (i_1 - i_P) & (j_1 - j_P) & (k_1 - k_P) \\ (i_2 - i_P) & (j_2 - j_P) & (k_2 - k_P) \\ & \vdots & \\ (i_{P-1} - i_P) & (j_{P-1} - j_P) & (k_{P-1} - k_P) \end{bmatrix}$$

and

$$V = \begin{bmatrix} \hat{d}_{Pq}^2 - \hat{d}_{1q}^2 + i_1^2 - i_P^2 + j_1^2 - i_P^2 + k_1^2 - k_P^2 \\ \hat{d}_{Pq}^2 - \hat{d}_{2q}^2 + i_2^2 - i_P^2 + j_2^2 - i_P^2 + k_2^2 - k_P^2 \\ \vdots \\ \hat{d}_{Pq}^2 - \hat{d}_{(P-1)q}^2 + i_{(P-1)}^2 - i_P^2 + j_{(P-1)}^2 - i_P^2 + k_{(P-1)}^2 - k_P^2 \end{bmatrix}$$

Then, the location of qth TN can be obtained from:

$$L_q = (U^T U)^{-1} U^T V \tag{17.6}$$

17.3 3D NODE LOCALIZATION USING MLP MODEL

RSS-based techniques rely on measuring the RSS of wireless signals transmitted by ANs. Based on LNSM, the RSS is used to calculate the separation between the TN and the AU. RSS-based techniques can achieve reasonable accuracy depending on factors like the quality of signal measurements, propagation models used, and calibration accuracy. However, they may suffer from limitations due to environmental factors, multipath effects, and signal interference. NN-based techniques, on the other hand, utilize ANNs to learn the mapping between RSS values and node positions. They leverage the power of machine learning to establish a relationship between signal measurements and node localization. Neural network–based techniques can potentially provide higher accuracy and precision compared to RSS-based techniques. By leveraging the learning capabilities of neural networks, they can capture complex relationships between signal measurements and node positions, leading to improved localization performance. A popular type of NN in machine learning and deep learning is the multi-layer perceptron (MLP) [29]. It is a feedforward neural network, which means that data moves without loops or feedback connections in a single direction, from the input layer to the output layer. The MLP is made up of numerous layers of artificial neurons or interconnected nodes. While the neurons in the output layer correspond to the expected outcomes, each neuron in the input layer represents an input feature. Hidden layers are those layers that exist between the input and output layers. The BP technique, which is widely utilized, can be effectively used to train the MLP weights [30]. Two essential procedures are involved in training an MLP: backpropagation and forward propagation. The anticipated outputs are computed when the input data is fed through the network during forward propagation. The calculated inaccuracy is then calculated by comparing the calculated outputs to the desired outputs. In order to reduce the discrepancy between expected and desired outputs, BP is used to adjust the weights of the links in the network based on the computed error. Up till the network achieves a suitable degree of accuracy, this iterative procedure is continued. MLPs have a number of benefits. They are suitable for a variety of tasks, including classification, regression, and pattern recognition, due to their ability to learn and represent complicated patterns in data. Given enough hidden units and the right activation functions, they can also approximate any non-linear function. MLPs can be used for both supervised and unsupervised learning and can handle both continuous and categorical input data. As a result, TN positions in a WSN are likewise estimated using the MLP model [26].

The MLP model utilized for localizing the UN in the UAV-assisted WSN is illustrated in Figure 17.3, where the MLP employs P AUs to localize the qth TN. This picture depicts the MLP model, which has P input elements, H hidden neurons, and a three-neuron output layer. The three output neurons calculated the qth TN's x, y, and z coordinates. The weights in an MLP determine the strength of connections between neurons, influencing their relatedness. Each neuron processes the weighted total of its inputs through an activation function to create an output, which is then transmitted to the next layer. The activation function in the network introduces non-linearity,

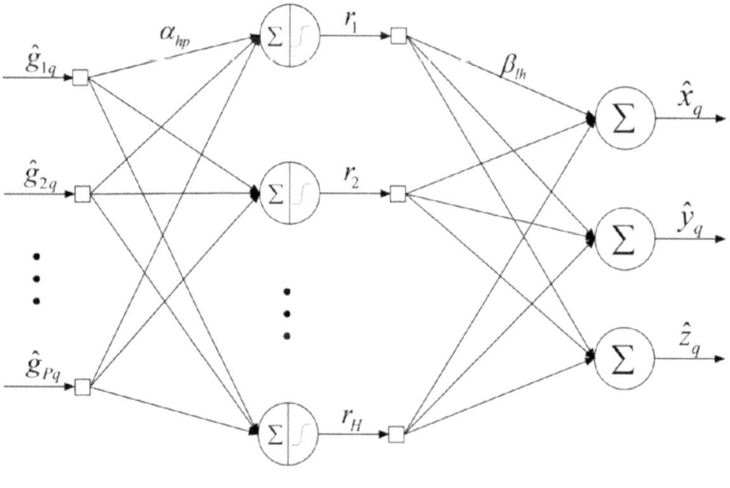

FIGURE 17.3 Architecture of MLP model for localizing qth TN.

allowing it to identify complex patterns and generate non-linear predictions. Consequently, the output of each hidden neuron can be represented as follows:

$$r_h = \psi\left(\sum_{p=1}^{P} \alpha_{hp}\, \hat{g}_{pq}\right), \quad h = 1,2,\ldots,H \tag{17.7}$$

where α_{hp} represents connection weight between the hth hidden node and pth input node. The sigmoid function $\psi(.)$ is employed as the non-linear activation function, defined as $\psi\ (t) = 1/(1 + e^{-t})$. The three output neurons act as simple summation nodes. Thus, the estimated x, y, and z coordinates of the qth TN are acquired from the outputs of these three neurons.

$$\hat{x}_q = \sum_{h=1}^{H} \beta_{1h} r_h,\ \hat{y}_q = \sum_{h=1}^{H} \beta_{2h} r_h,\ \hat{z}_q = \sum_{h=1}^{H} \beta_{3h} r_h \tag{17.8}$$

where β_{1h}, β_{2h}, and β_{3h}, respectively, are the connection weights between first, second, and third output nodes and hth hidden node.

17.4 3D MODEL LOCALIZATION USING PROPOSED RNN MODEL

The MLPs also have some limitations. To generalize successfully and avoid overfitting, they need a lot of labelled training data. They can be sensitive to the initial weights and architecture choices, making them prone to getting stuck in local optima. Training large MLPs can be computationally expensive, especially when dealing with high-dimensional data. To address some of these limitations, variations of MLPs have been developed, such as RNNs for sequential data. These variations leverage specific properties of the data to improve performance and efficiency. RNNs are more suited for tasks involving sequential or time-dependent data because they feature feedback connections, which, in contrast to feedforward neural networks, allow information to remain over time steps. The ability of an RNN to preserve an internal state, or memory, that records data from earlier inputs is its important feature [31]. This memory is passed along with the current input to produce an output and update the state for the next time step. This recurrent nature allows RNNs to capture temporal dependencies and context in the data. Because of its built-in self-feedback structure, the RNN model might outperform the MLP model. Also, since it avoids using a hidden layer, its complexity is also smaller than the MLP network. Training an RNN involves a process called real-time recurrent learning (RTRL) [32]. It is similar to backpropagation in feedforward neural networks but takes into account the sequential nature of the data. RTRL propagates the error from the output back through each time step, allowing the network to learn from its past mistakes and adjust its weights accordingly. Thus, RNN is considered to be efficient for node localization in WSNs.

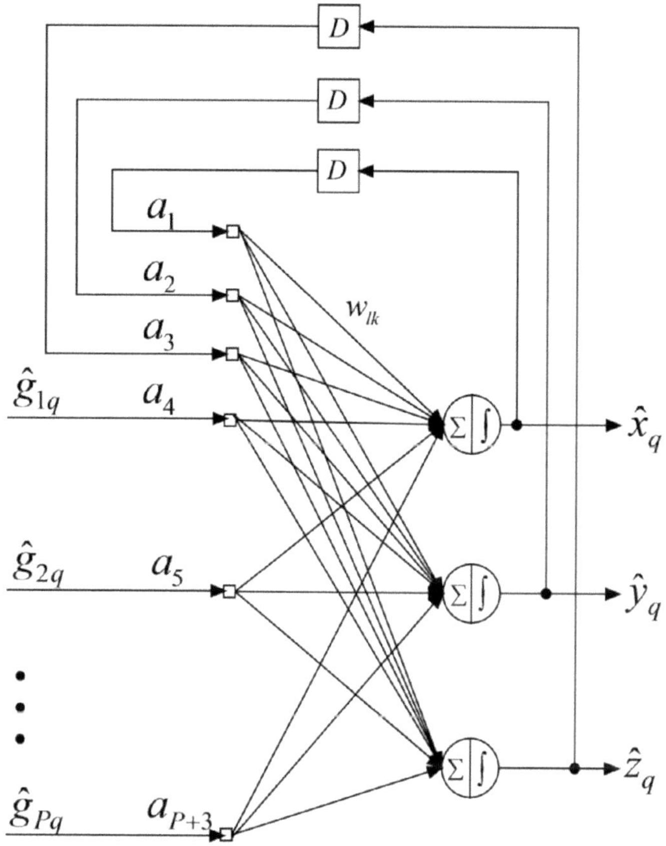

FIGURE 17.4 Architecture of RNN model for localizing qth TN.

The architecture of RNN model used for localizing qth TN is illustrated in Figure 17.4. The RNN model shown in this figure has an input layer of P + 3 elements. Out of these P + 3 elements, P elements contain planar distances of qth TN from P AUs, and three elements contain estimated x, y, and z position of the qth TN in the previous iteration that are fed back from the output layer. The output layer consists of three neurons, and these neurons estimates x, y, and z position of the qth TN. The external input vector $\boldsymbol{I} = \left[\hat{g}_{1q}, \hat{g}_{2q}, \ldots, \hat{g}_{Pq}\right]^T$ and the output vector with a one-time instant delay and zero initial state $\boldsymbol{O} = \left[D(\hat{x}_q), D(\hat{y}_q), D(\hat{z}_q)\right]^T$ are merged to create an input vector, denoted as $\boldsymbol{a}$. Subsequently, the kth element of the input vector can be expressed as:

$$a_k = \begin{cases} O_k, & \text{if } k \in [1,2,3] \\ I_{k-3}, & \text{if } k \in [4,5,\ldots,3+P] \end{cases} \tag{17.9}$$

Each output neuron consists of a summer and a non-linear activation. Hence, the estimated x, y, and z coordinate of qth TN are obtained from the three output neurons as:

$$\hat{x}_q = \psi\left(\sum_{k=1}^{P+3} w_{1k} a_k\right), \hat{y}_q = \psi\left(\sum_{k=1}^{P+3} w_{2k} a_k\right), \hat{z}_q = \psi\left(\sum_{k=1}^{P+3} w_{3k} a_k\right) \tag{17.10}$$

where w_{1k}, w_{2k}, and w_{3k}, respectively, are the connection weights between the first, second, and third output nodes and the kth input element. The activation function $\psi(.)$ is the same sigmoid activation function that we used for MLP.

17.5 SIMULATION ANALYSIS

In this section, the efficacy of the RNN-based node localization is tested and compared with MLP- and RSSI-based node localization through simulation analysis. In these simulations, we have assumed at least four UAVs are used to localize each TN. The Monto Carlo simulations are performed by averaging 100 times to locate each TN to obtain true locations. Table 17.1 provides an overview of the remaining simulation parameters [33]. The average localization error (AvgLE) that is used to evaluate the accuracy of various localization techniques is defined as:

$$\text{AvgLE} = \frac{1}{Q}\sum_{q=1}^{Q}\sqrt{(\hat{x}_q - x_q)^2 + (\hat{y}_q - y_q)^2} \tag{17.11}$$

TABLE 17.1
Simulation Parameters

Parameter	Value
Sensor field dimension	$100 \times 100 \times 35\ m$
Number of AUs (P)	4
Number of TNs (Q)	25
Height of the UAV (h_p)	40 m
Height of the TNs (h_q)	0–35 m
Path-loss constant (γ)	3.1058
Standard deviation (σ)	2.8189
Reference distance (d_0)	1 m
Transmitted power at reference distance ($P(d_0)$)	−27.143 dB
Number of output nodes of MLP and RNN	3
Number of input nodes of MLP (P)	4
Number of input nodes of RNN ($P + 3$)	7
Number of hidden neurons of the MLP (H_N)	8
Learning rate of MLP and RNN	0.08

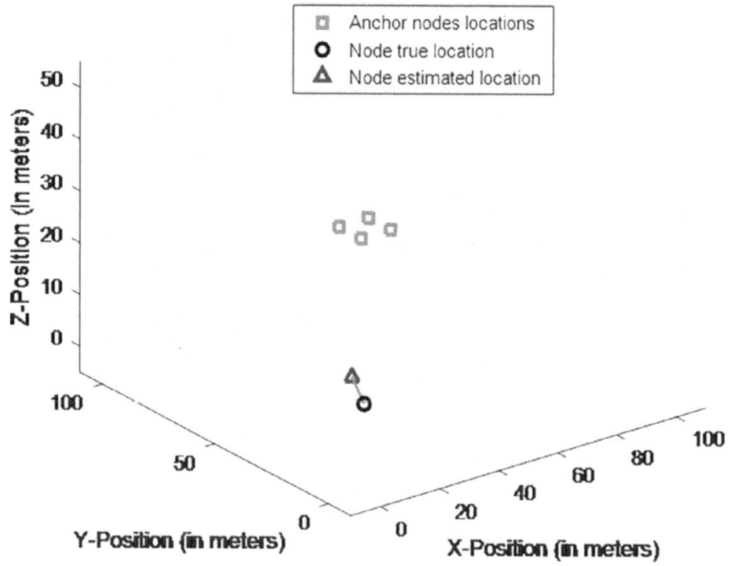

FIGURE 17.5 Localization of qth TN using four AUs.

Figure 17.5 shows actual position and estimated position of the qth TN using RSS with the help of four AUs. The four AUs are located at a height of 40 m from the ground. The RSS-based localization with increased number of AUs is depicted in Figure 17.6. Figure 17.6(a), Figure 17.6(b), and Figure 17.6(c) illustrate the estimated and actual positions of Q TNs when the number of AUs is 4, 5, and 6, respectively. It is observed that as the number of AUs increases, the actual and estimated positions converge, indicating improved accuracy. With more anchor nodes, the algorithm can incorporate additional measurements and use triangulation or trilateration techniques to estimate the target node's position more accurately. As illustrated in Table 17.2, the localization error decreases with an increasing number of AUs. This indicates that a higher number of anchor nodes contributes to minimizing the localization error, leading to improved accuracy in estimating the positions of the target nodes.

Figure 17.7(a), Figure 17.7(b), and Figure 17.7(c) illustrate the actual and estimated positions of Q TNs with four AUs using RSS-, MLP-, and the proposed RNN-based localization, respectively. RSS-based techniques can achieve reasonable accuracy depending on factors like the quality of signal measurements, propagation models used, and calibration accuracy. However, they may suffer from limitations due to environmental factors, multipath effects, and signal interference. By contrast, NN-based techniques can potentially provide higher accuracy and precision compared to RSS-based techniques. By leveraging the learning capabilities of neural networks, they can capture complex relationships between signal measurements and node positions, leading to improved localization performance, as shown

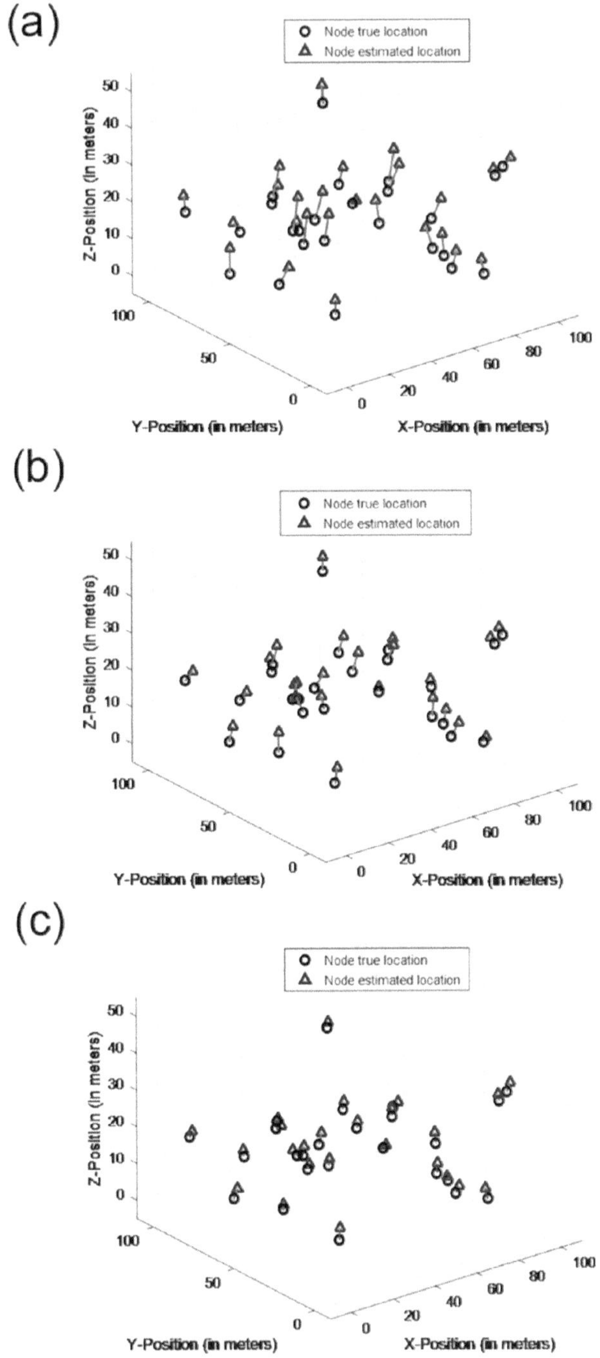

FIGURE 17.6 RSS localization of Q TNs with (a) $P = 4$ AUs, (b) $P = 5$ AUs, and (c) $P = 6$ AUs.

TABLE 17.2
AvgLE of the RSS Algorithm with Increased Number of AUs

P	AvgLE (in *meters*)
4	4.1273
5	3.2424
6	2.8573

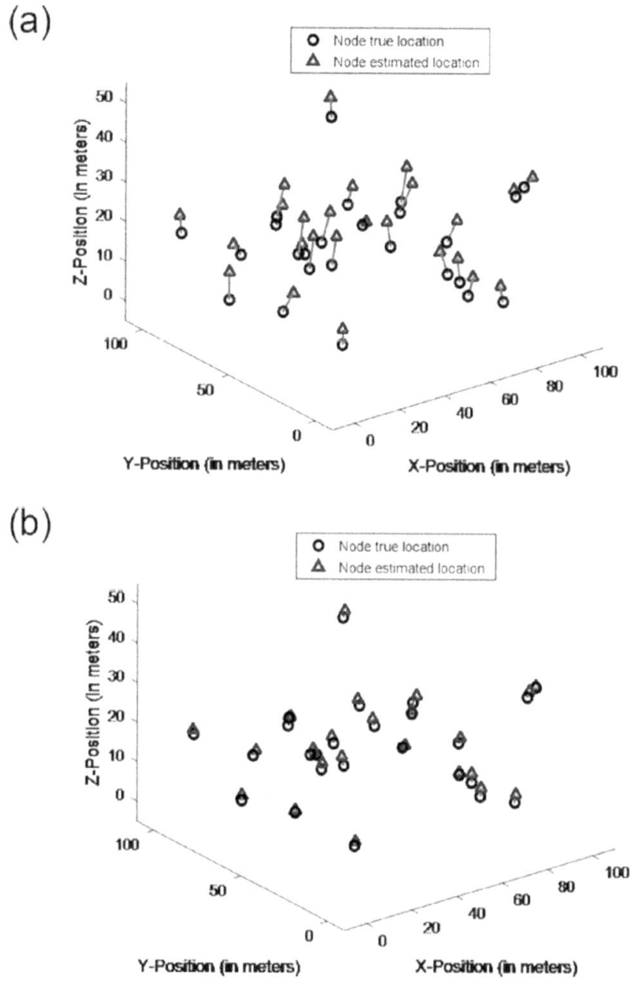

FIGURE 17.7 Localization of Q TNs with four AUs using (a) RSS, (b) MLP, and (c) RNN.

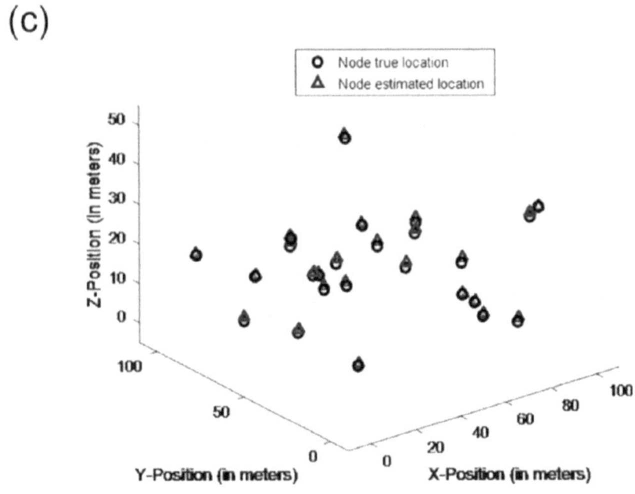

FIGURE 17.7 Continued

TABLE 17.3
AvgLE of Various Localization Techniques

Technique	AvgLE (in *meters*)
RSS	4.1273
MLP	1.1035
RNN	0.8462

in Figure 17.7(b) and Figure 17.7(c), where the actual estimated positions almost coincide. Further, RNN-based localization has an edge over MLP-based localization because the RNN also considers the previous outputs while training the network. This may lead to further lower AvgLE over ML, as presented in Table 17.3.

Figure 17.8(a), Figure 17.8(b), and Figure 17.8(c) illustrate the localization errors acquired for each TN using RSS, MLP, and RNN techniques, respectively. The choice of localization algorithm and techniques used in the system can influence the localization error of individual nodes. Different algorithms may have varying levels of accuracy and robustness, and certain algorithms may perform better in specific scenarios. The variance in the localization errors while using RSS technique is more, as we see in Figure 17.8(a), whereas the variance of localization errors while using MLP and RNN techniques is less due to their inherent capabilities. Thus, NN-based techniques try to localize all nodes with nearly equal precision compared to the RSS technique, as we see in Figure 17.8(b) and Figure 17.8(c).

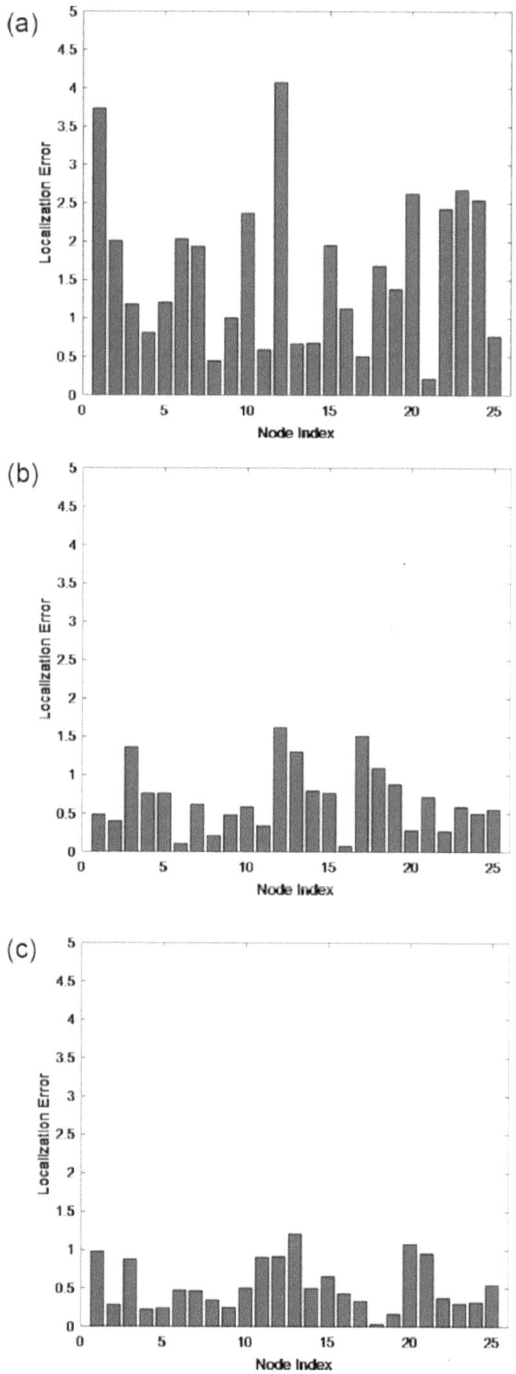

FIGURE 17.8 Localization error of each TN using (a) RSS, (b) MLP, and (c) RNN.

17.6 CONCLUSION

In conclusion, our research introduces a pioneering paradigm for node localization in wireless sensor networks through the innovative use of UAVs and deep learning techniques. The classical challenges associated with RSS-based localization methods, including susceptibility to environmental factors and signal distortions, are effectively addressed by our proposed approach. By harnessing the power of NNs, specifically the MLP and RNN architectures, our method exhibits a remarkable ability to learn complex patterns and relationships within received signals. This enables the system to adapt dynamically to environmental variations, ensuring robust and accurate node localization. The inclusion of UAVs in the localization process leverages air-to-ground communication links, facilitating consistent and reliable data exchange with sensor nodes. Our extensive simulations validate the superiority of the proposed UAV-assisted node localization using NNs over traditional methods. Notably, the recurrent nature of the RNN model, incorporating previous network outputs during training, contributes to its enhanced performance compared to the versatile MLP. The achieved localization accuracy surpasses existing benchmarks, emphasizing the efficacy of our approach in real-world scenarios. The implications of this research extend beyond improved localization accuracy, impacting resource management in UAV-assisted wireless networks. The demonstrated efficiency and adaptability of our method underscore its potential for applications such as environmental monitoring, disaster management, and asset tracking, where precise localization is imperative.

As we move forward, further exploration of deep learning models and optimization strategies promises to enhance the scalability and applicability of our approach. The integration of real-world experiments will validate the robustness of our findings and pave the way for practical implementation. In essence, our work contributes to the evolving landscape of localization techniques, offering a transformative solution for resource management in UAV-assisted wireless networks.

REFERENCES

[1] S. R. Jino Ramson and D. J. Moni, "Applications of Wireless Sensor Networks—A Survey," *2017 International Conference on Innovations in Electrical, Electronics, Instrumentation and Media Technology (ICEEIMT)*, Coimbatore, pp. 325–329, 2017, doi: 10.1109/ICIEEIMT.2017.8116858.

[2] A. Paul and T. Sato, "Localization in Wireless Sensor Networks: A Survey on Algorithms, Measurement Techniques, Applications and Challenges," *Journal of Sensor and Actuator Networks*, vol. 6, no. 4, p. 24, 2017, doi: 10.3390/jsan6040024.

[3] Y. Xu, G. Zhenguo and N. Qiang, "Unmanned Aerial Vehicle–Assisted Node Localization for Wireless Sensor Networks," *International Journal of Distributed Sensor Networks*, vol. 13, no. 12, 2017.

[4] V. Annepu, D. R. Sona, C. V. Ravikumar, K. Bagadi, M. Alibakhshikenari, A. A. Althuwayb, B. Alali, C. S. Virdee, G. Pau, I. Dayoub, C. H. See and F. Falcone, "Review on Unmanned Aerial Vehicle Assisted Sensor Node Localization in Wireless Networks: Soft Computing Approaches," *IEEE Access*, vol. 10, pp. 132875–132894, 2022, doi: 10.1109/ACCESS.2022.3230661.

[5] V. Annepu, A. Rajesh and K. Bagadi, "Radial Basis Function-Based Node Localization for Unmanned Aerial Vehicle-Assisted 5G Wireless Sensor Networks," *Neural Computing and Applications*, vol. 33, pp. 12333–12346, 2021, doi: 10.1007/s00521-021-05835-9.
[6] M. T. Nguyen, C. V. Nguyen, H. T. Do, H. T. Hua, T. A. Tran, A. D. Nguyen, G. Ala and F. Viola, "UAV-Assisted Data Collection in Wireless Sensor Networks: A Comprehensive Survey," *Electronics*, vol. 10, no. 21, p. 2603, 2021, doi: 10.3390/electronics 10212603.
[7] J. M. Li, D. M. Yan, G. Wang and L. Zhang, "An Improved SIFT Algorithm for Unmanned Aerial Vehicle Imagery," *35th International Symposium on Remote Sensing of Environment*, p. 17, 2014, doi: 10.1088/1755-1315/17/1/012187.
[8] H. Shakhatreh, A. H. Sawalmeh, A. Al-Fuqaha, Z. Dou, E. Almaita, I. Khalil, N. S. Othman, A. Khreishah and M. Guizani, "Unmanned Aerial Vehicles (UAVs): A Survey on Civil Applications and Key Research Challenges," *IEEE Access*, vol. 7, pp. 48572–48634, 2019, doi: 10.1109/ACCESS.2019.2909530.
[9] N. Vlad, P. Daniele, M. Michele and B. Luca, "Fly, Wake-up, Find: UAV-based Energy-Efficient Localization for Distributed Sensor Nodes," *Sustainable Computing: Informatics and Systems*, vol. 34, 2022, doi: 10.1016/j.suscom.2022.100666.
[10] P. Wang, M. Xu, D. Han and X. Zhou, "UAV Imagery Assisted Localization for Wireless Sensor Networks," *IEEE Transactions on Mobile Computing*, vol. 19, no. 7, pp. 1706–1719, 2020.
[11] J. Yang, Z. Tian, J. Tang and J. Ni, "A Comprehensive Survey of UAV-Assisted Localization Techniques in Wireless Sensor Networks," *IEEE Communications Surveys & Tutorials*, vol. 23, no. 2, pp. 1600–1625, 2021.
[12] N. K. Vaegae, K. K. Pulluri, K. Bagadi and O. O. Oyerinde, "Design of an Efficient Distracted Driver Detection System: Deep Learning Approaches," *IEEE Access*, vol. 10, pp. 116087–116097, 2022.
[13] K. P. Bagadi, V. Annepu and S. Das, "Recent Trends in Multiuser Detection Techniques for SDMA–OFDM Communication System," *Physical Communication*, vol. 20, pp. 93–108, 2016.
[14] K. P. Bagadi and S. Das, "Multiuser Detection in SDMA-OFDM Wireless Communication System Using Complex Multilayer Perceptron Neural Network," *Wireless Personal Communications*, vol. 77, pp. 21–39, 2014.
[15] K. P. Bagadi and S. Das, "Efficient Complex Radial Basis Function Model for Multiuser Detection in a Space Division Multiple Access/Multiple-Input Multiple-Output-Orthogonal Frequency Division Multiplexing System," *IET Communications*, vol. 7, pp. 1394–1404, 2013.
[16] K. Bagadi, T. Abrao and F. Benedetto, "A Novel Machine Learning Approach for Intelligent Spectrum Management in Cognitive Radio Networks," *IEEE Networking Letters*, vol. 5, pp. 232–236, doi: 10.1109/LNET.2023.3300274.
[17] K. P. Bagadi and S. Das, "Minimum Symbol Error Rate Multiuser Detection Using an Effective Invasive Weed Optimization for MIMO/SDMA–OFDM System," *International Journal of Communication Systems*, vol. 27, no. 12, pp. 3837–3854, 2014, doi: 10.1002/dac.2579
[18] K. Bagadi, C. V. Ravikumar, M. Alibakhshikenari, Ch. Nagaraj, A. Rajesh, S. Aïssa, I. Dayoub, F. Falcone and E. Limiti, "Precoded Large Scale Multi-User-MIMO System Using Likelihood Ascent Search for Signal Detection," *Radio Science*, vol. 57, no. 12, 2022, doi: 10.1029/2022RS007573.
[19] N. R. Challa and K. Bagadi, "Likelihood Ascent Search Detection for Coded Massive MU – MIMO Systems to Mitigate IAI and MUI," *Radioelectronics and Communication Systems*, vol. 63, no. 5, pp. 223–234, 2020, doi: 10.3103/S0735272720050015.

[20] N. R. Challa and K. Bagadi, "Design of Near-Optimal Local Likelihood Search-Based Detection Algorithm for Coded Large-Scale MU-MIMO System," *International Journal of Communication Systems*, vol. 33, no. 12, p. e4436, 2020, doi: 10.1002/dac.4436.
[21] K. P. Bagadi and S. Das, "Neural Network Based Multiuser Detection Techniques in SDMA-OFDM System," *2011 Annual IEEE India Conference*, BITS Pilani Campus, Hyderabad, 16–18 December 2011.
[22] S.-A. Hassan, T. Rahim and S.-Y. Shin, "An Improved Deep Convolutional Neural Network-Based Autonomous Road Inspection Scheme Using Unmanned Aerial Vehicles," *Electronics*, vol. 10, no. 22, p. 2764, 2021, doi: 10.3390/electronics10222764.
[23] H. Li, W. Zhou, W. Li and J. Cao, "A UAV-Based Localization Method for Wireless Sensor Networks Using Deep Learning," *IEEE Internet of Things Journal*, vol. 6, no. 1, pp. 364–375, 2019.
[24] X. Wang, L. Fu, N. Cheng, R. Sun, T. Luan, W. Quan and K. Aldunaikhy, "Joint Flying Relay Location and Routing Optimization for 6G UAV–IoT Networks: A Graph Neural Network-Based Approach," *Remote Sensing*, vol. 14, no. 17, p. 4377, 2022, doi: 10.3390/rs14174377.
[25] C. V. Ravi Kumar and K. P. Bagadi, "MC–CDMA Receiver Design Using Recurrent Neural Networks for Eliminating Multiple Access Interference and Nonlinear Distortion," *International Journal of Communication System*, vol. 30, no. 16, pp. 1–13, 2017, doi: 10.1002/dac.3328.
[26] V. Annepu and A. Rajesh, "An Unmanned Aerial Vehicle-Aided Node Localization Using an Efficient Multilayer Perceptron Neural Network in Wireless Sensor Networks," *Neural Computing & Applications*, vol. 32, pp. 11651–11663, 2020, doi: 10.1007/s00521-019-04653-4.
[27] A. Al-Hourani, S. Kandeepan and S. Lardner, "Optimal LAP Altitude for Maximum Coverage," *IEEE Wireless Communications Letters*, vol. 3, no. 6, pp. 569–572, 2014.
[28] Y. Xu, J. Zhou and P. Zhang, "RSS-Based Source Localization When Path-Loss Model Parameters Are Unknown," *IEEE Communications Letters*, vol. 18, no. 6, pp. 1055–1058, 2014.
[29] J. Naskath, G. Sivakamasundari and A. A. S. Begum, "A Study on Different Deep Learning Algorithms Used in Deep Neural Nets: MLP SOM and DBN," *Wireless Personal Communications*, vol. 128, pp. 2913–2936, 2023, doi: 10.1007/s11277-022-10079-4.
[30] A. Singh, S. Kushwaha, M. Alarfaj and M. Singh, "Comprehensive Overview of Backpropagation Algorithm for Digital Image Denoising," *Electronics*, vol. 11, no. 10, p. 1590, 2022, doi: 10.3390/electronics11101590.
[31] A. Lazar, G. Pipa and J. Triesch, "SORN: A Self-Organizing Recurrent Neural Network," *Frontiers in Computational Neuroscience*, vol. 3, 2009.
[32] R. J. Williams and D. Zipser, "Experimental Analysis of the Real-time Recurrent Learning Algorithm," *Connection Science*, vol. 1, no. 1, pp. 87–111, 2007.
[33] B. Yuan, R. He, B. Ai, R. Chen, G. Wang, J. Ding and Z. Zhong, "A UAV-Assisted Search and Localization Strategy in Non-Line-of-Sight Scenarios," *IEEE Internet of Things Journal*, vol. 9, no. 23, 2022.

18 A Smartphone-Based Context-Aware Driver Rating System for Cities in Developing Economies

Gayatree Parbat, Munshi Yusuf Alam, and Subrata Nandi

18.1 INTRODUCTION

In today's world, transportation systems play a crucial role in both developed and developing nations, with significant impacts on their economies, societies, and overall quality of life. But transportation systems in developing countries face a range of complex issues and challenges. Developing countries may have inadequate or poorly maintained transportation infrastructure, including roads, bridges, and public transmit, due to economic instability, lack of monitoring systems, rapid urbanization, high population density, etc. The number of two- and four-wheel drive vehicles is growing daily as a result of the rise in commercial services such as product delivery and pick-and-drop. Instead of using public transportation like buses and trains, people are choosing to reserve a cab and a two-wheeler in order to arrive promptly, safely, and comfortably. Thus, driving has become one of the sources of employment for many individuals in developing nations.[1] In this context, driver rating systems are an essential component of modern ride-sharing and delivery platforms, promoting safety, quality, accountability, and customer satisfaction. Nowadays, many ride-sharing and delivery platforms predominantly use incentive-based or token-based driver rating systems. An incentive-based driver rating system typically rewards drivers with incentives, such as bonuses, discounts, or other perks, for achieving high ratings to motivate drivers to provide excellent service to maintain or increase

FIGURE 18.1 Certain characteristics of the road surface (poor lighting, different speed breaker profiles, potholes, damaged road patches, road construction) which affect smooth and safe drive.

DOI: 10.1201/9781032625034-18

their income. On the other hand, a token-based driver rating system could involve drivers earning tokens or credits for good performance [1, 2]. These tokens might be redeemable for various benefits, like priority dispatch, better shifts, or additional opportunities. Both systems aim to motivate drivers to provide better service. But sometimes drivers may receive poor ratings in certain circumstances where they are not entirely responsible for the negative feedback. According to some recent surveys, certain external or environmental factors, like street lighting and road conditions, can actually have an impact on drivers' ratings in certain situations. For example, a lot of potholes, uneven roads, unmarked speed breakers, or construction zones shown in Figure 18.1 can make for an uncomfortable and unhealthy ride for passengers. Roads in disrepair can lead to slower travel times, which might frustrate passengers particularly if they are in a hurry. Similarly insufficient street lighting can make passengers feel less safe, especially during night-time rides, which can lead to lower ratings. Poor visibility may make it difficult for passengers as well as unknown drivers to locate the location, which also leads to misunderstanding and lower ratings. To mitigate the effects of poor road conditions and inadequate lighting, some fair and effective driver rating systems are facilitated through mobile device sensing, which is a transformative concept poised to revolutionize road safety and infrastructure monitoring, because smartphones are widely available and owned by many individuals, making them a cost-effective option for data collection. Deploying a dedicated sensor device can be expensive and logistically challenging in the context of many urban and suburban cities in developing countries. In an era where the reliability of transportation systems is paramount, this innovative system harnesses the power of modern technology to address two critical issues, that is, road anomalies and non-functioning street lights. By leveraging the capabilities of mobile devices, users can contribute up-to-date road information from their devices, by which we can create a comprehensive solution that not only assesses proper driver rating but also contributes to the enhancement of road conditions and public safety. Here, we explore the development, implementation, and potential benefits of such a system, offering a glimpse into the future of smart and accountable transportation networks.

18.1.1 Related Work

Research indicates that crowdsensing, as opposed to upgrading the current infrastructure, is the most practical, affordable, and easily deployable method for large-scale public infrastructure monitoring in developing and underdeveloped regions of various countries [3, 4]. Few studies have been done in the literature that use crowdsensing to monitor and control urban street lights. According to Poulsen et al.'s research [5], street lighting colour can be dynamically controlled by smartphones based on public occupancy. In order to automatically gather lighting data for street light monitoring, the authors of [3] have created a car-top sensor platform with a light sensor array. However, these methods are still obtrusive for in-the-wild deployments because they require specific hardware components to be mounted on top of the vehicle. On the other hand, we investigate in this work whether crowdsensing from smartphones can be used for efficient street light monitoring. A handy platform with a clear view of the lighting conditions for the smartphone light sensor is necessary for this kind of approach. We take advantage of the navigation apps that are used for two-wheeler commuting in order to achieve this [6, 7]. In fact, bike-sharing programs, a relatively new form of public transportation, have seen great success in a number of major cities across the world [7]. In many parts of the world, motorcycles (such as bikes, scooters, etc.) and bicycles are regarded as some of the most practical forms of urban transportation [8–11]. One notable research that looks into the issue of whether or not there is an adequate supply of light is conducted by Kumar et al. [12]. They processed the photos after using external cameras mounted on top of the vehicle. In this case, it might be challenging to filter out the ambient lighting from surrounding buildings and shops. Moreover, establishing cameras on the vehicle's roof in order to gather data might not be feasible for developing nations. Furthermore, image processing methods require more memory than information from location and ambient light sensors. The goal of Ashraf Zatari et al. [13] is to create a flexible and automated system for measuring road lighting parameters (such as glare, luminance, and illumination) across the entire city. They have employed CCD cameras with digital storage capabilities that are more accurate. Two or more pre-calibrated cameras were installed in order to measure lighting parameters and determine the instantaneous observation position on the road. The digitally saved photos underwent offline analysis. This also comes at a great cost. Using luminance and threshold increment measurements, Anne-Mari Ylinen [14] et al. concentrated on examining the impact of mesopic dimensioning on energy consumption and life cycle costs of the LED installations for street lighting. For the study, they employed a Techno-Team LMK Mobile Advanced digital luminance camera. Installing a sensor at each lamp is the way that Chunguo Jing et al. [15] suggested controlling and monitoring street lights. The control centre, RTUs, and GPRS module make up the wireless sensor network technology that this system uses to aggregate sensed data. It requires sophisticated software and hardware design, which developing nations cannot afford.

18.1.2 System Model

An outline of the suggested scheme in Figure 18.2 shows the **fair driver rating system (FDRS)**. The modules in each component are then briefly described. The graphic displays the different FDRS components. Our primary focus is on identifying three characteristics of the road surface: (a) damaged road patches, (b) potholes, and (c) speed breakers. We also keep an eye on the street light illumination level and identify and locate any malfunctioning light posts. For the driver rating system, we refer to these characteristics as the point of concerns (PoC). The modules are arranged into two main components: phone processing and server processing, depending on where and when they will be implemented.

Phone processing. In order to identify potential candidates for primary PoCs using geotagged sensor data, we first gather raw sensor data using smartphone sensors that are mounted, pocketed, or placed on the dashboard for various vehicle modalities (two, three, and four wheels). We then apply an auto-thresholding mechanism to the sensor data. The smartphone's phone

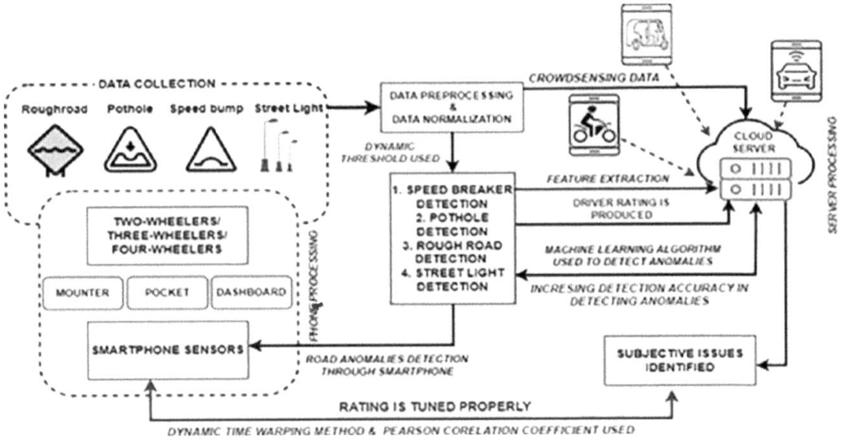

FIGURE 18.2 System architecture of FDRS.

processing module, also known as the smartphone sensing module, is where this procedure is carried out. Subsequently, the crowdsourced sensor data is uploaded to the server for additional processing.

Server processing. This step involves sending each user's geotagged primary PoC candidates to the server, where we use a decision tree–based classification technique to identify the primary PoCs (potholes and speed bumps) and secondary PoCs (broken road patches), based on data from individual users. It should be mentioned that this step is required since noise from the vehicle's suspension and other external elements could have caused a sensor signature to be formed. Additionally, the extraction of secondary PoCs – potholes that occur repeatedly throughout a road segment from a damaged road patch – depends on the original PoCs. Subsequently, we employ a clustering-based technique to aggregate the PoC data from several trails obtained from crowdsourcing data in order to determine the precise positions of the PoCs. GPS only provides an estimated location in this case, and there is a chance that user data from different trails is not precisely aligned in space. Finally, data is processed via a server, which we refer to as the server processing module. Lastly, we create an aural alarm system to help drivers, a recommendation system for a safe and comfortable alternative route, a recommended system for affordable road maintenance, etc., utilizing the data shown in the dashboard.

18.1.3 Issues and Challenges

- **Phone related issues.** Mostly, the issues are related to the placement and positioning of the phones and also related to different brands of phones. We see different sensor signatures in Figure 18.3 when the phone is kept in the pocket, as opposed to when it is installed over the vehicle. This is mainly because the body vibration of the pocketed phone cancels out the vibration

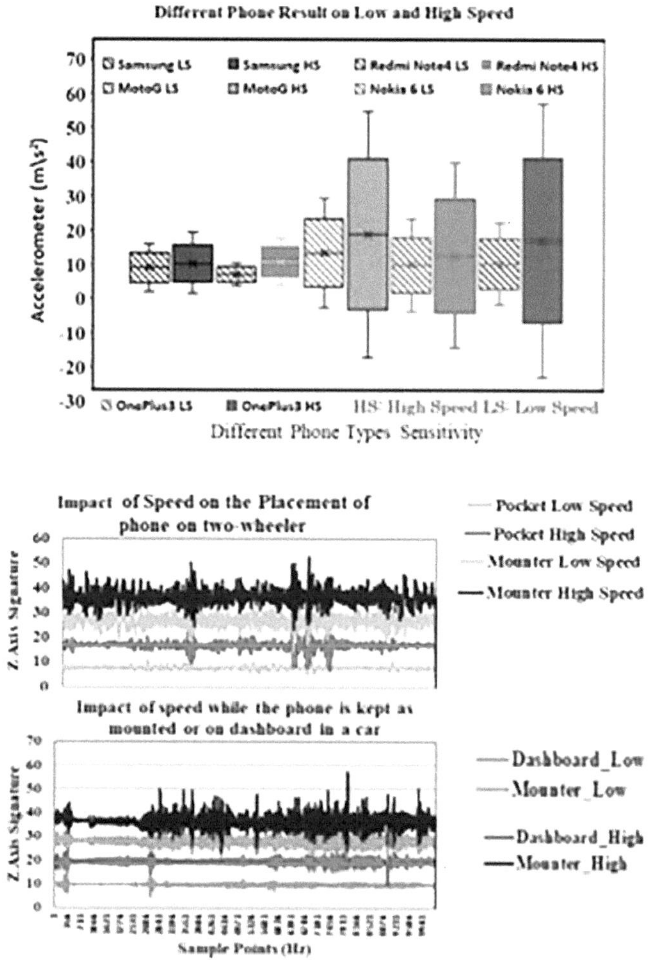

FIGURE 18.3 Comparative analysis of different phone sensitivity, placement of phone on speed.

of the vehicle when it drives over an uneven section of road. This makes mobile crowdsensing for road surface monitoring challenging, because the position of the smartphone is a key factor in the choice. The effect of high and low speed on various phone types is significant. For instance, MotoG exhibits a rather large influence in this Figure 18.3 because of its low-cost phones. All phone models exhibit this variation, with the exception of the HTC Desire Pro, and it is more apparent when users move at higher speeds than at slower ones. On the other hand, sensor signatures change significantly at a comparatively fast rate. Figure 18.3 shows phone sensitivity on high and low speed on different brands of smartphone and while the placement of mobile phones varies in cars.

- **Road-related issues.** Roads in developing countries like India can vary greatly in quality. While major highways are generally well-maintained, rural roads and some city streets may be in disrepair, with potholes, uneven surfaces, and insufficient signage. Potholes are a common problem on Indian roads, and they tend to worsen during and after heavy rainfall. Poor drainage systems in many Indian cities can result in water-logging on roads. Hence, it can reduce visibility on the roads, making it more challenging for drivers to navigate. Speed breakers are often built with varying heights, widths, and profiles, as shown in Figure 18.1. This inconsistency can lead to confusion and makes it challenging for drivers to anticipate their effect. In many cases, speed breakers are not adequately marked or painted, which can make them difficult to spot, especially at night or during adverse weather conditions. Though many existing road anomaly detection systems perform nicely on smooth roads for four-wheelers and six-wheelers in detecting speed breakers and potholes, they fail to detect speed breakers and potholes fairly in uneven roads using two-wheelers and three-wheelers, as shown in Figure 18.4. Invisibility of these anomalies on this type of rough road produces more ambiguity in the signature of speed breakers and potholes captured by smartphone sensors.

  Our tests reveal that while the current systems can detect speed breakers and potholes with a good accuracy of 80–85% and 65–77%, respectively, on smooth road segments, their performance dramatically decreases on rough roads (46–68% for potholes and 56–75% for speed breakers). In summary, we find that existing systems are unable to detect potholes and speed breakers on uneven roads with an accuracy of more than 80%. Furthermore, since broken road patches can be mistaken for potholes and speed breakers, they need to be identified independently, something that none of the current systems do. Thus, algorithms that can estimate a dynamic

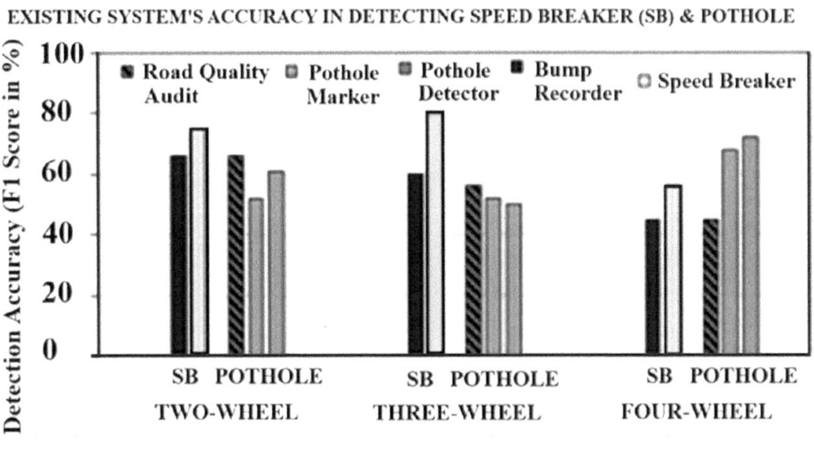

FIGURE 18.4 Comparative study of existing systems in detecting road anomalies.

threshold for each occurrence without requiring input from users and that take into consideration a wide range of variables, including the phone's location and orientation, driving manoeuvres, and phone types, must be developed.

- **Vehicle-related issues.** On a 400 m broken road stretch, the accelerometer signature (Z-Axis, which is mostly used to detect a speed breaker or a pothole) is displayed in Figure 18.5. Because the vehicle suspension level varies depending on the type of vehicle, it has been noticed that the accelerometer sensor signature differs as well. The accelerometer signature over a car 18.5(a) is smoother than the accelerometer signature over other vehicles, as seen in Figure 18.5(b), 18.5(c), and 18.5(d), mainly because a car's suspension is superior to that of other vehicles. A variation of this kind in the sensor signature typically affects the quality of road surface monitoring, and the methodology needs to be tuned based on the vehicle type.
- **Driver-related issues.** An interesting scenario has been observed in the given figure (Figure 18.6), where the sensor signature differs with differ-

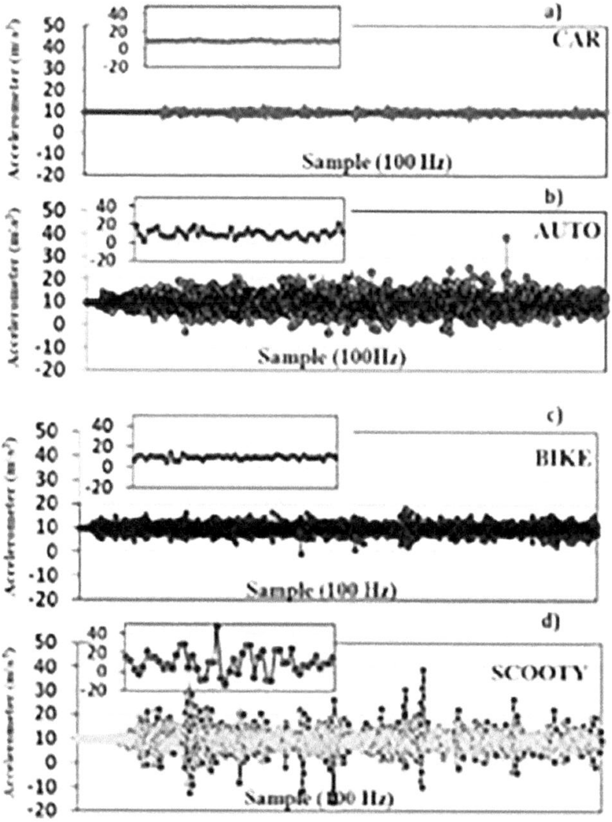

FIGURE 18.5 Impact of accelerometer sensors based on different vehicle.

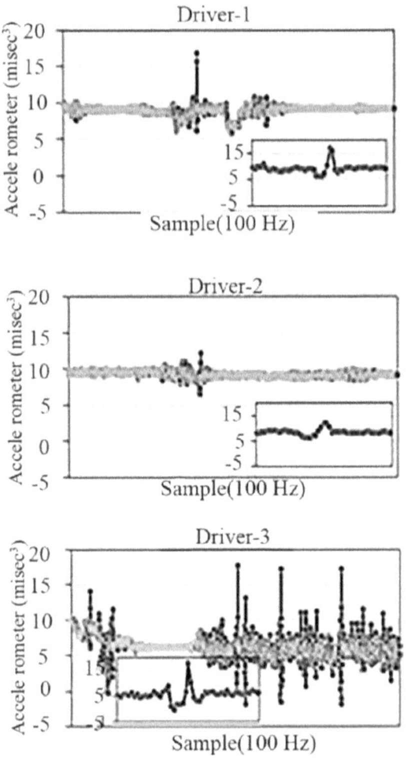

FIGURE 18.6 Comparative study of driving signature among three drivers on the same road, the same vehicle.

ent drivers on the same road while driving a two-wheeler (Scooty) in the controlled environment. On an experimental road with three speed breakers, various driving manoeuvres resulted in various road signatures. We see that over the data from Driver 3, all three speed breaker signatures are apparent, but over the data from the other two drivers, we only see one speed breaker signature. Inspection revealed that if a two-wheeler is travelling extremely slowly over a speed breaker, the sensor data does not generate a good signature due to disbalance of the driver.

- **Street light issues.** Street lighting is especially critical for night-time driving in developing nations, where road infrastructure can vary greatly. Many drivers rely on well-lit highways to travel safely at night. Adequate street lighting improves visibility on the road, allowing drivers to detect potential risks, such as speed breakers, potholes, pedestrians, bicycles, and other vehicles.

  Drivers fail to properly assess the height and location of the speed breakers, especially if the road surface is extensively damaged during the rainy season, with several waterlogged potholes that are not visible at night, owing to inadequate road lighting. A sudden high peak relative to the average peak can be formed while crossing a speed breaker at night, which may grade the driver poorly, but poor sight owing to insufficient illumination may be the true cause. Another issue is that the street lighting policies[2] in Figure 18.8 often vary based on the types of roads and the specific needs of those areas shown in Figure 18.7. Lighting levels in residential areas are often lower, and lighting fixtures are commonly shorter and tailored to offer appropriate illumination. The distance between light posts can be increased, ensuring enough lighting at junctions and crosswalks. Lighting built for both safety and energy efficiency, such as LED lights, may be used on roads with moderate traffic instead of halogen lights. As a result of the changing distance between poles at different locations, determining the number of poles in a roadway section may be difficult. In the latter instance, identifying a broken light post becomes difficult because light from the opposite side may contribute to the brightness as ambient light, resulting in surprising findings. Aside from that, the lighting level fluctuates greatly while detecting halogen, LED, and CFL technology-based street lights using smartphone sensor.

- **Subjective issues.** We believe that the subjective related issues which are not mentioned by any authors yet before are discussed thoroughly. The issues which are involved in the system design are as follows. First, we have tried to classify all those road surface anomalies like speed breakers, potholes, and broken patches through smartphone sensors based on their severity, which may be the actual cause of affecting the driver's rating system. Because of inexperience on the road, good drivers are sometimes forced to drive poorly. Drivers are unable to regulate their speed because of the lack of sufficient road illumination and varying profiles of unmarked speed

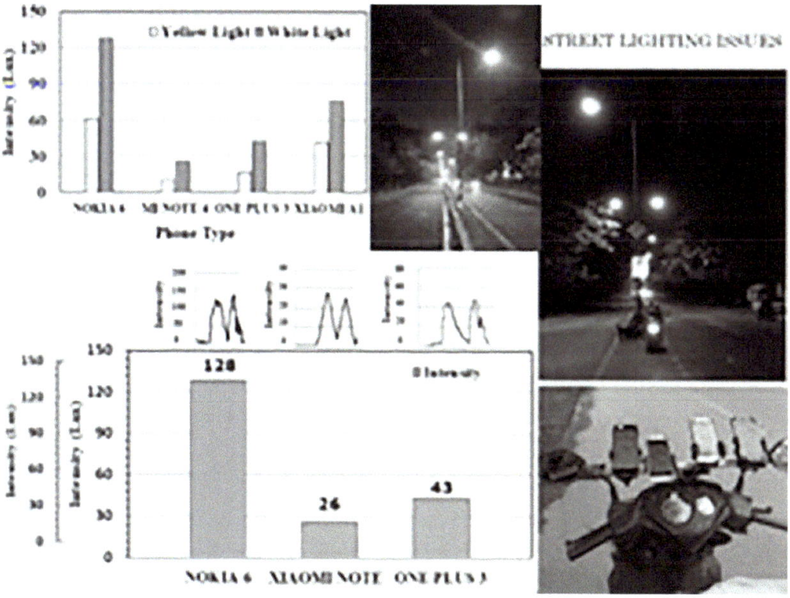

FIGURE 18.7 Different street light effects based on different phones.

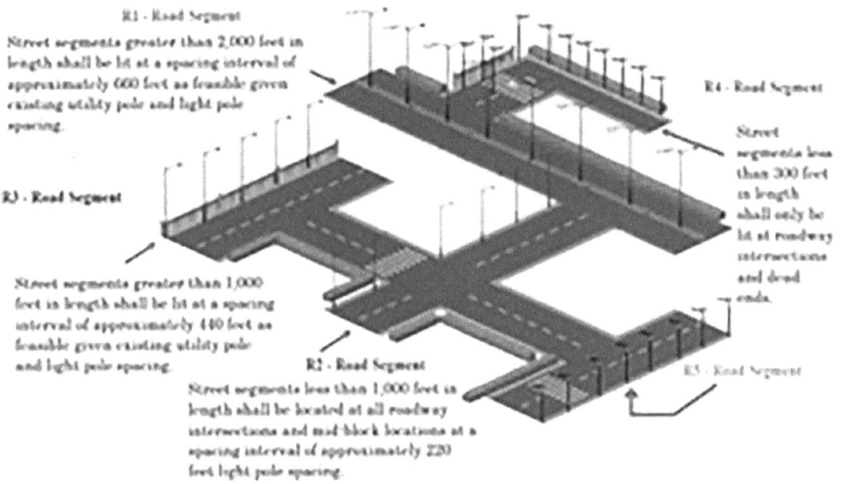

FIGURE 18.8 Different street light effects based on different types of roads.

> breakers, and they occasionally indicate their immediate break or cross the speed breaker and pothole with high speed. Because of this type of driving manoeuvre, the smartphone sensor signature indicates an abrupt result, and both commuters and the system give the driver a low rating. In this scenario, our system delivers a fair rating to the driver on smooth roads where the vehicle has a good suspension level. However, due to ambiguity in recognition, the same technique gives the driver a low rating on rough roads, where the road surfaces are entirely cracked over a broad span.

To overcome all these subjective concerns in driver profiling, ride-sharing and delivery companies can establish clear rating rules and encourage drivers to use the rating system fairly and consistently. Monitor and analyze rating data to detect and rectify potential bias and inconsistency issues. Use data analytics to uncover trends and patterns in driver evaluations, which can aid in distinguishing between driver-related and external factors influencing ratings.

18.2 METHODOLOGY

In this section, first, we present the methodology used to develop the fair driver rating system (FDRS). Next, we split our work between a remote server and a smartphone. Smartphones are only used to collect sensor data. The remote server receives the data and handles all additional processing. Thus, the remote server itself handles all the calculation. Consequently, this reduces the phone's computational burden. At last, we present the details of the techniques used to provide a fair driver rating after identifying an appropriate maximum-intensity level for a working light post and calculate the distance between the poles, which all together are considered as subjective issues.

18.2.1 Data Collection and Analysis

The gathering of data has been placed over three different types of roads, namely, smooth roads, rough roads, and broken patches, within and around the suburban city of Durgapur, covering almost 12,000 km of road. A smooth 10 to 12 km road with four potholes, seven speed breakers, and no broken road patches has been evaluated for system training. Three damaged road patches, 12 potholes, and 15 speed breakers can be found along 5 km of uneven road patch. The following table provides specifics on the data we collected during the course of our study.

18.2.2 Smartphone Sensing for Detecting Speed Breaker, Pothole, Broken Patch, and Street Light Using Dynamic Threshold

A number of experiments have been carried out to find out the factors that primarily affect the threshold values used to separate out different anomalies. A sample of observations from the controlled experiments is illustrated in Figure 18.9, where we adjusted the threshold values and observed that for which threshold value yielded the highest accuracy result for detecting a PoC. The graph depicts the raw Z-axis acceleration readings, the moving average of the speed over time, and the appropriate speed breaker and pothole detection thresholds.

The appropriate threshold value for speed bump and pothole detection may increase as the average speed increases. The positive Z-axis acceleration over a speed bump grows as speed increases, while the negative Z-axis acceleration over a pothole increases as speed increases. As a result, vehicle speed determines both acceleration over an anomaly and the threshold for anomaly detection. And thus we consider vehicle speed to be a confusing element in auto threshold generation. Our starting threshold is estimated under 20 km/h for our experiments, which is the speed at which that threshold can identify a pothole or a speed breaker using a vehicle. To determine the initial two-wheeler threshold, we asked five separate two-wheeler drivers to set their smartphone on a mounter and cross a speed bump at a regulated, low speed of 20 km/h. A speed breaker has a signature that ranges from high to low. We took the maximum and minimum values from each generated signature and used the lowest among the highest values and the highest among the lowest values to determine as initial thresholds. A dynamic threshold value is calculated to eliminate false positives when a two-wheeler encounters a speed bump at a higher speed of more than 20 km/h, applying its instantaneous speed and similar to detect potholes. Similar studies were carried out for three-wheeler and four-wheeler vehicles under various placement possibilities to determine the initial thresholds. Based on the preceding facts, we use an automatic regressive time series model to estimate the threshold value based on vehicle speed measurements.

For a certain type of vehicle, the smartphone's mount point, etc. Let $\boldsymbol{Vt}$ be the vehicle's instantaneous velocity at time $\boldsymbol{t}$, expressed in km/h, and let $\boldsymbol{S_T}$ represent the base point up to which the static threshold operates. Beyond this, the system switches to the dynamic threshold mode. At time $t = 0$, let $\boldsymbol{T_0}$ be the starting threshold value. Furthermore, $\boldsymbol{D}$ is the threshold's starting point for rise when the system is in dynamic threshold mode. A scaling factor known as $\boldsymbol{S}$ determines the pace at which

TABLE 18.1
Summary of Data Collection for Fair Driving Rating System (FDRS)

Data collection for FDRS	Smartphone/ mobile phone sensing events	No. of samples for each event	Types of phones and its no.	Phone position and placement	Drivers' specification, and no.	Types of vehicles and nos.	Types of roads and span
Road surface anomalies: speed breaker, pothole, broken road patch	Speed breaker: 37 Pothole:18 Broken patch: 4 Smartphone sensors: accelerometer, GPS gyroscope, GPS bearing	Total: 994 samples	HTC Desire Pro: (1) OnePlus: (3) Nokia 6: (1) Redmi Note 4/(Xiaomi): (2) Samsung Note 4: (1) Moto G: (1) **Total: 9**	**2-wheeler:** mounter, pocket **3-wheeler:** windshield mounter, pocket **4-wheeler:** windshield mounter, dashboard, pocket **Position: horizontal vertical, vertically tilted**	Total: 10 Known driver: 6 Unknown driver: 4	**2-wheeler** 1. Honda Dio 2. Aviator 3. Honda Achiever 4. FZS 5. Yamaha **Total: 5** **3-wheeler:** auto rickshaw **Total: 7** **4-wheeler:** 1. i10 2. Dzire, 3. Etios. **Total: 3**	Rough road: 425 km Smooth road: 248 km Total: **673**
Street light monitoring: mal functioning street lights, light intensity monitoring	Smartphone GPS, IMU light sensor, Lux Meter (YOKOGAWA 51001-Illuminanc e Meter)	**Total: 10,000** sample trips Light Poles: 11,769 **Effective lights:** 13,670 **Defective lights:** 1,213	1. OnePlus3 2. Nokia 6 3. Redmi 4. ote 6 Pro, 5. RealMe U1 **Total: 5**	Position: Horizontal	All known drivers **Total: 4**	2-wheeler 1. Scooty: Honda Dio 2. Bike: Hero Achiever) 3-wheelers/ **4-wheelers: NA**	Total: 414 km Nonfunctioning street lights road (local/market road) Functioning street lights road (local/GT road/NH)
Driver rating	Smartphone sensor: accelerometer, GPS, light sensor	Bad drive sample: 56 (4–6 km each) Good drive sample: 68 (3–5 km each)	Smartphone app SafeDrive App/ RoADApp	**Phone position:** Horizontal, vertical tilting, **Phone placement:** Pocket, mounted	Total: 8 Known driver: 2 Unknown driver: 6	2-wheeler 1. Scooty: Honda Dio 2. Bike: Hero Achiever) 3-wheelers/4-wheelers: NA	Total: 650 km Rough road: 370 km Smooth road: 210 km Broken patch: 70 km

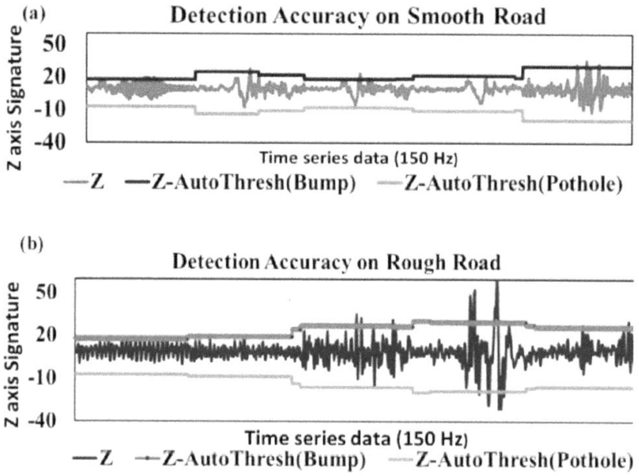

FIGURE 18.9 Based on vehicle speed and different road types, threshold varies.

the threshold rises for each unit increase in speed above the lower limit. T_t is the automated threshold, which is established at t depending on T_0 and Vt, two independent parameters. **Tt** is calculated as follows:

$$\mathrm{T_t} = \mathrm{T_0} + \left(\frac{1}{t} \sum_{j=0}^{t} V_j - D \right) * S_T \text{ if } \sum_{j=0}^{t} V_j - D > S_{T'} \text{ Otherwise } \mathrm{T_0} \tag{18.1}$$

The base point S_T is employed in this approach because the lowest peak should be present in the signal signature even at low vehicle speeds. D is utilized in order to change the lowest value at which the threshold should start to adjust dynamically. In almost all situations, D should equal S_T. D is greater than S_T in situations where the vehicle shakes a lot at lower speeds and then reduces vibration above a particular speed typically for a two-wheeler. D is less than S_T in circumstances where the vehicle begins to vibrate more at higher speeds after a specific speed value. We derive T_0 empirically based on the trials. The moving average of the vehicle speed is used as a confounding variable in the proposed automated thresholding strategy to create a dynamic threshold value rather than a static one. We discovered that while the previous algorithm correctly identifies the existence of events like speed bumps and potholes with average accuracy of 83% and 75%, respectively, it fails to accurately identify potholes in a significant proportion of situations. Furthermore, the prior approach fails to classify broken road patches, which is depicted as a series of sequential speed bumps and potholes. To overcome the ambiguity caused by the occurrences, we will require new features in the future.

We choose to use machine learning techniques to classify the events precisely. To reduce processing overhead on smartphones, we send data regarding geotagged PoC candidates and their signatures to a server, where we try to significantly enhance our system's performance by working on the outcomes of this part.

18.2.3 Server Processing for Increasing Detection Accuracy and Removing Anomalies, Signature Ambiguity Using Machine Learning Algorithms

We created RoADApp [16], an Android application, to detect PoC candidates whose locations are geotagged. After all PoCs are collected, they are transferred to a cloud server for additional processing, which may reduce the system's processing cost while also providing comprehensive information regarding road surface conditions. At the first level of classification, we use a decision tree–based classifier, shown in Figure 18.10, to distinguish different types of primary PoCs (speed breakers and potholes), and then at the second level of classification, we separate secondary anomalies (broken road patches) from the combinations of potholes and speed breakers.

- **Decision tree.** Training phase in machine learning consists of known dataset gathered from event samples from several trips covering 425 km over 15 km training road. The relevant occurrences (speed breaker/pothole/broken patch) have been manually labelled in order to label the data. Training is carried out in order to determine the various computational values that would accurately classify the various events. During the 'testing phase', event samples gathered from multiple trips covering 248 km, spanning 11 km test road, are tested. Z_t is the measured Z-axis acceleration at time t while detecting PoC based on the temporal threshold.

T_t is given by equation (1). Features Z_t^{Next} and Z_t^{Prev} are used to remove the ambiguity nature from detected PoCs as speed breaker and potholes. We have developed a decision tree–based classifier based on these features, discussed here.

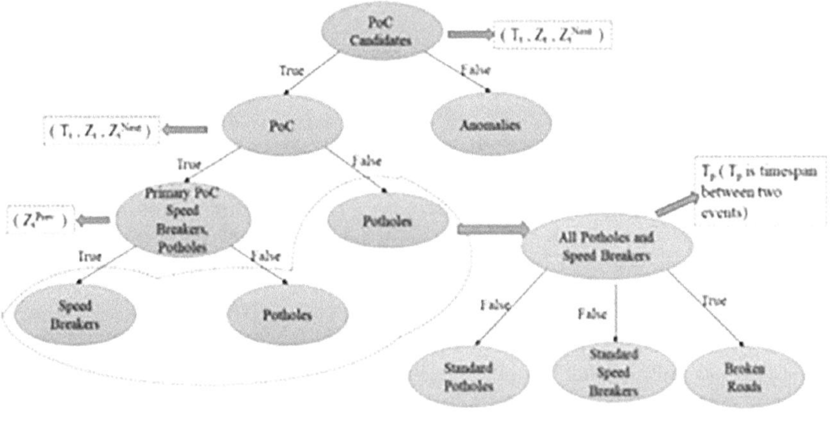

FIGURE 18.10 Removing ambiguities and classifying the road surface anomalies based on decision tree.

For the presence of a PoC, or due to some anomalies, the root node predicts events registered with the system that have truly occurred for. In the next level, nodes classified the events into speed breakers, potholes, as well as broken patches. If the pothole signature is unclear and recognized as speed breaker, the classifier generates another pair of branches that classify the speed breaker as speed breaker or pothole. The accelerometer signatures are employed at various levels of the classifier. If a real event such as the speed breaker is detected, Z_t^{Next} will be the negative local minima just after Z_t due to the decrease in acceleration when the vehicle approaches the speed breaker, followed by a sudden peaking of the increased negative acceleration in the direction of gravity when the vehicle just crosses the speed breaker. However, in the instance of the pothole, we observe a positive local maxima due to a positive acceleration when the vehicle crosses the side edge of the pothole soon after Z_t. When numerous subsequent potholes come one after another inside the small-time window Δ, a pothole signature might occasionally be misinterpreted with a speed breaker. In this situation, we additionally use the feature Z_t^{Prev} to see if another pothole was spotted recently, within the same window Δ time. At the next level of classification for finding the broken patch, multiple potholes and speed breakers are combined to check whether they are encountered very frequently. Thus, the features $\boldsymbol{T_p}$ and $\boldsymbol{S_p}$, where T_P is the duration between the time of detection for a PoC and the immediate next PoC **p**. $\boldsymbol{S_P}$ is the instantaneous speed of the vehicle at the detected PoC **p,** which has been considered to check distance and time taken by the system to detect all PoCs. Generally, it has been observed that to cross a speed breaker or even a small pothole, 2 to 3 sec are required in normal conditions by the vehicle having a rapid speed of 5 km/h to 10 km/h. Taking the duration between the two events into account, if we notice that they are spaced at very close intervals, we group them together and identify the new event as a broken road patch.

- **K-medoids clustering method to localize the road anomalies.** The trail gathered by every single user is subjected to the decision tree–based [17] classifier. Lastly, we merge the data from several trails (gathered via crowdsourcing from various users) in order to (1) correctly locate the PoCs and (2) improve the detection confidence of PoCs. To do this, we first combine the PoCs identified from multiple trails and subsequently employ unsupervised k-medoids clustering [17, 18]. This method is predicated on minimizing the total of the dissimilarities between points on the dataset of n objects that are labelled as corresponding to a cluster. The centre of that cluster is identified as a certain spot. The k-medoids algorithm is a method for splitting data into k-groups. We employ a silhouette [19, 20] technique to find $\boldsymbol{k}$ a priori in order to ascertain its value. As shown in Figure 18.11, the centre of the cluster is considered as the location of the PoC. Further, we eliminate a detected PoC if the number of member points in a cluster less than, $[M/3] + 1$, where M_T represents the local number of trails in the crowdsourced data. This means that for a PoC to be deemed valid, at least one-third of the trails must be involved in its detection.

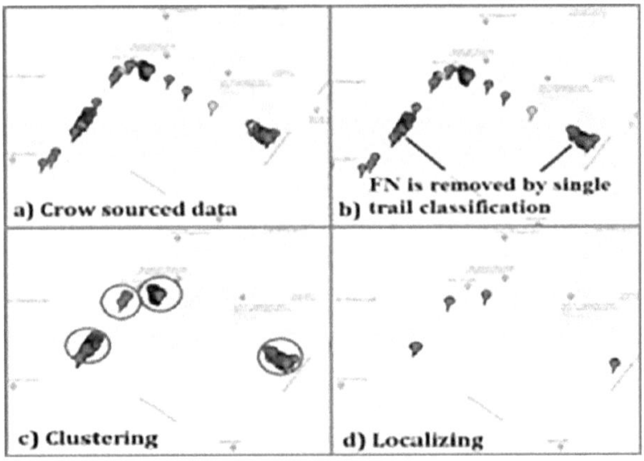

FIGURE 18.11 K-means clustering method: from crowdsourced data > decision tree is applied to remove the ambiguity > k-means clustering > centroids are selected as road anomalies > localizing the road anomalies to the graph.

18.2.4 Street Light Detection Using K-Means Clustering/ Elbow Method

Below the light pole, the light intensity is always at its highest. Thus, we just need to concentrate on the local maxima in order to determine the precise location of a light pole [21]. To do this, we create a local maxima filter that removes the remaining data while preserving all the local maxima that were present in the original data. Even after eliminating some of the superfluous data with this filter, we will still obtain some additional maxima that are present at extremely close distances. The ambient light is to blame for this. We employ the second filter, the distance-based noise filter, to get rid of those. Any road has light posts that are always spaced at a fixed distance – let us say β. This constant distance may contain a few small mistakes and fluctuates depending on the type of road. Light poles are the maxima at distance β, while noise that needs to be filtered out is the noise present at very little distance ('β'). Only the valuable maxima are retained by the distance-based noise filter, which discards the other data. Thus, our search continues from $(\beta - \alpha)$ meters to $(\beta + \alpha)$ meters. This was added to make up for the inaccuracies brought about by both non-uniform pole distance $(\beta > \alpha)$ and GPS. We treat any maxima in this distance range that have a value higher than the automated threshold as working poles, or else, we consider them non-functional poles [22]. The next step is to combine the local points across the peak values to determine the precise location of the light poles that are in operation. This can be achieved by merging the local peaks, which are spatially close to one another, using a clustering-based technique. The number of operational light poles on the road section should be the same as the optimal value of the number of clusters, k. However, we employed an unsupervised k-means clustering approach because the number of operational poles (clusters) is unknown beforehand. Heuristic to estimate desired (k): it is not easy to figure out what k is worth. According to Indian norms,[3] the pole spacing ranges from 20 to 50 m and depends on the type of road. Therefore, we calculate that k_{min} = L 50 and k_{max} = L 20, where L is the road segment's length. After performing k-means clustering on the determined local peaks, we compute k using the elbow approach [15], changing the value of k between k_{min} and k_{max}. We determine the optimal value of k with nearly identical results using both the standard and the silhouette coefficient as score/metric. Furthermore, we note that both metrics typically yield a value of k that is lower than the true number of functioning poles, suggesting the existence of false negatives (FN), or instances in which a functioning light pole that is present in ground truth are missed. Figure 18.12(b) displays the number of such clusters found for the sample road segment; to facilitate easy differentiation, intensity peaks in successive clusters are displayed in alternating colours. Figure 18.12(c) depicts the elbow [22]. The non-working light poles shown in Figure 18.12(d) do not have an intensity signature recorded, in contrast to the functioning poles. As a result, it is necessary to estimate the number of non-functional poles and their geolocations indirectly using the pole separation data that is taken from the geolocation of functioning poles.

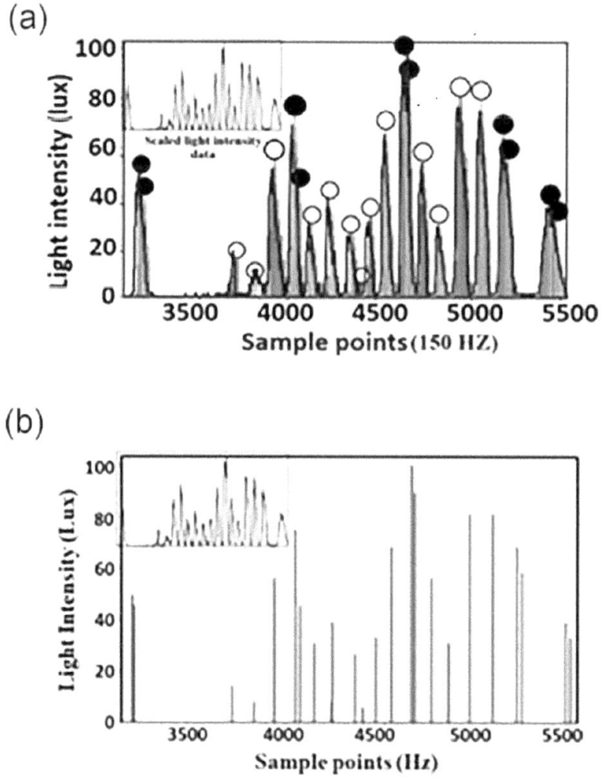

FIGURE 18.12 Detection of functioning and non-functioning street lights using k-medoids clustering algorithm.

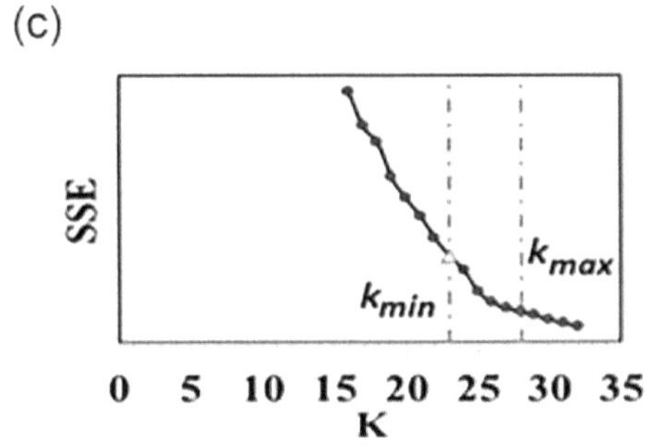

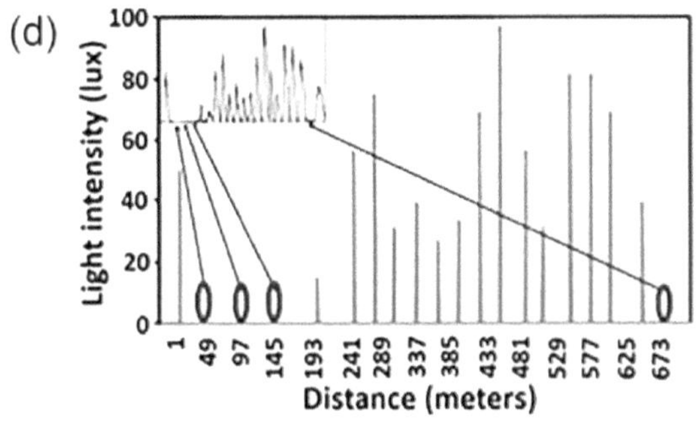

FIGURE 18.12 Continued

1. Scaled light intensity inset figure as our sample data. White circle shows single cluster head, whereas overlapping cluster heads are shown in black colour.
2. Overlap light poles are selected and removed in the next using k-medoids algorithm.
3. Elbow method to find the value of *k*.
4. Missing light poles are marked as elliptical shape based on the distance between two poles.

18.2.5 Driver Riding Pattern Matching Using Dynamic Time Warping Method on Time Series Data

- **Dynamic time warping method.** Though dynamic time warping (DTW) has been employed in partial shape matching in a variety of other application settings, we have seen in our investigation that DTW [23–25] is a good method that fits our objective. We are able to recognize the user-identified nice rides and the various kinds of bumps the vehicle has encountered during its journey.

Let $R = \{r_1, r_2, \cdots, r_m\}$ represent the time series signal corresponding to it. Let the time series signal for any sample ride to be rated be represented by $Q = \{q_1, q_2, \cdots,$

q_n}. Here p_t is the sample of Z-axis readings at time t, and s_t is a particular event type at time t. The square root of the cost of the best warping path is used to calculate the warping distance in DTW. In order to determine the best match between these two sequences, namely P and S, the optimal path has been found through the matrix (m × n), where (i^{th} and j^{th}) elements of the matrix correspond to the squared distance:

$$\mathrm{d}\left(r_i, q_j\right) = (r_i - q_j)^2 \tag{18.2}$$

The relationship between points p_i and s_j is indicated by this equation. The best speed breaker matching path, or optimal warping path, is generated using an equation based on the shortest warping distance, which can be expressed as follows.

$$\mathrm{DTW}\left(\mathrm{R}, \mathrm{Q}\right) = \min\left\{\sqrt{\sum_{i=0}^{k} W_i}\right\} \tag{18.3}$$

Here, W_i is the matrix element $(i, j)_k$, as it belongs to the kth element of a warping path W, a contiguous set of matrix elements which depicts a mapping between the time series R and Q.

18.3 RESULT ANALYSIS

18.3.1 Speed Breaker, Pothole, and Broken Patch Detection Accuracy Based on Smartphone Processing and Server Processing

We start by examining the smartphone's automated-tune thresholding mechanism's performance in terms of detection. We provide the best accuracy achieved by applying a dynamic threshold using the automated-tune threshold module. We compare the performance of the static and dynamic thresholding approaches. In order to do this, we tested with various static threshold settings and reported the F1-score for various smartphones (Figure 18.13) and vehicles (Figure 18.14), with the automated-tune threshold providing a better F1-score. We can see from the two figures that dynamic thresholds greatly increase the detection accuracy across all smartphone models and vehicle kinds. Using the automated-tune threshold instead of a static threshold, we observe an average increase of 44% in the accuracy of speed breakers and 40% in the accuracy of potholes on smooth roads, and an average increase of 37% in the accuracy of speed breakers and 46% in the accuracy of potholes on rough roads. As illustrated in Figure 18.13, the automated threshold generation in the very inexpensive Xiaomi phones results in the greatest improvement in the speed breaker and the pothole detection accuracy. Our tests have shown that some phones, such as the Xiaomi and MotoG, have a wider accelerometer Z-axis fluctuation range than others. So while using static threshold, a slight jerk has frequently also been identified as a speed breaker. The outcome is an increase in false positive rates, which lowers both phones' precision scores.

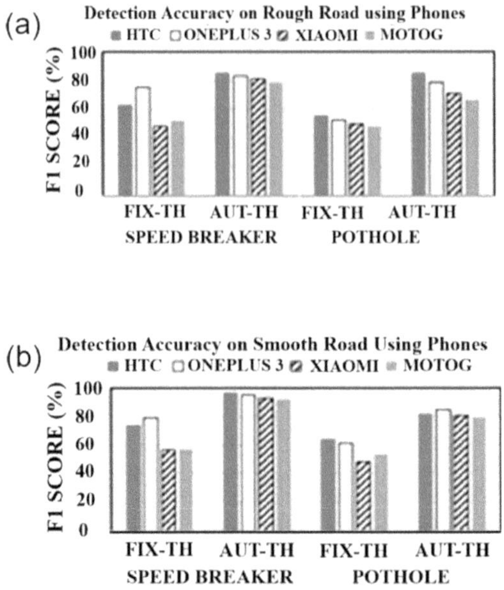

FIGURE 18.13 Anomalies detection using fixed and automated threshold using different phones.

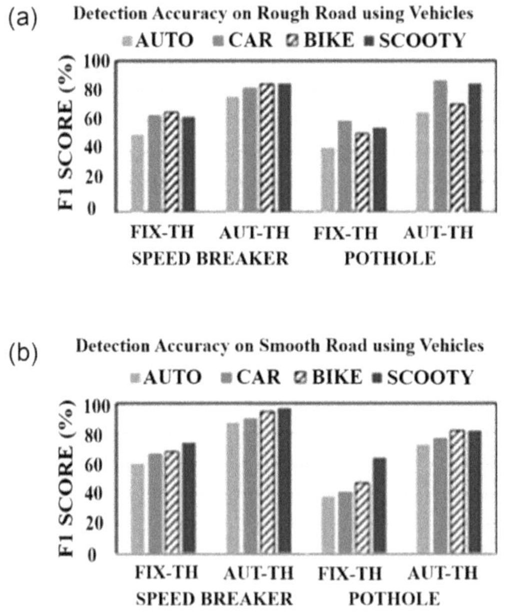

FIGURE 18.14 Anomalies detection using fixed and automated threshold using different vehicles.

However, when auto-thresholding is employed, as demonstrated in Figure 18.14, the threshold dynamically adjusts, that is, it rises or falls in response to changes in speed. As a result, minor jerks are not recorded as points of failure. This reduces the false positive rate and boosts precision. Consequently, we find that auto-threshold improves system performance more in Xiaomi and MotoG phones – which also happen to be the relatively inexpensive phones we tested in our research.

18.3.2 Automated Threshold Results in Phone Processing and Tuned Result in Server Processing

Despite the fact that Figure 18.13 and Figure 18.14 show that the F1-score for speed breakers on a smooth road is 96%, for potholes, it is 84%. Because smooth roads usually create less vibration in vehicles, especially for two-wheelers. Increased vibration on poorly maintained roads causes more noise to be added to the accelerometer signature, decreasing detection accuracy to 85% on poorly maintained roads and 75% on uneven roads. Our recommended approach uses automated-tune threshold to detect potholes and speed breakers in a reasonable manner when broken patches were not taken into consideration. We made multiple trips during our experiment, but in order to support our assertion, we have included three separate, random trips in the following. Two-wheelers (scooters) and the HTC Desire phone model were used on the three trips depicted in Figure 18.15(a) as a trail for a controlled environment test. The system has an average accuracy of 80–85% for identifying speed breakers and 55–75% for identifying potholes on uneven terrain. The ambiguous signature of events on uneven roadways causes detection accuracy to decrease, leading to an increase in false positive and false negative rates. Further server processing is carried out to eliminate these ambiguities, and the accuracy is fairly raised, as shown in Figure 18.15(b). By eliminating the ambiguity from the improperly identified PoCs using cell phones, the decision tree classifier performs admirably on our dataset.

18.3.3 Comparative Results on Crowdsensing Data Based on Different ML Classifiers

We examine the results using several classifiers, as Figure 18.16 summarizes. As illustrated by Figure 18.10, we note that the linearity of the separation plane utilizing the features Z_t, Z_t^{Next}, and Z_t^{Prev} allows the k-nearest neighbour (KNN) and J48 [26] decision tree to outperform other classifiers. For this specific challenge, overfitting of the data results in poor performance of a support vector machine (SVM)-based classifier under different hyperparameters. This explains why we decided to categorize different road surface anomalies from the raw sensor data using a J48 decision tree.

18.3.4 Street Light Detection Accuracy

1. **Street light detection and their localization on the map.** Using its built-in light sensor, the smartphone may capture images of several peaks that are

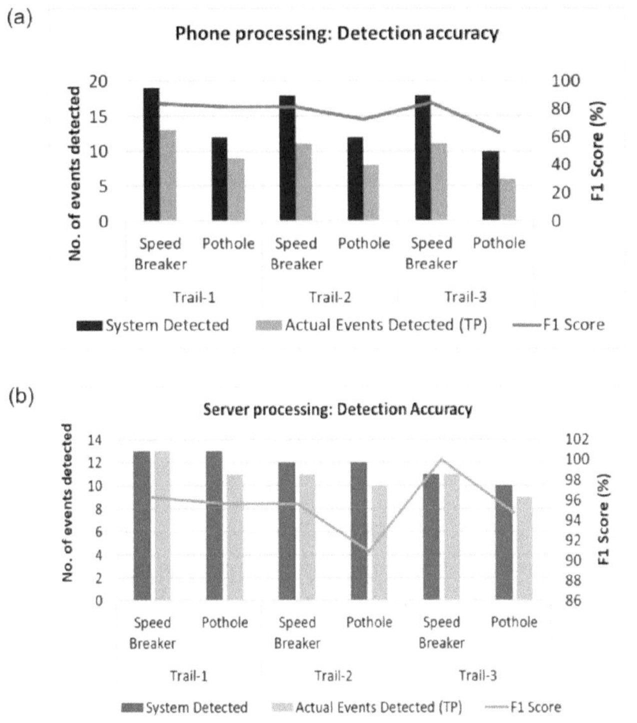

FIGURE 18.15 (a) Phone processing: detection accuracy. (b) Server processing: increasing detection accuracy.

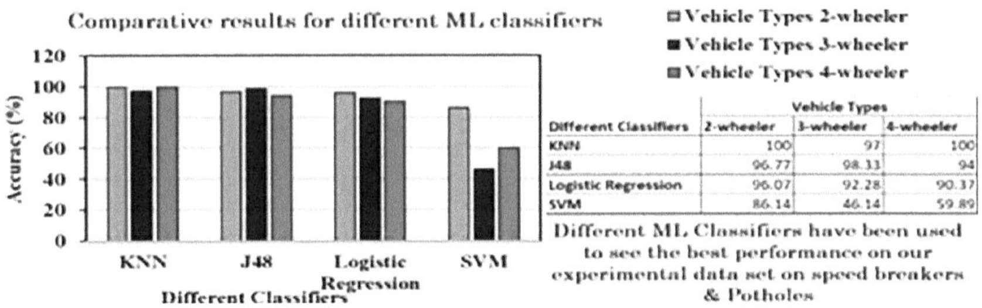

Different Classifiers	Vehicle Types		
	2-wheeler	3-wheeler	4-wheeler
KNN	100	97	100
J48	96.77	98.33	94
Logistic Regression	96.07	92.28	90.37
SVM	86.14	46.14	59.89

FIGURE 18.16 Comparative result analysis of road anomalies using different classifiers.

just below the light pole. In order to demonstrate our approach, we use data taken at a sampling rate of 150 Hz while travelling across a 700 m long section of road that has 16 working and 4 non-working light poles as ground truth. Two tasks are necessary for light pole detection to function in our proposed system: first, identify the abrupt shift in light intensity, or peak, and secondly, combine the local points across the peak values in order to determine the precise location of the light poles.

The following step will be to combine the local points across the peak values in order to determine the precise location of the operational light poles. This was accomplished by merging the local peaks, which are close to one another in space, using a clustering-based method. The k-means clustering with elbow approach was used to achieve the goal of having the ideal number of clusters, *k*, equal to the number of operational light poles on the road stretch.

We tested how well our proposed system worked on various phones. The results of the comparison are shown in Figure 18.17. The phones' overall performance has been noted to be good, with >0.95. However, because of their poor light sensors, inexpensive phones like the Redmi Note 4 and Realme U1 perform worse when it comes to identifying working light poles. As a result, it can be said that our algorithm performs well and is phone-invariant when it comes to identifying light poles of all kinds. A 1.2 km stretch of road was examined with a OnePlus 3 smartphone mounted on a Scooty, where 11 non-working street light poles were found and 2 light poles were flagged by the system as false positives, shown in Figure 18.17.

2. **Road categorizing based on light intensity of street lights.** We found in the study that the illumination strength of street lights varies depending on which segments of the road they are in. High-intensity street lights are found on bypasses and urban roads, medium-intensity lights are found on local urban roads, and low-intensity lights are found on rural roads. Local urban roads, however, could also have dim lighting. We reduce the intensity of these three types of street lights to one-third, two-thirds, and one-third, respectively. Based on our trained model, Figure 18.17 illustrates how the experimental route can be divided. With nearly flawless precision, the segmentation according to light intensity is completed. Red, yellow, and green, respectively, represent low-, medium-, and high-intensity lights on the map. A later classification of the route will be made in accordance with the safety concern – discussed next. Here in Figure 18.17, we get only low- and high-intensity lights.

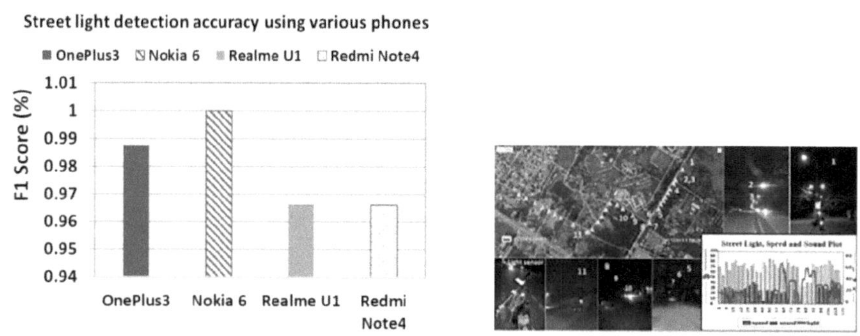

FIGURE 18.17 F1-score for detecting street lights using different phone covering almost 1.2 km road.

18.3.5 DTW Result for Driver-Tuned Rating Based on Subjective Issues

We can detect the user-identified signatures of the many kinds of speed breakers the vehicle has to navigate in addition to the smooth journeys. From the collection of rides, we chose a decent ride signature with the lowest spike height. It should be noted that the specified FDRS system records various speed breaker signatures from both known (Kn) and unknown (Un) drivers, in addition to volunteer manual ratings, GPS logs, and light illumination data collected both during the day and at night. In this module, we use the DTW (dynamic time warping) approach to compare the data with the specified baseline (excellent ride bump signature). Initially, data on bump signatures on our experimental 2 km road patch were gathered from knowledgeable and known drivers. Second, the dynamic time warping method is used to compare the collected sample data with baseline data. It generates a certain warping distance, which has been further examined in order to distinguish between the various kinds of speed breakers that the drivers ride over. The FDRS system ranks the driver as a good driver if the produced warping distance is closer to the mean value of the typical speed breakers; if the produced value is significantly higher than the mean value, the driver is rated as a bad driver. The pillion rider marked every single road feature – speed breaker, pothole, broken patch – during the trial and gave a five-scale Likert rating – pleasant, good, average, unsafe, risky – for the driving behaviour on each one. Figure 18.18 displays sample plots of the accelerometer signatures for daytime and night-time speed breaker crossing events with and without brakes. It should be highlighted that various intriguing scenarios are produced when the presence of a street light is detected by light sensors and the accelerometer signature of the speed breaker is combined. The accelerometer sensor's Z-component can serve as a reliable signature, which has been thoroughly described in the decision tree analysis section. We can detect the user-identified signatures of the many kinds of speed breakers the vehicle has to navigate in addition to the smooth journeys. Therefore, the most important thing to do is figure out what a good riding signature is. From the collection of good rides, we choose a decent ride signature with the lowest spike height. Table 18.2 summarizes these observations for various types of speed breakers and daytime or night-time riding conditions. It should be noted that the specified FDRS system records various speed breaker signatures from both known (Kn) and unknown (Un) drivers, in addition to volunteer manual ratings, GPS logs, and light illumination data collected both during the day and at night.

Let $S_B = \{S_1, S_2, S_3, \cdots, S_N\}$ represent the set of events for crossing speed breakers along a path, where each S_i refers to the event of crossing a high peak, low peak, or multiple peaks. Consider $R(S_B)$ as an accumulated ranking across road anomalies along the trip; the final score of the driver may be calculated as follows:

$$\mathrm{R}\left(\mathrm{S_B}\right) = \frac{\sum_{i=1}^{N} R\left(S_i\right)}{\left|\left(S_B\right)\right|} \tag{18.4}$$

Here, $R(S_i)$ is the rating of road features crossed by the driver. Using the Pearson correlation technique, we have demonstrated a positive correlation between the

TABLE 18.2
DTW Distances for Various Known (Kn) and Unknown (Un) Driver Journeys During the Day and Night

Driver	SB_1	SB_2	SB_3	SB_4	SB_5	SB_6	Time (min)	SR_scale (5)
Mean_ (DTW)	**1.4217 9**	**1.5835 9**	**1.4674 7**	**1.4243 8**	**1.2769 9**	**1.4461 9**	**3.00**	**Score**
D1_(Kn)	1.1744	1.0743	1.5787	1.4033	1.4721	1.1560	2.17	4.80
D1_(Kn)	1.4547	1.3924	1.9957	1.4720	0.9203	0.5976	3.36	3.51
D2_(Kn)	1.3563	1.5777	1.4420	1.1109	0.5462	1.4791	2.57	4.78
D2_(Kn)	1.8643	2.0016	1.4619	1.7317	1.6339	1.5598	2.10	2.69
D3_(Un)	1.0343	1.7361	1.5076	1.6063	0.9938	0.5925	3.12	4.01
D4_(Un)	1.6587	1.2980	0.8560	1.4296	1.8489	2.0401	3.02	3.31
D5_(Un)	3.0538	1.0638	1.2559	0.5205	1.7523	0.9641	4.03	3.78
D6_(Un)	1.5790	2.1455	0.3718	1.7968	1.0592	1.8913	2.39	3.00
D7_(Un)	2.2803	1.6937	1.1956	1.2776	1.8389	1.6726	3.13	2.68
D8_(Un)	1.7991	1.2858	1.4702	1.5863	1.7831	1.5873	2.48	3.69

Note: The mean values are derived from 50 trials. SB: *speed breaker*; SR: *system rating*.

TABLE 18.3
Pearson Relation Coefficient (PRCT) of Manual Rating (MRT) and System Rating (SRT)

	SB 1		SB 2		SB 3		SB 4		SB 5		SB 6		
Drivers	**MRT**	**SRT**	**MRT**	**SRT**	**MRT**	**SRT**	**MRT**	**SRT**	**MRT**	**SRT**	**MRT**	**SRT**	**PRCT**
D1	5	1	5	1	3	5	4	3	3	2	2	4	0.71
D1	4	5	4	4	5	5	5	5	4	5	4	5	0.32
D2	2	2	3	5	3	4	4	5	4	5	4	5	0.44
D2	4	5	5	5	2	1	5	5	5	5	3	5	0.77
D3	3	1	5	3	4	5	2	1	4	3	2	1	0.73
D4	4	2	4	5	5	5	3	2	3	4	4	4	0.38
D5	4	3	3	4	2	5	2	1	5	5	3	1	0.32
D6	3	1	5	5	3	5	4	2	4	5	4	5	0.75
D7	5	4	2	1	4	5	3	5	1	2	3	1	0.60

manual rating (MR) and the rating provided by our system in real-world scenarios, denoted as system rating (SR), as shown in Table 18.3. If we take into account the unknown driver, we might not determine that the overall system rating coincides with the manual user rating to a respectable extent. As a result, it is necessary to adjust the total rating by adding two elements that are crucial for measuring driving performance. The α factor is used to describe day and night vision, and the β factor is used to describe the characteristics of a bumpy road surface. As a result, equation (4) is changed to:

$$R(S_B) = \frac{\sum_{i=1}^{N} R(S_i)}{|(S_B)|} + \alpha + \beta \quad (18.5)$$

Only the β is considered during the day, while both the α and β variables are considered at night. Based on the varying warping distance values of the bumps that drivers must navigate during the day and at night, a specific value is assigned to each factor. The computation of α will equal (5 – SR)/2, while the computation of β will take into account distinct bump features. For high peak, low peak, and numerous peaks, the corresponding values are $(\beta_1) = (5 - SR)/2$, $(\beta_2) = (5 - SR)/3$, and $(\beta_3) = (5 - SR)/4$, respectively. To determine the driver's final score – which might be either good or bad – all ratings are computed and added together. After tuning, we might receive the driver D1's score for the high peak bump of 5 and the low peak bump of 4.3, respectively, which could have an effect on the total score. It yields a useful result about driver D1, who takes less time (2.10 min) than driver D4 (3.12 min) but has a score of 4.4, which is respectable. On the other hand, driver D4 receives 3.4, indicating a medium score and rating equivalent to a usual drive.

18.3.6 A Decent Rating Provided by FDRS

GPS is used in the server to locate road surface irregularities, which are measured in terms of how they affect driving. We have just taken into account the abnormalities in the road surface that affect commuters at night. The accelerometer signature, vehicle speed, and frequency of events are used to categorize the risk of those irregularities in the road surface. We converted the entire road trip into a Likert scale by multiplying the light intensity and the state of the road surface shown in

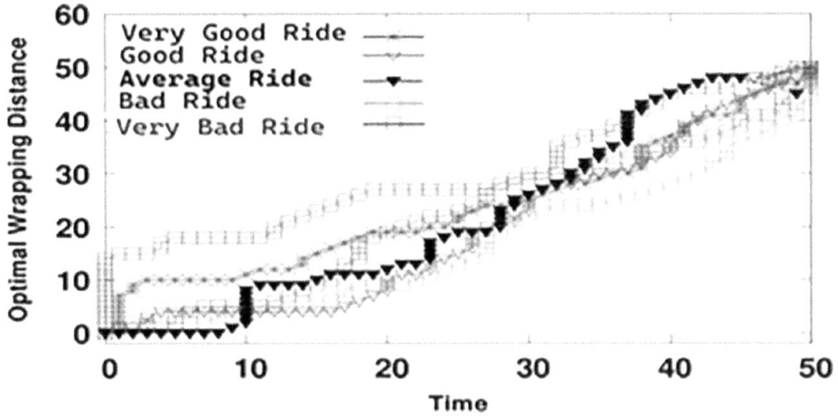

FIGURE 18.18 Study of warping distances associated with excellent and bad ride time series (DTW result).

Figure 18.20: strongly agree (pleasant road = 5), agree (good road = 4), neutral (average/bad road = 3), disagree (unsafe road = 2), and strongly disagree (risky road = 1) are the assigned Likert scale values. The system may designate a place as a dangerous zone if it automatically identifies a sizable, unavoidable pothole and simultaneously measures the street light intensity using GPS tagging. If the intensity is too low or the street lights are malfunctioning. Similarly, the algorithm might flag a road as dangerous if it has a significant area of fractured road and no street lights for many hundred meters. Because broken road patches absorb more light than smooth roads, the algorithm may classify a road as average or worse if it has broken road surface for several hundred meters yet high street light intensity overall. The system may designate a road as dangerous if it has massive, unavoidable potholes, high-peak speed breakers, shown in the middle of Figure 18.19, or

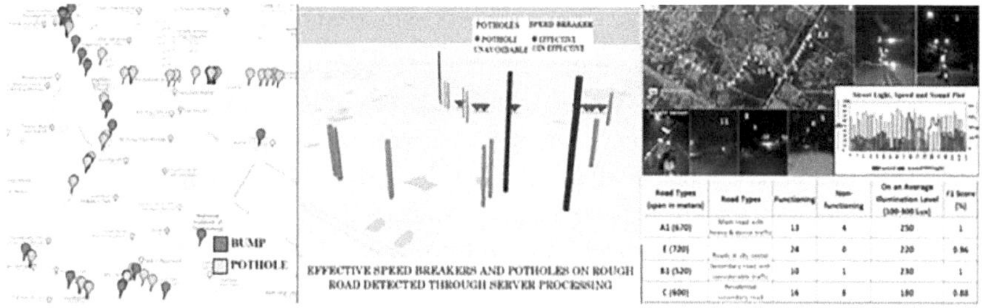

FIGURE 18.19 The effectiveness of road surface anomalies and the intensity of the street lights are subjective matters.

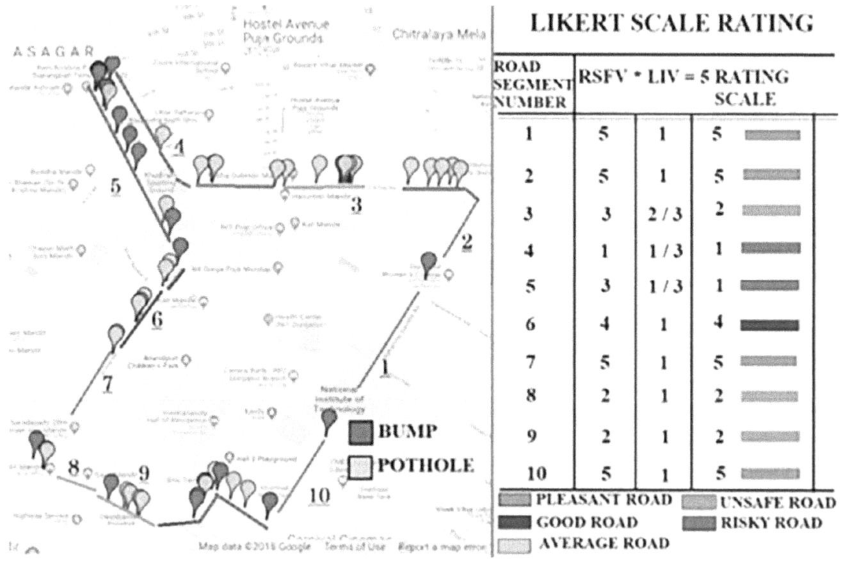

ROAD SEGMENT NUMBER	RSFV	* LIV	= 5 RATING SCALE
1	5	1	5
2	5	1	5
3	3	2 / 3	2
4	1	1 / 3	1
5	3	1 / 3	1
6	4	1	4
7	5	1	5
8	2	1	2
9	2	1	2
10	5	1	5

FIGURE 18.20 A decent driver rating system using Likert scale.

cracked road patches but no street lights near them. If the appropriate information arrives at the appropriate moment with its ground truth, the civil authority will have an easier job prioritizing taking action. By providing a proper alarm before commuter's travel, the risk will be reduced. By visualizing into the map with accurate information about the road journey during the night, the civil authority may be able to repair or replace on time.

18.4 CONCLUSION

In order to assess drivers' performance on normal road conditions in developing nations, we have created a smartphone app–based system and suggested a rating system. The entire application system can quickly identify the many road conditions that a driver may encounter during the day or even at night, such as high and low peaks and numerous bumps. The approach is readily adaptable to additional road characteristics, such as broken patch and potholes. With regard to each individual road event, our suggested system demonstrates an exceptional result of the driver's aggressiveness, which can be further accumulated to provide the driver with a final rating. According to our research, judging a driver's performance beyond a specific checkpoint is a personal decision. Because of this, a reliable method for rating driving behaviour in emerging regions needs to consider other factors, such as traffic in real time, the time of day, vehicle conditions, etc. It is crucial to locate these areas' markers on traffic GIS maps. We are currently experimenting with various vehicle modes and other features in an effort to give drivers warning systems before approaching such hazardous road conditions during the day or night. We also hope to notify other drivers who may be nearby to take extra precautions to ensure a safe drive by communicating with the devices in their proximity.

NOTES

1 https://www.ibef.org/industry/india-automobiles
2 https://me-lewiston.civicplus.com/DocumentCenter/View/549/082-StreetLighting Policy?bidId.
3 https://niua.in/csc/assets/pdf/key-documents/phase-2/Energy-Green-Building/Energy-Efficient-Street-Lighting-Guidelines-BEE.pdf.

REFERENCES

[1] G. Castignani, T. Derrmann, R. Frank, and T. Engel, "Driver behavior profiling using smartphones: A low-cost platform for driver monitoring," IEEE Intelligent Transportation Systems Magazine, vol. 7, pp. 91–102, 2015. doi: 10.1109/MITS.2014.2328673.
[2] D. A. Johnson and M. M. Trivedi, "Driving style recognition using a smartphone as a sensor platform," 2011 14th International IEEE Conference on Intelligent Transportation Systems (ITSC), pp. 1609–1615, 2011. IEEE.
[3] F. Restuccia, N. Ghosh, S. Bhattacharjee, S. K. Das, and T. Melodia, "Quality of information in mobile-crowdsensing: Survey and research challenges," ACM Transactions on Sensor Networks, vol. 13, no. 4, p. 34, 2017.

[4] J. Liu, H. Shen, H. S. Narman, W. Chung, and Z. Lin, "A survey of mobile crowdsensing techniques: A critical component for the internet of things," ACM Transactions on Cyber-Physical Systems, vol. 2, no. 3, p. 18, 2018.
[5] E. S. Poulsen, A. Morrison, H. J. Andersen, and O. B. Jensen, "Responsive lighting: The city becomes alive," ACM Mobile HCI, pp. 217–226, 2013.
[6] F. Wu, C. Wen, Y. Guo, J. Wang, Y. Yu, C. Wang, and J. Li, "Rapid localization and extraction of street light poles in mobile lidar point clouds: A supervoxel-based approach," IEEE Transactions on Intelligent Transportation Systems, vol. 18, no. 2, pp. 292–305, 2016.
[7] S. Sabatini, M. Corno, S. Fiorenti, and S. M. Savaresi, "Vision-based pole-like obstacle detection and localization for urban mobile robots," IEEE IV, pp. 1209–1214, 2018.
[8] W. Gu, Z. Zhou, Y. Zhou, H. Zou, Y. Liu, C. J. Spanos, and L. Zhang, "Bikemate: Bike riding behavior monitoring with smartphones," ACM MobiQuitious, pp. 313–322, 2017.
[9] D. Tomaras, I. Boutsis, and V. Kalogeraki, "Modeling and predicting bike demand in large city situations," IEEE PerCom, pp. 1–10, 2018.
[10] L. Abolhassani, A. P. Afghari, and H. M. Borzadaran, "Public preferences towards bicycle sharing system in developing countries: The case of Mashhad, Iran," Sustainable Cities and Society, vol. 44, pp. 763–773, 2019.
[11] H. Han, B. Meng, and W. Kim, "Bike-traveling as a growing phenomenon: Role of attributes, value, satisfaction, desire, and gender in developing loyalty," Tourism Management, vol. 59, pp. 91–103, 2017.
[12] S. Kumar, A. Deshpande, S. S. Ho, J. S. Ku, and S. E. Sarma, "Urban street lighting infrastructure monitoring using a mobile sensor platform," IEEE Sensors Journal, vol. 16, no. 12, pp. 4981–4994, 2016.
[13] A. Zatari, G. Dodds, K. Rafferty, and R. Robinson, "Glare, luminance, and illuminance measurements of road lighting using vehicle mounted CCD cameras," LEUKOS: The Journal of the Illuminating Engineering Society of North America, vol. 1, pp. 85–106, 2013. doi: 10.1582/leukos.2004.01.02.005.
[14] A-M. Ylinen, L. Tähkämö, M. Puolakka, and L. Halonen, "Road Lighting Quality, Energy Efficiency, and Mesopic Design – LED Street Lighting Case Study," LEUKOS: The Journal of the Illuminating Engineering Society of North America, vol. 8, 2011. doi: 10.1582/LEUKOS.2011.08.01.001.
[15] C. Jing, D. Shu, and D. Gu, "Design of streetlight monitoring and control system based on wireless sensor networks," 2nd IEEE Conference on Industrial Electronics and Applications, pp. 57–62, 2007. IEEE. Doi: 10.1109/ICIEA.2007.4318369.
[16] M. Y. Alam, A. Nandi, A. Kumar, S. Saha, M. Saha, S. Nandi, and S. Chakraborty, "Crowdsourcing from the True crowd: Device, vehicle, road-surface and driving independent road profiling from smartphone sensors," Pervasive and Mobile Computing, vol. 61, p. 101103, 2020, doi: 10.1016/j.pmcj.2019.101103.
[17] J. Kim, M. Hwang, D.-H. Jeong, and H. Jung, "Technology trends analysis and forecasting application based on decision tree and statistical feature analysis," Expert Systems with Applications, vol. 39, no. 16, pp. 12618–12625, 2012.
[18] A. Batra, "Analysis and approach: K-means and K-medoids data mining algorithms," ICACCT, 5th IEEE International Conference on Advanced Computing & Communication Technologies, pp. 274–279, 2011.
[19] N. Ailon, R. Jaiswal, and C. Monteleoni, "Streaming k-means approximation," Advances in Neural Information Processing Systems, pp. 10–18, 2009.
[20] S. Aranganayagi and K. Thangavel, "Clustering categorical data using silhouette coefficient as a relocating measure," International Conference on Computational Intelligence and Multimedia Applications, ICCIMA 2007, vol. 2, pp. 13–17, 2007. IEEE.

[21] Y. Yu, J. Li, H. Guan, C. Wang, and J. Yu, "Semiautomated extraction of street light poles from mobile LiDAR point-clouds," IEEE Transactions on Geoscience and Remote Sensing, vol. 53, no. 3, pp. 1374–1386, 2014.

[22] M. Y. Alam, S. Imam, H. Anurag, S. Saha, S. Nandi, and M. Saha, "LiSense: Monitoring city street lighting during night using smartphone sensors," 2018 IEEE International Conference on Data Mining Workshops (ICDMW), Singapore, pp. 596–603, 2018, doi: 10.1109/ICDMW.2018.00092.

[23] D. J. Berndt and J. Clifford, "Using dynamic time warping to find patterns in time series," Proceedings of the 3rd International Conference on Knowledge Discovery and Data Mining, pp. 359–370, 1994. AAAI.

[24] M. Alam, H. Anurag, Md. Imam, S. Saha, M. Saha, S. Nandi, and S. Chakraborty. "Urban safety as a service during bike navigation: My smartphone can monitor my street-lights," *2020 IEEE International Conference on Smart Computing (SMARTCOMP),* Bologna, Italy, 2020, pp. 114–121. doi: 10.1109/SMARTCOMP50058.2020.00035.

[25] S. Chu, E. Keogh, and D. Hart, "Iterative deepening dynamic time warping for time series," Proceedings of the 2nd SIAM International Conference on Data Mining.

[26] N. Bhargava, G. Sharma, R. Bhargava, and M. Mathuria, "Decision tree analysis on J48 algorithm for data mining," International Journal of Advanced Research in Computer Science and Software Engineering, vol. 3, no. 6, pp. 1114–1119, 2013.

Index

M

N

O

P

Q

R

S

T

U

V

W

Y

Z